T0291692

CAMBRIDGE LIBRARY COLLECTION

Books of enduring scholarly value

Mathematical Sciences

From its pre-historic roots in simple counting to the algorithms powering modern desktop computers, from the genius of Archimedes to the genius of Einstein, advances in mathematical understanding and numerical techniques have been directly responsible for creating the modern world as we know it. This series will provide a library of the most influential publications and writers on mathematics in its broadest sense. As such, it will show not only the deep roots from which modern science and technology have grown, but also the astonishing breadth of application of mathematical techniques in the humanities and social sciences, and in everyday life.

Elements of the Mathematical Theory of Electricity and Magnetism

The British physicist Sir Joseph John Thomson published the first edition of his Elements of the Mathematical Theory of Electricity and Magnetism in 1895 and this fourth edition in 1909, three years after he was awarded the Nobel Prize in Physics for his theoretical and experimental investigations on the conduction of electricity by gases. In this book for students his intention is to give 'an account of the fundamental principles of the Mathematical theory of Electricity and Magnetism and their more important applications, using only simple mathematics'.

Cambridge University Press has long been a pioneer in the reissuing of out-of-print titles from its own backlist, producing digital reprints of books that are still sought after by scholars and students but could not be reprinted economically using traditional technology. The Cambridge Library Collection extends this activity to a wider range of books which are still of importance to researchers and professionals, either for the source material they contain, or as landmarks in the history of their academic discipline.

Drawing from the world-renowned collections in the Cambridge University Library, and guided by the advice of experts in each subject area, Cambridge University Press is using state-of-the-art scanning machines in its own Printing House to capture the content of each book selected for inclusion. The files are processed to give a consistently clear, crisp image, and the books finished to the high quality standard for which the Press is recognised around the world. The latest print-on-demand technology ensures that the books will remain available indefinitely, and that orders for single or multiple copies can quickly be supplied.

The Cambridge Library Collection will bring back to life books of enduring scholarly value across a wide range of disciplines in the humanities and social sciences and in science and technology.

Elements of the Mathematical Theory of Electricity and Magnetism

JOHN JOSEPH THOMSON

CAMBRIDGE UNIVERSITY PRESS

Cambridge New York Melbourne Madrid Cape Town Singapore São Paolo Delhi

Published in the United States of America by Cambridge University Press, New York

www.cambridge.org
Information on this title: www.cambridge.org/9781108004909

© in this compilation Cambridge University Press 2009

This edition first published 1909
This digitally printed version 2009

ISBN 978-1-108-00490-9

ELEMENTS

OF THE

MATHEMATICAL THEORY

OF

ELECTRICITY AND MAGNETISM

CAMBRIDGE UNIVERSITY PRESS

London: FETTER LANE, E.C.

C. F. CLAY, Manager

Edinburgh: 100, PRINCES STREET
Berlin: A. ASHER AND CO.
Leipzig: F. A. BROCKHAUS
New York: G. P. PUTNAM'S SONS
Bombay and Calcutta: MACMILLAN AND CO., Ltd.

ELEMENTS

OF THE

MATHEMATICAL THEORY

OF

ELECTRICITY AND MAGNETISM

BY

Sir J. J. THOMSON, M.A., D.Sc., LL.D., Ph.D., F.R.S.,

FELLOW OF TRINITY COLLEGE, CAMBRIDGE;
CAVENDISH PROFESSOR OF EXPERIMENTAL PHYSICS IN THE
UNIVERSITY OF CAMBRIDGE;
PROFESSOR OF NATURAL PHILOSOPHY IN THE
ROYAL INSTITUTION, LONDON

FOURTH EDITION

CAMBRIDGE:
AT THE UNIVERSITY PRESS
1909

First   Edition 1895.
Second   Edition 1897.
Third   Edition 1904.
Fourth   Edition 1909.

PREFACE TO FIRST EDITION

IN the following work I have endeavoured to give an account of the fundamental principles of the Mathematical theory of Electricity and Magnetism and their more important applications, using only simple mathematics. With the exception of a few paragraphs no more advanced mathematical knowledge is required from the reader than an acquaintance with the Elementary principles of the Differential Calculus.

It is not at all necessary to make use of advanced analysis to establish the existence of some of the most important electromagnetic phenomena. There are always some cases which will yield to very simple mathematical treatment and yet which establish and illustrate the physical phenomena as well as the solution by the most elaborate analysis of the most general cases which could be given.

The study of these simple cases would, I think, often be of advantage even to students whose mathematical attainments are sufficient to enable them to follow the solution of the more general cases. For in these simple cases the absence of analytical difficulties allows attention to be more easily concentrated on the physical aspects of the question, and thus gives the student a more vivid idea and a more manageable grasp of the subject than he would be likely to attain if he merely regarded electrical phenomena through a cloud of analytical symbols.

I have received many valuable suggestions and much help in the preparation of this book from my friends Mr H. F. Newall of Trinity College and Mr G. F. C. Searle of Peterhouse who have been kind enough to read the proofs. I have also to thank Mr W. Hayles of the Cavendish Laboratory who has prepared many of the illustrations.

J. J. THOMSON.

CAVENDISH LABORATORY,
CAMBRIDGE.
September 3, 1895.

PREFACE TO THE SECOND EDITION

IN this Edition I have through the kindness of several correspondents been able to correct a considerable number of misprints. I have also made a few verbal alterations in the hope of making the argument clearer in places where experience has shown that students found unusual difficulties.

J. J. THOMSON.

CAVENDISH LABORATORY,
CAMBRIDGE.
November, 1897.

PREFACE TO THE THIRD EDITION

THE most important of the alterations made in this Edition is a new chapter on the properties of moving electrified bodies; many of these properties may be proved in a simple way, and the important part played by moving charges in Modern Physics seems to warrant a discussion of their properties in even an Elementary Treatise.

I have much pleasure in thanking Mr G. F. C. Searle of Peterhouse for many valuable suggestions, and for his kindness in reading the proof sheets of the first five chapters; to Mr P. V. Bevan of Trinity College I am indebted for similar assistance with the subsequent chapters.

J. J. THOMSON.

CAVENDISH LABORATORY,
CAMBRIDGE.
    October 4, 1904.

PREFACE TO THE FOURTH EDITION

IN this Edition a few additions and corrections have been made.

J. J. THOMSON.

CAVENDISH LABORATORY,
CAMBRIDGE.
    April 26, 1909.

TABLE OF CONTENTS

CHAP.   PAGES

I.   General Principles of Electrostatics .   .   .   1— 59

II.   Lines of Force .   .   .   .   60— 83

III.   Capacity of Conductors. Condensers   .   .   84—119

IV.   Specific Inductive Capacity   .   .   .   .   120—144

V.   Electrical Images and Inversion   .   .   .   145—190

VI.   Magnetism .   .   .   .   191—231

VII.   Terrestrial Magnetism   .   .   .   .   232—245

VIII.   Magnetic Induction .   .   .   .   .   246—282

IX.   Electric Currents   .   .   .   .   .   283— 328

X.   Magnetic Force due to Currents   .   .   .   329—386

XI.   Electromagnetic Induction   .   .   .   .   387—456

XII.   Electrical Units : Dimensions of Electrical
        Quantities .   .   .   .   457—479

XIII.   Dielectric Currents and the Electromagnetic
        Theory of Light .   .   .   .   .   .   480—505

XIV.   Thermoelectric Currents .   .   .   .   506—518

XV.   The Properties of Moving Electric Charges .   519—546

INDEX   .   .   .   .   547—550

ELEMENTS OF THE MATHEMATICAL THEORY OF ELECTRICITY AND MAGNETISM

CHAPTER I

GENERAL PRINCIPLES OF ELECTROSTATICS

1. Example of Electric Phenomena. Electrification. Electric Field. A stick of sealing-wax after being rubbed with a well dried piece of flannel attracts light bodies such as small pieces of paper or pith balls covered with gold leaf. If such a ball be suspended by a silk thread, it will be attracted towards the sealing-wax, and, if the silk thread is long enough, the ball will move towards the wax until it strikes against it. When it has done this, however, it immediately flies away from the wax; and the pith ball is now repelled from the wax instead of being attracted towards it as it was before the two had been in contact. The piece of flannel used to rub the sealing-wax also exhibits similar attractions for the pith balls, and these attractions are also changed into repulsions after the balls have been in contact with the flannel.

The effects we have described are called 'electric' phenomena, a title which as we shall see includes an enormous number of effects of the most varied kinds. The example we have selected, where electrical effects are produced by rubbing two dissimilar bodies against each other, is the oldest electrical experiment known to science.

The sealing-wax and the flannel are said to be *electrified,* or to be in a state of *electrification,* or to be charged with *electricity;* and the region in which the attractions and repulsions are observed is called the *electric field.*

2. Positive and Negative Electrification. If we take two pith balls A and B, coated with gold leaf and suspended by silk threads, and let them strike against the stick of sealing-wax which has been rubbed with a piece of flannel, they will be found to be repelled, not merely from the sealing-wax but also from each other. To observe this most conveniently remove the pith balls to such a distance from the sealing-wax and the flannel that the effects due to these are inappreciable. Now take another pair of similar balls, C and D, and let them strike against the flannel; C and D will be found to be repelled from each other when they are placed close together. Now take the ball A and place it near C; A and C will be found to be *attracted* towards each other. Thus, a ball which has touched the sealing-wax is repelled from another ball which has been similarly treated, but is attracted towards a ball which has been in contact with the flannel. The electricity on the balls A and B is thus of a kind different from that on the balls C and D, for while the ball A is repelled from B it is attracted towards D, while the ball C is attracted towards B and repelled from D; thus when the ball A is attracted the ball C is repelled and *vice versâ.*

The state of the ball which has touched the flannel is said to be one of *positive electrification,* or the ball is said to be *positively electrified*; the state of the ball which has touched the sealing-wax is said to be one of *negative electrification,* or the ball is said to be *negatively electrified.*

We may for the present regard 'positive' and 'negative' as conventional terms, which when applied to electric phenomena denote nothing more than the two states of electrification described above. As we proceed in the subject, however, we shall see that the choice of these terms is justified, since the properties of positive and negative electrification are, over a wide range of phenomena, contrasted like the properties of the signs *plus* and *minus* in Algebra.

The two balls *A* and *B* must be in similar states of electrification since they have been similarly treated; the two balls *C* and *D* will also for the same reason be in similar states of electrification. Now *A* and *B* are repelled from each other, as are also *C* and *D*; hence we see that *two bodies in similar states of electrification are repelled from each other*: while, since one of the pair *A, B* is attracted towards either of the pair *C, D*, we see that *two bodies, one in a positive state of electrification, the other in a negative state, are attracted towards each other.*

In whatever way a state of electrification is produced on a body, it is found to be one or other of the preceding kinds; i.e. the ball *A* is either repelled from the electrified body or attracted towards it. In the former case the electrification is negative, in the latter positive.

A method, which is sometimes convenient, of detecting whether the electrification of a body is positive or negative is to dust it with a mixture of powdered red lead and yellow sulphur which has been well shaken; the friction of the one powder against the other electrifies both powders, the sulphur becoming negatively, the red lead positively electrified. If now we dust a negatively electrified surface with this mixture, the positively electrified red lead will stick to the surface, while the negatively electrified sulphur will be easily detached, so that if we blow on the powdered surface the sulphur will come off while the red lead will remain, and thus the surface will be coloured red : if a positively electrified surface is treated in this way it will become yellow in consequence of the sulphur sticking to it.

3. **Electrification by Induction.** If the negatively electrified stick of sealing-wax used in the preceding experiments is held near to, but not touching, one end of an elongated piece of metal supported entirely on glass or ebonite stems, and if the metal is dusted over with the mixture of red lead and sulphur, it will be found, after blowing off the loose powder, that the end of the metal nearest to the sealing-wax is covered with the yellow sulphur, while the end furthest away is covered with red lead, showing that the end of the metal nearest the negatively electrified stick of sealing-wax is positively, the end remote from it negatively, electrified. In this experiment the metal, which has neither been rubbed nor been in contact with an electrified body, is said to be electrified by *induction*; the electrification on the metal is said to be *induced* by the electrification on the stick of sealing-wax. The electrification on the part of the metal nearest the wax is of the kind opposite to that on the wax, while the electrification on the more remote

parts of the metal is of the same kind as that on the wax. The electrification on the metal disappears as soon as the stick of sealing-wax is removed.

4. Electroscope. An instrument by which the presence of electrification can be detected is called an *electroscope*. All electroscopes give some indication of the amount of the electrification, but if accurate measurements are required a special form of electroscope or a more elaborate instrument, called an electrometer (Art. 60), is generally used.

A simple form of electroscope, called the gold leaf electroscope, is represented in Fig. 1. It consists of a

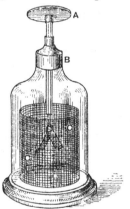

Fig. 1.

glass vessel fitting into a stand; a metal rod, with a disc of metal at the top and terminating below in two strips of gold leaf, passes through the neck of the vessel the rod passing through a glass tube covered inside and out with sealing-wax or shellac varnish and fitting tightly into a plug in the mouth of the vessel.

When the gold leaves are electrified they are repelled from each other and diverge, the amount of the divergence giving some indication of the degree of electrification. It is desirable to protect the gold leaves from the influence of electrified bodies which may happen to be near the electroscope, and from any electrification there may be on the surface of the glass. To do this we take advantage of the property of electrical action (proved in Art. 33), that a closed metallic vessel completely protects bodies inside it from the electrical action of bodies outside. Thus if the gold leaves could be completely surrounded by a metal vessel, they would be perfectly shielded from extraneous electrical influence: this however is not practicable, as the metal case would hide the gold leaves from observation. In practice, sufficient protection is afforded by a cylinder of metal gauze connected to earth, such as is shown in Fig. 1, care being taken that the top of the gauze cylinder reaches above the gold leaves.

If the disc of the electroscope is touched by an electrified body, part of the electrification will go to the gold leaves; these will be electrified in the same way, and therefore will be repelled from each other. In this case the electrification on the gold leaves is of the same sign as that on the electrified body. When the electrified body does not touch the disc but is held near to it, the metal parts of the electroscope will be electrified by induction; the disc, being the part nearest the electrified body, will have electrification opposite to that of the body, while the gold leaves, being the parts furthest from the electrified body, will have the same kind of electrification as the body, and will repel each other. This repulsion will cease as soon as the electrified body is removed.

If, when the electrified body is near the electroscope, the disc is connected to the ground by a metal wire, then the metal of the electroscope, the wire and the ground, will correspond to the elongated piece of metal in the experiment described in Art. 3. Thus, supposing the body to be negatively electrified, the positive electrification will be on the disc, while the negative will go to the most remote part of the system consisting of the metal of the electroscope, the wire and the ground, i.e. the negative electrification will go to the ground and the gold leaves will be free from electrification. They cease then to repel each other and remain closed. If the wire is removed from the disc while the electrified body remains in the neighbourhood, the gold leaves will remain closed as long as the electrified body remains stationary, but if this is removed far away from the electroscope the gold leaves diverge. The positive electrification, which, when the electrified body was close to the electroscope, concentrated itself on the disc so as to be as near the electrified body as possible, when this body is removed spreads to the gold leaves and causes them to diverge.

If, when the electroscope is charged, we wish to determine whether the charge is positive or negative, all we have to do is to bring near to the disc of the electroscope a stick of sealing-wax, which has been negatively electrified by friction with flannel; the proximity of the negatively electrified wax, in consequence of the induction (Art. 3), increases the negative electrification on the gold leaves. Hence, if the presence of the sealing-wax increases the divergence of the leaves, the original electrification was negative, but if it diminishes the divergence the original electrification was positive.

5. Charge on an electrified body. Definition of equal charges. Place on the disc of the electroscope a metal vessel as nearly closed as possible, the opening being only just wide enough to allow electrified

Fig. 2.

bodies to be placed inside. Then introduce into this vessel a charged body suspended by a silk thread, and let it sink well below the opening. The gold leaves of the electroscope will diverge, since they will be electrified by induction (see Art. 3), but the divergence will remain the same however the body is moved about in the vessel. If two or more electrified bodies are placed in the vessel the divergence of the gold leaves is the same however the electrified bodies are moved about relatively to each other or to the vessel. The divergence of the gold leaves thus measures some property of the electrified body which remains constant however the body is moved about within the vessel. This property is called the charge on the body, and two bodies, A and B, have equal charges when the divergence of the gold leaves is the same when A is inside the vessel placed on the disc of the electroscope and B far away, as when B is inside and A far away. A and B are each supposed to be suspended by dry silk threads, for such threads do not allow the electricity to escape along them; see Art. 6. Again, the charge on a body C is twice that on A if, when C is introduced into the vessel, it produces the same effect on the electroscope as that produced by A and B when introduced together. B is a body whose charge has been proved equal to that on A in the way just described. Proceeding in this way we can test what multiple the charge on any given electrified body is of the charge on another body, so that if we take the latter charge as the unit charge we can express any charge in terms of this unit.

Two bodies have equal and opposite charges if when introduced simultaneously into the metal vessel they produce no effect on the divergence of the gold leaves.

6. Insulators and Conductors. Introduce into the vessel described in the preceding experiment an electrified pith ball coated with gold leaf and suspended by a dry silk thread: this will cause the gold leaves to diverge. If now the electrified pith ball is touched with a stick of sealing-wax, an ebonite rod or a dry piece of glass tube, no effect is produced on the electroscope, the divergence of the gold leaves is the same after the pith ball has been touched as it was before. If, however, the pith ball is touched with a metal wire held in the hand or by the hand itself, the gold leaves of the electroscope immediately fall together and remain closed after the wire has been withdrawn from the ball. Thus the pith ball loses its charge when touched with a metal wire, though not when touched with a piece of sealing-wax. We may thus divide bodies into two classes, (1) those which, when placed in contact with a charged body, can discharge the electrification, these are called *conductors*; (2) those which can not discharge the electrification of a charged body with which they are in contact, these are called *insulators*. The metals, the human body, solutions of salts or acids are examples of conductors, while the air, dry silk threads, dry glass, ebonite, sulphur, paraffin wax, sealing-wax, shellac are examples of insulators.

When a body is entirely surrounded by insulators it is said to be *insulated*.

7. *When electrification is excited by friction or by any other process, equal charges of positive and negative electricity are always produced.* To show this, when the electrification is excited by friction, take a piece of sealing-wax and electrify it by friction with a piece of flannel; then, though both the wax and the flannel are charged with electricity, they will, if introduced together into the metal vessel on the disc of the electroscope (Art. 5), produce no effect on the electroscope, thus showing that the charge of negative electricity on the wax is equal to the charge of positive electricity on the flannel. This can be shown in a more striking way by working a frictional electrical machine, insulated and placed inside a large insulated metal vessel in metallic connexion with the disc of an electroscope; then, although the most vigorous electrical effects can be observed near the machine inside the vessel, the leaves of the electroscope remain unaffected.

showing that the total charge inside the vessel connected with the disc has not been altered though the machine has been in action.

To show that, when a body is electrified by induction, equal charges of positive and negative electrification are produced, take an electrified body suspended by a silk thread, lower it into the metal vessel on the top of the electroscope and observe the divergence of the gold leaves; then take a piece of metal suspended by a silk thread and lower it into the vessel near to but not in contact with the electrified body; no alteration in the divergence of the gold leaves will take place, showing that the total charge on the piece of metal introduced into the vessel is zero. This piece of metal is, however, electrified by induction, so that its charge of positive electrification excited by this process is equal to its charge of negative electrification.

Again, when two charged bodies are connected by a conductor, the sum of the charges on the bodies is unaltered, i.e. the amount of positive electrification gained by one is equal to the amount of positive electrification lost by the other. To show this, take two electrified metallic bodies, A and B, suspended from silk threads, and introduce A into the metal vessel, noting the divergence of the gold leaves; then introduce B into the vessel and observe the divergence when the two bodies are in the vessel together: now take a piece of wire wound round one end of a dry glass rod and, holding the rod by the other end, place the wire so that it is in contact with A and B simultaneously; no alteration in the divergence of the gold leaves will be produced by this process, showing that the sum of the charges on A and B is unaltered. Take away the wire and remove

B from the vessel, and now again observe the divergence of the gold leaves; it will not (except in very special cases) be the same as it was before B was put into the vessel, thus proving that, though a transference of electrification between A and B has taken place, the sum of the charges on A and B has not changed.

8. Force between bodies charged with electricity. *When two charged bodies are at a distance r apart, r being very large compared with the greatest linear dimension of either of the bodies, the repulsion between them is proportional to the product of their charges and inversely proportional to the square of the distance between them.*

This law was first proved by Coulomb by direct measurement of the force between electrified bodies; there are, however, other methods by which the law can be much more rigorously established; as these can be most conveniently considered when we have investigated the properties of this law of force, we shall begin by assuming the truth of this law and proceed to investigate some of its consequences.

9. Unit charge. We have seen in Art. 5 how the charges on electrified bodies can be compared with each other; in order, however, to express the numerical value of any charge it is necessary to have a definite unit of charge with which the charge can be compared.

The unit charge of electricity is defined to be such that when two bodies each have this charge, and are separated by unit distance in air they are repelled from each other with unit force. The dimensions of the charged bodies are assumed to be very small compared with the unit distance.

It follows from this definition and the law of force previously enunciated that the repulsion between two small bodies with charges e and e' placed in air at a distance r apart is equal to

$$\frac{ee'}{r^2}.$$

The expression ee'/r^2 will express the force between two charged bodies, whatever the signs of their electrifications, if we agree that, when the expression is positive, it indicates that the force between the bodies is a repulsion, and that, when this expression is negative, it indicates that the force is an attraction. When the charges on the bodies are of the same kind ee' is positive, the force is then repulsive; when the charges are of opposite sign ee' is negative, the force between the bodies is then attractive.

Electric Intensity. The electric intensity at any point is the force acting on a small body charged with unit positive charge when placed at the point, the electrification of the rest of the system being supposed to be undisturbed by the presence of this unit charge.

Total Normal Electric Induction over a Surface. Imagine a surface drawn anywhere in the electric field, and let this surface be completely divided up as in the figure, into a network of meshes, each mesh being so small that the electric intensity at any point in a mesh may be regarded as constant over the mesh. Take a point in each of these meshes and find the component of the electric intensity at that point in the direction of the

normal drawn from the outside of the surface at that
point, and multiply this normal component by the area

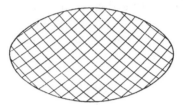

Fig. 3.

of the mesh; the sum of these products for all the meshes
on the surface is defined to be the total normal electric
induction over the surface.   This is algebraically expressed
by the relation

$$I = \Sigma N\omega,$$

where I is the total normal electric induction, N the component of the electric intensity resolved along the normal
drawn from the outside of the surface at a point in a
mesh, and ω is the area of the mesh: the symbol Σ denotes
that the sum of the products $N\omega$ is to be taken for all the
meshes drawn on the surface.

With the notation of the integral calculus

$$I = \int N dS,$$

where dS is an element of the surface, the integration
extending all over the surface.

10. Gauss's Theorem.   We can prove all the propositions about the forces between electrified bodies, which
we shall require in the following discussion of Electrostatics, by the aid of a theorem due to Gauss. This
theorem may be stated thus: *the total normal electric
induction over any closed surface drawn in the electric*

field is equal to 4π times the total charge of electricity inside the closed surface.

We shall first prove this theorem when the electric field is that due to a single charged body.

Let *O* (Fig. 4) be the charged body, whose dimensions are supposed to be so small, compared with its distances

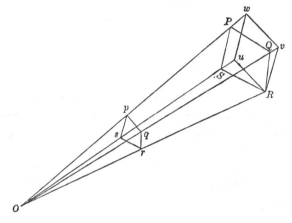

Fig. 4.

from the points at which the electric intensity is measured, that it may be regarded as a point. Let *e* be the charge on this body.

Let *PQRS* be one of the small meshes drawn on the surface, the area being so small that *PQRS* may be regarded as plane: join *O* to *P, Q, R, S,* and let a plane through *R* at right angles to *OR* cut *OS, OQ, OP* respectively in *u, v, w*: with centre *O* describe a sphere of unit radius, and let the lines *OP, OQ, OR, OS* cut the surface of this sphere in the points *p, q, r, s* respectively. The area *PQRS* is assumed to be so small that the electric intensity may be regarded as constant over it; we may take as the value of the electric intensity e/OR^2, which is the value it has at R.

The contribution of this mesh to the total normal induction is, by definition, equal to

$$\text{area } PQRS \times N,$$

where N is the normal component of the electric intensity at R.

Now $$N = \frac{e}{OR^2} \times \cos\theta,$$

where θ is the angle between the outward normal to the surface at R, and OR the direction of the electric intensity. The normal to the surface is at right angles to $PQRS$, and OR is at right angles to the area $Ruvw$, and hence the angle between the normal to the surface and OR is equal to the angles between the planes $PQRS$ and $Ruvw$.

Hence

area $PQRS \times \cos\theta = $ the area of the projection of the area $PQRS$ on the plane $Ruvw$

$$= \text{area } Ruvw \dots\dots\dots\dots\dots(1).$$

Consider the figures $Ruwv$ and $rspq$. Ru is parallel to rs since they are in the same plane and both at right angles to OR, and for similar reasons Rv is parallel to rq, vw to pq, uw to sp. The figure $Ruwv$ is thus similar to $rspq$: and the areas of similar figures are proportional to the squares of their homologous sides. Hence

$$\text{area } Ruwv : \text{area } rspq = Ru^2 : rs^2$$
$$= OR^2 : Or^2,$$

so that
$$\frac{\text{area } Ruwv}{OR^2} = \frac{\text{area } pqrs}{Or^2}$$

$$= \text{area } pqrs \ \dots\dots\dots\dots(2),$$

since Or is equal to unity by construction.

The contribution of the mesh $PQRS$ to the total normal induction is equal to

$$\text{area } PQRS \times \frac{e}{OR^2} \times \cos\theta$$

$$= e \times \frac{\text{area } Ruvw}{OR^2} \text{ by equation (1)}$$

$$= e \times \text{area } pqrs \text{ by equation (2).}$$

Thus the contribution of the mesh to the total normal induction is equal to e times the area cut off a sphere of unit radius with its centre at O by a cone having the mesh for a base and its vertex at O.

By dividing up any finite portion of the surface into meshes and taking the sum of the contributions of each mesh, we see that the total normal induction over the surface is equal to e times the area cut off a sphere of unit radius with its centre at O by a cone having the boundary of the surface as base and its vertex at O.

Let us now apply the results we have obtained to the case of a closed surface.

First take the case where O is inside the surface. The total normal induction over the surface is equal to e times the sum of the areas cut off the unit sphere by cones with their bases on the meshes and their vertices at O, and since the meshes completely fill up the closed surface the sum of the areas cut off the unit sphere by the cones will be the area of the sphere, which is equal to 4π, since its radius is unity. Thus the total normal induction over the closed surface is $4\pi e$.

Next consider the case when O is outside the closed surface.

Draw a cone with its vertex at O cutting the closed surface in the areas $PQRS$, $P'Q'R'S'$. Then the magnitude of the total normal induction over the area $PQRS$

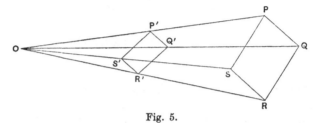

Fig. 5.

is equal to that over the area $P'Q'R'S'$, since they are each equal to e times the area cut off by this cone from a sphere whose radius is unity and centre at O. But over the surface $PQRS$ the electric intensity points along the outward drawn normal so that the sign of the component resolved along the outward drawn normal is positive; while over the surface $P'Q'R'S'$ the electric intensity is in the direction of the inward drawn normal so that the sign of its component along the outward drawn normal is negative. Thus the total normal induction over $PQRS$ is of opposite sign to that over $P'Q'R'S'$, and since they are equal in magnitude they will annul each other as far as the total normal induction is concerned. Since the whole of the closed surface can be divided up in this way by cones with their vertices at O, and since the two sections of each of these cones neutralize each other, the total normal induction over the closed surface will be zero.

We thus see that when the electric field is due to a small body with a charge e, the total normal induction over any closed surface enclosing the charge is $4\pi e$, while it is equal to zero over any closed surface not enclosing the charge. We have therefore proved Gauss's theorem when the field is due to a single small electrified body.

We can easily extend it to the general case when the field is due to any distribution of electrification. For we may regard this as arising from a number of small bodies having charges e_1, e_2, e_3... &c. Let N be the component along the outward drawn normal to the surface of the resultant electric intensity, N_1 the component in the same direction due to e_1, N_2 that due to e_2 and so on; then

$$N = N_1 + N_2 + N_3 + \ldots.$$

If ω is the area of the mesh at which the normal electric intensity is N, the total normal induction over the surface is

$$\Sigma N \omega$$
$$= \Sigma (N_1 + N_2 + N_3 + \ldots)\, \omega$$
$$= \Sigma N_1 \omega + \Sigma N_2 \omega + \Sigma N_3 \omega + \ldots,$$

that is, the total normal electric induction over the surface due to the electrical system is equal to the sum of the normal inductions due to the small charged bodies of which the system is supposed to be built up. But we have just seen that the total normal induction over a closed surface due to any one of these is equal to 4π times its charge if the body is inside the surface, and is zero if the body is outside the surface. Hence the sum of the total normal inductions due to the several charged bodies, i.e. that due to the actual field, is 4π times the charge of electricity inside the closed surface over which the normal induction is taken.

11. Electric intensity at a point outside a uniformly charged sphere. Let us now apply the theorem to find the electric intensity at any point in the region outside a sphere uniformly charged with electricity.

Let O be the centre of the sphere, P a point outside the sphere at which the electric intensity is required.

Through P draw a spherical surface with its centre at O. Let R be the electric intensity at P. Since the charged sphere is uniformly electrified, the direction of the intensity will be OP, and it will have the same value R at any point on the spherical surface through P. Hence since at each point on this surface the normal electric intensity is equal to R, the total normal induction over the sphere through P is equal to $R \times$ (surface of the sphere), i.e. $R \times 4\pi . OP^2$. By Gauss's theorem this is equal to 4π times the charge enclosed by the spherical surface, that is to 4π times the charge on the inner sphere. If e is this charge we have therefore

$$R \times 4\pi OP^2 = 4\pi e,$$

$$R = \frac{e}{OP^2}.$$

Hence the intensity at a point outside a uniformly electrified sphere is the same as if the charge on the sphere were concentrated at the centre.

12. Electric intensity at a point inside a uniformly electrified spherical shell. Let Q be a point inside the shell, R the electric intensity at that point. Through Q draw a spherical surface, centre O; then as before, the normal electric intensity will be constant all over this surface. The total normal induction over this sphere is therefore $R \times$ area of sphere, i.e.

$$R \times 4\pi . OQ^2.$$

By Gauss's theorem this is equal to 4π times the charge of electricity inside the spherical surface passing through Q; hence as there is no charge inside this surface,

$$4\pi R \times OQ^2 = 0,$$

or $\qquad\qquad R = 0.$

Hence the electric intensity vanishes at any point inside a uniformly electrified spherical shell.

13. Infinite Cylinder uniformly electrified. We shall next consider the case of an infinitely long circular cylinder uniformly electrified. Let P be a point outside the cylinder at which we wish to find the electric intensity. Through P describe a circular cylinder coaxial with the electrified one, draw two planes at right angles to the axis of the cylinder at unit distance apart, and consider the total normal induction over the closed surface formed by the curved surface of the cylinder through P and the two plane ends. Since the electrified cylinder is infinitely long and is symmetrical about its axis, the electric intensity at all points at the same distance from the axis of the cylinder will be the same, and the electric intensity at P will by symmetry be along a radius drawn through P at right angles to the axis of the cylinder.

Thus the electric intensity at any point on either plane end of the cylinder will be in the plane of that end, and will therefore have no component at right angles to it; the plane ends will therefore contribute nothing to the total normal induction over the surface. At each point of the cylindrical surface the electric intensity is at right angles to the curved surface and is equal to R. The total normal induction over the surface is therefore

$R \times$ (area of the curved surface of the cylinder).

But since the length of the curved surface is unity its area is equal to $2\pi r$, where r is the distance of P from the axis of the cylinder. If E is the charge per unit length on the electrified cylinder, then by Gauss's theorem the total normal induction over the surface is equal to $4\pi E$. The total normal induction is however equal to $R \times 2\pi r$, hence

$$R \times 2\pi r = 4\pi E,$$

$$R = \frac{2E}{r}.$$

Thus, in the case of the cylinder, the electric intensity varies inversely as the distance from the axis of the cylinder.

We can prove in the same way as for the uniformly electrified spherical shell that the electric intensity vanishes at any point inside a uniformly electrified cylindrical shell.

14. Uniformly electrified infinite plane. In this case we see by symmetry (1) that the electric intensity will be normal to the plane, (2) that the electric intensity will be constant at all points in a plane parallel to the electrified one. Draw a cylinder $PQRS$, Fig. 6, the axis of the cylinder being at right angles to the plane, the ends of the cylinder being planes at right angles to the axis. Since this cylinder encloses no electrification the total normal induction over its surface is zero by Gauss's

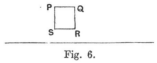

Fig. 6.

theorem. But since the electric intensity is parallel to the axis of the cylinder the normal intensity vanishes over the curved surface of the cylinder. Let F be the electric intensity at a point on the face PQ—this is along the outward drawn normal if the electrification on the plane is positive—F' the electric intensity at a point on the face RS, ω the area of either of the faces PQ or RS, then the total normal induction over the surface $PQRS$ is equal to

$$F\omega - F'\omega;$$

and since this vanishes by Gauss's theorem

$$F = F',$$

or the electric intensity at any point, due to the infinite uniformly charged plane, is independent of the distance of the point from the plane. It is, therefore, constant in magnitude at all points in the field, acting upwards in the region above the plane, downwards in the region below it.

To find the magnitude of the intensity at P. Draw through P (Fig. 7) a line at right angles to the plane and prolong it to Q, so that Q is as far below the plane as P is above it. With PQ as axis describe a right circular cylinder bounded by planes through P and Q parallel to the electrified plane. Consider now the total normal induction over the surface of this cylinder. The electric intensity is everywhere parallel to the axis of the cylinder, and has, therefore, no normal component over the curved

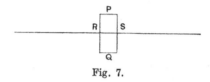

Fig. 7.

surface of the cylinder, the total normal induction over the surface thus arises entirely from the flat ends. Let R be the magnitude of the electric intensity at any point in the field, ω the area of either of the flat ends of the cylindrical surface. Then the part of the total normal induction over the surface $PQRS$ due to the flat end through P is $R\omega$. The part due to the flat end through Q will also be equal to this and will be of the same sign, since the intensity at Q is along the outward drawn normal. Thus since the normal intensity vanishes over the curved surface of $PQRS$ the total normal induction over the closed surface is $2R\omega$. If σ is the quantity of electricity per unit area of the plane the charge of electricity inside the closed surface is $\sigma\omega$; hence by Gauss's theorem

$$2R\omega = 4\pi\sigma\omega,$$

or $\hspace{3em} R = 2\pi\sigma.$

By comparing this with the results given in Arts. 11 and 13 the student may easily prove that the intensity due to the charged plane surface is half that just outside a charged spherical or cylindrical surface having the same charge of electricity per unit area.

15. Lines of Force. A line of force is a curve drawn in the electric field, such that its tangent at any point is parallel to the electric intensity at that point.

16. Electric Potential. This is defined as follows: the electric potential at a point P exceeds that at Q by the work done *by the electric field* on a body charged with unit of electricity when the latter passes from P to Q. The path by which the unit of electricity travels from P to Q is immaterial, as the work done will be the same whatever the nature of the path. To prove this suppose that the work done on the unit charge when it travels along the path PAQ is greater than when it travels along the path

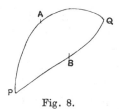

Fig. 8.

PBQ. Since the work done by the field on the unit of electricity when it goes from P to Q along the path PBQ is equal to the work which must be done by applied mechanical forces to bring the unit from Q to P along QBP, we see that if we make the unit travel round the closed curve $PAQBP$ the work done by the field on the unit when it travels along PAQ is greater than the work spent by the applied forces in bringing it back from Q to P along the path QBP. Thus though the unit of electricity is back at the point from which it started, and if the field is entirely due to charges of electricity, everything is the same as when it started, we have, if our hypothesis is correct, gained work. This is not in accordance with the principle of the Conservation of Energy, and we therefore conclude that the hypothesis on which it is founded, i.e. that the work done on unit electric charge when it travels from P to Q depends on the path by which it travels, is incorrect.

Since electric phenomena only depend upon differences of potential it is immaterial what point we take as the one at which we call the potential zero. In mathematical investigations it simplifies the expression for the potential to assume as the point of zero potential one at an infinite distance from all the electrified bodies.

If P and Q are two points so near together that the electric intensity may be regarded as constant over the distance PQ, then the work done by the field on unit charge when it travels from P to Q is $F \times PQ$, if F is the electric intensity resolved in the direction PQ. If V_P, V_Q denote the potentials at P and Q respectively, then since by definition $V_P - V_Q$ is the work done by the field on unit charge when it goes from P to Q we have

$$V_P - V_Q = F \times PQ,$$

hence
$$F = \frac{V_P - V_Q}{PQ} \dots\dots\dots\dots\dots(1),$$

thus the electric intensity in any direction is equal to the rate of diminution of the potential in that direction.

Hence if we draw a surface such that the potential is constant over the surface (a surface of this kind is called an equipotential surface) the electric intensity at any point on the surface must be along the normal. For since the potential does not vary as we move along the surface, we see by equation (1) that the component of the electric intensity tangential to the surface vanishes.

Conversely a surface over which the tangential component of the intensity is everywhere zero will be an equipotential surface, for since there is no tangential intensity no work is done when the unit charge moves along the surface from one point to another; that is, there is no difference of potential between points on the surface.

The surface of a conductor placed in an electric field must be an equipotential surface when the field is in equilibrium, for there can be no tangential electric intensity, otherwise the electricity on the surface would move along the surface and there could not be equilibrium. It is this fact that makes the conception of the potential so important in electrostatics, for the surfaces of all bodies made of metal are equipotential surfaces.

17. Potential due to a uniformly charged sphere. The potential at P is the work done by the electric field when unit charge is taken from P to an infinite distance. Let us suppose that the unit charge travels from P to an infinite distance along a straight line passing through the centre of the sphere. Let $QRST$ be a series of points

Fig. 9.

very near together along this line. If e is the charge on the sphere, O its centre, the electric intensity at Q is e/OQ^2, while that at R is e/OR^2; as Q and R are very near together these quantities are very nearly equal, and we may take

the average electric intensity between Q and R as equal
to $e/OQ.OR$, the geometric mean of the intensities at Q
and R. Hence the work done by the field as the unit
charge goes from Q to R is equal to

$$\frac{e}{OQ.OR}\,QR$$

$$=\frac{e}{OQ.OR}\,(OR-OQ)$$

$$=\frac{e}{OQ}-\frac{e}{OR}.$$

Similarly the work done by the field as the charge goes
from R to S is

$$\frac{e}{OR}-\frac{e}{OS},$$

as it goes from S to T

$$\frac{e}{OS}-\frac{e}{OT},$$

and so on. The work done by the field as the charge goes
from Q to T is the sum of these expressions, and this
sum is equal to

$$\frac{e}{OQ}-\frac{e}{OT},$$

and we see, by dividing up the distance between the points
into a number of small intervals and repeating the above
process that this expression will be true when Q and T are
a finite distance apart, and that it always represents the
work done by the field on the unit charge as long as
Q and T are two points on a radius of the sphere. The
potential at P is the work done by the field when the

unit charge goes from P to an infinite distance, and is therefore by the preceding result equal to

$$\frac{e}{OP}.$$

This is also evidently the potential at P of a charge e placed on a small body enclosing O if the dimensions of the body over which the charge is spread are infinitesimal in comparison with OP.

18.   The electric intensity vanishes at any point inside a closed equipotential surface which does not enclose any electric charge.   We shall first prove that the potential is constant throughout the volume enclosed by the surface; then it will follow by equation (1), Art. 16, that the electric intensity vanishes throughout this volume.

For if the potential is not constant it will be possible to draw a series of equipotential surfaces inside the given one; let us consider the equipotential surface for which the potential is very nearly, but not quite, the same as for the given surface. As the difference of potential between this and the outer surface is very small the two surfaces will be close together, and they cannot cut each other, for if they did, any point in their intersection would have two different potentials.

Suppose for a moment that the potential at the inner surface is greater than that at the outer.

Let P be a point on the inner surface, Q the point where the normal at P drawn outwards to the inner surface cuts the outer surface. Then, since the electric intensity from P to Q is equal to $(V_P - V_Q)/PQ$ and since by hypothesis $V_P - V_Q$ is positive, we see that the normal electric intensity over the second surface is everywhere in the direction of the outward drawn normal to the surface, and therefore that the total normal electric induction over the surface will be *positive*. Hence there must be a positive charge inside the surface, as the total normal induction over the surface is, by Gauss's theorem, proportional to the charge enclosed by the surface. Hence, as by hypothesis there is no charge inside the surface, we see that the potential over the inner surface cannot be greater than that at the outer surface. If the potential at the inner surface were less than that at the outer, then the normal electric intensity would be everywhere in the direction of the inward normal, and, as before, we can show by Gauss's theorem that this would require a negative charge inside the surface. Hence, as there is no charge either positive or negative the potential at the inner surface can neither be greater nor less than at the outer surface, and must therefore be equal to it. In this way we see that the potential at all points inside the surface must have the same value as at the surface, and since the potential is constant the electric intensity will vanish inside the surface.

19. It follows from this that if we have a closed hollow conductor there will be no electrification on its inner surface unless there are electrified bodies inside the hollow. Let Fig. 10 represent the conductor with a cavity inside it. To prove that there is no electrification at P a point on the inner surface, take any closed surface enclosing a small portion α of the inner surface near P; by Gauss's theorem the charge on α is proportional to the total normal electric induction over the surface surrounding α. The electric intensity is however zero everywhere over this surface. It is zero over the part of the closed surface which is in the material of the shell because this part of the surface is in a conductor, and when there is equilibrium the electric intensity is zero

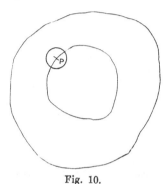

Fig. 10.

at any point in a conductor. The electric intensity is zero over the part of the closed surface which is inside the cavity because the surface of the cavity being the surface of a conductor is an equipotential surface, and as we have just seen the electric intensity inside such a surface is zero unless it encloses electric charges. Thus since the electric intensity vanishes at each point on the closed surface surrounding α, the charge at α must vanish; in this way we can see that there is no electrification at any point on the inner cavity. The electrification is all on the outer surface of the conductor.

20. Cavendish Experiment. The result proved in Art. 18 that when the force between two charged bodies varies inversely as the square of the distance between them the electric intensity vanishes throughout the interior of an electrified conductor enclosing no charge, leads to the most rigorous experimental proof of the truth of this law.

Let us for simplicity confine our attention to the case when the electrified conductor is a sphere positively electrified.

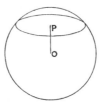

Fig. 11.

Consider the state of things at a point P inside a sphere whose centre is O, Fig. 11: through P draw a plane at right angles to OP. The electrification on the portion of the sphere above this plane produces an electric intensity in the direction PO, while the electrification on the portion of the sphere below the plane produces an electric intensity in the direction OP. When the law of force is that of the inverse square these two intensities balance each other, the greater distance from P of the electrification below the plane being compensated by the larger electrified area.

Now suppose that intensity varies as r^{-p}, then if p is greater than 2 the intensity diminishes more quickly as the distance increases than when the law of force is that of the inverse square, so that if the larger area below the plane was just sufficient to compensate for the greater distance when the law of force was that of the inverse square it will not be sufficient to do so when p is greater than 2; thus the electrification on the portion of the sphere above the plane will gain the upper hand and the resultant electric intensity will be in the direction PO. Again, if p is less than 2 the intensity will not diminish so rapidly when the distance increases, as it does when p is equal to 2, so that, if the greater area below the plane is sufficient to compensate for the increased distance when the law of force is that of the inverse square, it will be more than sufficient to do so when p is less than 2; in this case the electrification below the plane will gain the upper hand, and the electric intensity at P will be in the direction OP.

Now suppose we have two concentric metal spheres connected by a wire, and that we electrify the outer sphere positively, then if $p = 2$ there will be no electric intensity inside the outer sphere, and therefore no movement of electricity to the inner sphere which will therefore remain unelectrified. If p is greater than 2 we have seen that the electric intensity due to the positive charge on the outer sphere will be towards the centre of the sphere, i.e. the force on a negative charge will be from the inner sphere towards the outer. Negative electricity will therefore flow from the inner sphere, which will be left with a positive charge.

If however p is less than 2, the electric intensity due to the charge on the outer sphere will be from the centre of the sphere, and the direction of the force acting on a positive charge will be from the inner sphere to the outer. Positive electricity will therefore flow from the inner sphere to the outer, so that the inner sphere will be left with a negative charge.

Thus, according as p is greater than, equal to or less than 2, the charge on the inner sphere will be positive, zero or negative. By testing the state of electrification on the inner sphere we can therefore test the law of force. This is what was done by Cavendish in an experiment made by him, and which goes by his name*. The following is a description of a slight modification, due to Maxwell, of Cavendish's original experiment.

The apparatus for the experiment is represented in Fig. 12.

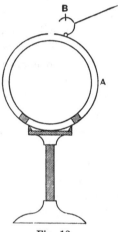

Fig. 12.

The outer sphere A, made up of two tightly fitting hemispheres, is fixed on an insulating stand, and the inner sphere is fixed concentrically with the outer one by means of an ebonite ring. Connection between the inner and outer spheres is made by a wire fastened to a small metal disc B which acts as a lid to a small hole in the outer sphere. When the wire and the disc are lifted up by a silk thread the electrical condition of the inner sphere can be tested by pushing an insulated wire connected to an electroscope (or preferably to a quadrant electrometer, see Art. 60) through the hole until it is in

---

* Mr Woodward (*Nature*, March 4, 1909) has pointed out that Priestley (*History of Electricity*, 2nd Edition, 1769, p. 711) anticipated Cavendish in this proof of the law of the inverse square.

contact with the inner sphere. The experiment is made
as follows: when the two spheres are in connection a
charge of electricity is communicated to the outer sphere,
the connection between the spheres is then broken by
lifting the disc by means of the silk thread; the outer
sphere is then discharged and kept connected to earth;
the testing wire is then introduced through the hole and
put into contact with the inner sphere. Not the slightest
effect on the electroscope can be detected, showing that
if there is any charge on the inner sphere it is too small
to affect the electroscope. To determine the sensitiveness
of the electroscope or electrometer, a small brass ball
suspended by a silk thread, is placed at a considerable
distance from the two spheres. After the outer sphere is
charged (suppose positively) the brass ball is touched and
then left insulated; in this way the ball gets by induction
a negative charge amounting to a calculable fraction, say α,
of the original charge communicated to the outer sphere.
Now when the outer sphere is connected to earth this
negative charge on the ball will induce a positive charge
on the outer sphere which is a calculable fraction, say β, of
the charge on the ball. If we disconnect the outer sphere
from the earth and discharge the ball this positive charge
on the outer sphere will be free to go to the electroscope
if this is connected to the sphere. When the ball is not
too far away from the sphere this charge is sufficient to
deflect the electroscope, i.e. a fraction $\alpha\beta$ of the original
charge on the sphere is sufficient to deflect the electroscope, showing that the charge on the inner sphere in
the Cavendish experiment could not have amounted to
$\alpha\beta$ of the charge communicated to the outer sphere*. If

* Since the electroscope is connected with the inner sphere in the
first part of the Cavendish experiment and with the outer sphere in the

the force between two charges is assumed to vary as r^{-p}, we can calculate the charge on the inner sphere and express it in terms of p, and then, knowing from the Cavendish experiment that this charge is less than $\alpha\beta$ of the original charge, we can calculate that p must differ from 2 by less than a certain quantity. In this way it has been shown that p differs from 2 by less than 1/20,000.

21. Definition of surface density. When the electrification is confined to the surface of a body, the charge per unit area is called the surface density of the electricity.

22. Coulomb's Law. The electric intensity at a point P close to the surface of a conductor surrounded by air is at right angles to the surface and is equal to $4\pi\sigma$ where σ is the surface density of the electrification.

The first part of this law follows from Art. 16, since the surface of a conductor is an equipotential surface.

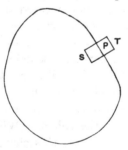

Fig. 13.

To prove the second part take on the surface a small area around P (Fig. 13) and through the boundary of this

second part, the capacity of the electroscope and its connections will not be the same in the two cases; in estimating the sensitiveness of the method a correction must be made on this account, this is easily done by the method of Art. 61.

area draw the cylinder whose generating lines are parallel
to the normal at P. Let this cylinder be truncated at T
and S by planes parallel to the tangent plane at P.

The total normal electric induction over this cylinder is
$R\omega$, where R is the normal electric intensity and ω the area
of the cross section. For $R\omega$ is the part of the total normal
induction due to the end T of the cylinder, and this is the
only part of the surface of the cylinder which contributes
anything to the total normal induction. For the intensity
along that part of the curved surface of the cylinder which
is in air is tangential to the surface and therefore has
no component along the normal, while since the electric
intensity vanishes inside the conductor the part of the
surface which is inside the conductor will not contribute
anything to the total induction. If σ is the surface density
of the electricity at P the charge inside the cylinder is
$\omega\sigma$; hence by Gauss's theorem

$$R\omega = 4\pi\omega\sigma$$

or $$R = 4\pi\sigma.$$

The result expressed by this equation is known as
Coulomb's Law. It requires modification when the conductor is not surrounded by air, but by some other insulator. See Art. 71.

23. Energy of an electrified system. If a number
of conductors are placed in an electric field, and if E_1 is
the charge on the first conductor, V_1 its potential, E_2 the
charge on the second conductor, V_2 its potential, and so
on, then we can show that the potential energy of this
system of conductors is equal to

$$\tfrac{1}{2}(E_1 V_1 + E_2 V_2 + \ldots).$$

To prove this we notice that the potentials of the

conductors will depend upon the charges of electricity on the conductors, in such a way that if the charge on every part of the system is increased m times, the potential at every point in the system will also be increased m times.

To find the energy of the system of conductors we shall suppose that each conductor is originally uncharged, and at potential zero, and that we bring a charge E_1/n from an infinite distance to the first conductor, a charge E_2/n from an infinite distance to the second conductor, a charge E_3/n to the third conductor, and so on. After this has been done, the potential of the first conductor will be V_1/n, that of the second V_2/n, and so on. Let us call this the first stage of the operation. Then bring from an infinite distance charges E_1/n to the first conductor, E_2/n to the second, and so on. When this has been done the potentials of the conductors will be $2V_1/n$, $2V_2/n$, ... Call this the second stage of the operation. Repeat this process until the first conductor has the charge E_1 and the potential V_1, the second conductor the charge E_2 and the potential V_2, and so on.

Then in the first stage the potential of the first conductor is zero at the beginning, and V_1/n at the end; the work done in bringing up to it the charge E_1/n is therefore greater than 0 but less than $\dfrac{E_1}{n} \cdot \dfrac{V_1}{n}$; similarly the work spent in bringing up the charge E_2/n to the second conductor is greater than zero but less than $\dfrac{E_2}{n} \cdot \dfrac{V_2}{n}$.

If $_1Q_1$ be the work spent in the first stage of the operations in charging the first conductor we have

$$_1Q_1 > 0, \quad _1Q_1 < \frac{1}{n^2} E_1 V_1.$$

In the second stage of the operations the potential of the first conductor is V_1/n at the beginning, and $2V_1/n$ at the end, so that the work spent in bringing up the charge E_1/n to the first conductor is greater than $\dfrac{E_1}{n}\cdot\dfrac{V_1}{n}$ but less than $\dfrac{E_1}{n}\cdot\dfrac{2V_1}{n}$; similarly the work spent in bringing up the charge E_2/n to the second conductor is greater than $\dfrac{E_2}{n}\cdot\dfrac{V_2}{n}$ but less than $\dfrac{E_2}{n}\cdot\dfrac{2V_2}{n}$. Thus if $_2Q_1$ is the work spent in this stage in charging the first conductor we have

$$_2Q_1 > \frac{1}{n^2}E_1V_1, \quad _2Q_1 < \frac{2}{n^2}E_1V_1.$$

Similarly if $_3Q_1$ is the work spent in the third stage in charging the first conductor we have

$$_3Q_1 > \frac{2}{n^2}E_1V_1, \quad _3Q_1 < \frac{3}{n^2}E,$$

and $_nQ_1$, the work spent in the last stage, is

$$> \frac{n-1}{n^2}E_1V_1,$$

and $\qquad\qquad\qquad < \dfrac{n}{n^2}E_1V_1.$

Now Q_1 the total amount of work spent in charging the first conductor is equal to $_1Q_1 + {}_2Q_1 + \ldots {}_nQ_1$, and is therefore

$$> \frac{1+2+3+\ldots(n-1)}{n^2}E_1V_1 < \frac{1+2+3+\ldots n}{n^2}E_1V_1,$$

i.e. $\qquad > \dfrac{(n-1)n}{2n^2}E_1V_1 < \dfrac{n\cdot(n+1)}{2n^2}E_1V_1,$

or $\qquad > \dfrac{1}{2}\left(1-\dfrac{1}{n}\right)E_1V_1 < \dfrac{1}{2}\left(1+\dfrac{1}{n}\right)E_1V_1.$

If however we make n exceedingly great the two limits coincide, and we see that Q_1 the total work spent in charging the first conductor is equal to $\frac{1}{2}E_1V_1$; and Q the work done in charging the whole system is given by the equation

$$Q = \frac{1}{2}(E_1V_1 + E_2V_2 + E_3V_3 + \ldots).$$

The work done in charging the conductors is stored up in the system as electrical energy, the potential energy of the system being equal to the work done in charging up the system; the energy only depends on the final state of the system and is independent of the way that state is arrived at. Hence we see from the above result that the energy of a system of conductors is one half the sum of the products obtained by multiplying the charge of each conductor by its potential.

24. Relation between the potentials and charges on the conductors. Superposition of electrical effects. Let V' be the potential at any point P when the first conductor has a charge E_1 and all the other conductors are without charge, and V'' the potential at P when the second conductor has the charge E_2 and all the other conductors are without charge; then when the first conductor has the charge E_1, the second the charge E_2, and all the other conductors are without charge, the potential at P will be $V' + V''$.

The conditions to be satisfied in this case are that the charges on the conductors should have the given values and that the surfaces of the conductors should be equipotential surfaces.

Now consider the distribution of electrification when the first conductor has the charge E_1 and the rest are without charge; this satisfies the conditions that the conductors are equipotential surfaces, that the charge on the first conductor is E_1, and that the charges on the other conductors are zero. The distribution of electrification when the second conductor is charged and the rest uncharged satisfies the conditions that the conductors are equipotential surfaces, that the charge on the first conductor is zero, that the charge on the second conductor is E_2, and that the charges on the other conductors are zero. If we take a new distribution formed by superposing the last two distributions, it will satisfy the conditions that the conductors are equipotential surfaces, that the charge on each conductor is the sum of the charges corresponding to the two solutions, i.e. that the charge on the first conductor is E_1, that on the second conductor E_2, and that on each of the other conductors zero. In other words, the new distribution will be that which occurs in the case when the first conductor has the charge E_1, the second the charge E_2, while the rest of the conductors are uncharged. But when two systems of electrification are superposed, the potential at P is the sum of the potentials due to the two systems separately, i.e. the potential at P is $V' + V''$, and hence the theorem is true.

25. We can extend this reasoning to the general case in which V' is the potential at P when the first conductor has the charge E_1, the other conductors being uncharged, V'' the potential at P when the second conductor has the charge E_2, the other conductors being uncharged, V''' the potential at P when the charge on the third conductor is

E_3, the other conductors being uncharged, and so on; and we then see that when the first conductor has the charge E_1, the second the charge E_2, the third the charge E_3, and so on, the potential at P is

$$V' + V'' + V''' + \dots.$$

26. When the first conductor has the charge E_1, the other conductors being uncharged and insulated, the potentials of the conductors will be proportional to E_1, that is, the potentials of the first, second, third, &c. conductors will be respectively

$$p_{11}E_1, \quad p_{12}E_1, \quad p_{13}E_1 \dots,$$

where p_{11}, p_{12}, p_{13} are quantities which do not depend upon the charges of the conductors or their potentials, but only upon their shapes and sizes and their positions with reference to each other. The quantities p_{11}, p_{12}, p_{13}, &c. are called *coefficients of potential*; their properties are further considered in Arts. 27—31. When the second conductor has the charge E_2, the other conductors being uncharged and insulated, the potentials of the conductors will be proportional to E_2, and the potentials of the first, second, third, &c. conductors will be

$$p_{21}E_2, \quad p_{22}E_2, \quad p_{23}E_2, \dots.$$

When the third conductor has the charge E_3, the other conductors being uncharged and insulated, the potentials of the first, second, third conductors will be

$$p_{31}E_3, \quad p_{32}E_3, \quad p_{33}E_3.$$

Hence by Art. 25, we see that when the first conductor has the charge E_1, the second the charge E_2, the third the charge E_3, and so on, V_1 the potential of the first conductor will be given by the equation

$$V_1 = p_{11}E_1 + p_{21}E_2 + p_{31}E_3 + \ldots,$$

V_2 the potential of the second conductor by the equation

$$V_2 = p_{12}E_1 + p_{22}E_2 + p_{32}E_3 + \ldots;$$

if V_3 is the potential of the third conductor

$$V_3 = p_{13}E_1 + p_{23}E_2 + p_{33}E_3 + \ldots,$$

$$\ldots\ldots\ldots\ldots\ldots\ldots\ldots\ldots\ldots$$

If we solve these equations we get

$$E_1 = q_{11}V_1 + q_{21}V_2 + q_{31}V_3 + \ldots,$$

$$E_2 = q_{12}V_1 + q_{22}V_2 + q_{32}V_3 + \ldots,$$

where the q's are functions of the p's and only depend upon the configuration of the system of conductors. The q's are called *coefficients of capacity* when the two suffixes are the same and *coefficients of induction* when the suffixes are different.

27. We shall now show that the coefficients which occur in these equations are not all independent, but that

$$p_{21} = p_{12}.$$

To prove this let us suppose that only the first and second conductors have any charges, the others being without charge and insulated. Then we may imagine the system charged, by first bringing up the charge E_1 from an infinite distance to the first conductor and leaving all the other conductors uncharged, and then when this has been done, bringing up the charge E_2 from an infinite distance to the second conductor. The work done in bringing the charge E_1 up to the first conductor will be the energy of the system, when the first conductor has the charge E_1 and the other conductors are without charge; the potential of the first conductor is in this case $p_{11}E_1$, so that by Art. 23 the work done is $\frac{1}{2}E_1 . p_{11}E_1$ or $\frac{1}{2}p_{11}E_1^2$. To find the work done in bringing up the charge E_2 to the second conductor let us suppose that this charge is brought up in instalments each equal to E_2/n. Then the potential of the second conductor before the first instalment is brought up is, by Art. 26, equal to $p_{12}E_1$, and after the first instalment has arrived it is $p_{12}E_1 + p_{22}\dfrac{E_2}{n}$.

Hence the work done in bringing up the first instalment will be between

$$p_{12}E_1 \frac{E_2}{n} \text{ and } \left(p_{12}E_1 + p_{22}\frac{E_2}{n}\right)\frac{E_2}{n}.$$

Similarly the work done in bringing up the second instalment E_2/n will be between

$$\left(p_{12}E_1 + p_{22}\frac{E_2}{n}\right)\frac{E_2}{n} \text{ and } \left(p_{12}E_1 + p_{22}\frac{2E_2}{n}\right)\frac{E_2}{n},$$

and the work done in bringing up the last instalment of the charge will be between

$$\left(p_{12}E_1 + p_{22}\frac{(n-1)E_2}{n}\right)\frac{E_2}{n} \text{ and } \left(p_{12}E_1 + p_{22}\frac{nE_2}{n}\right)\frac{E_2}{n}.$$

Thus the total amount of work done in bringing up the charge E_2 will be between

$$p_{12}E_1E_2 + \frac{1+2+3 \ldots +n-1}{n^2}p_{22}E_2^2$$

and
$$p_{12}E_1E_2 + \frac{1+2+3 \ldots +n}{n^2}p_{22}E_2^2,$$

that is, between

$$p_{12}E_1E_2 + \frac{1}{2}\left(1 - \frac{1}{n}\right)p_{22}E_2{}^2 \text{ and } p_{12}E_1E_2 + \frac{1}{2}\left(1 + \frac{1}{n}\right)p_{22}E_2{}^2,$$

but if n is very great these two expressions become equal to $p_{12}E_1E_2 + \frac{1}{2}p_{22}E_2{}^2$, which is therefore the work done in bringing up the charge E_2 to the second conductor when the first conductor has already received the charge E_1. Hence the work done in bringing up first the charge E_1 and then E_2 is

$$\tfrac{1}{2}p_{11}E_1{}^2 + p_{12}E_1E_2 + \tfrac{1}{2}p_{22}E_2{}^2.$$

It follows in the same way that the work done when the charge E_2 is first brought to the second conductor and then the charge E_1 to the first is

$$\tfrac{1}{2}p_{22}E_2{}^2 + p_{21}E_1E_2 + \tfrac{1}{2}p_{11}E_1{}^2,$$

but since the final state is the same in the two cases, the work required to charge the conductors must be the same; hence

$$\tfrac{1}{2}p_{11}E_1{}^2 + p_{12}E_1E_2 + \tfrac{1}{2}p_{11}E_1{}^2 = \tfrac{1}{2}p_{22}E_2{}^2 + p_{21}E_1E_2 + \tfrac{1}{2}p_{11}E_1{}^2,$$

i.e. $$p_{21} = p_{12}.$$

It follows from the way in which the q's can be expressed in terms of the p's, that $q_{21} = q_{12}$.

28. Now p_{12} is the potential of the second conductor when unit charge is given to the first, the other conductors being insulated and without charge, and p_{21} is the potential of the first conductor when unit charge is given to the second. But we have just seen that $p_{21} = p_{12}$, hence the potential of the second conductor when insulated and without charge due to unit charge on the first is equal to the potential of the first when insulated and without charge due to unit charge on the second, the remaining conductors being in each case insulated and without charge.

29. Let us consider some examples of this theorem. Let us suppose that the first conductor is a sphere with its centre at O, and that the second conductor is very small and placed at P, then if P is outside the sphere we know by Art. 17 that if unit charge is given to the sphere the potential at P is increased by $1/OP$. It follows from the preceding article that if unit charge be placed at P the potential of the sphere when insulated is increased by $1/OP$.

If P is inside the sphere then when unit charge is given to the sphere the potential at P is increased by $1/a$ where a is the radius of the sphere. Hence if the sphere is insulated and a unit charge placed at P the potential of the sphere is increased by $1/a$. Thus the increase in the potential of the sphere is independent of the position of P as long as it is inside the sphere.

Since the potential inside any closed conductor which does not include any charged bodies is constant, by Art. 18, we see by taking as our first conductor a closed surface, and as our second conductor a small body placed at a point P anywhere inside this surface, that since the potential at P due to unit charge on the conductor is independent of the position of P, the potential of the conductor when insulated due to a charge at P is independent of the position of P. Thus however a charged body is moved about inside a closed insulated conductor the potential of the conductor will remain constant. An example of this is afforded by the experiment described in Art. 5; the deflection of the electroscope is independent of the position of the charged bodies inside the insulated closed conductor.

30. Again, take the case when the first conductor is charged, the others insulated and uncharged; then

$$V_1 = p_{11}E_1,$$
$$V_2 = p_{12}E_1,$$

so that
$$\frac{V_1}{V_2} = \frac{p_{11}}{p_{12}}.$$

Now suppose that the first conductor is connected to earth while a charge E_2 is given to the second conductor, all the other conductors being uncharged; then since $V_1 = 0$ we have

$$0 = p_{11}E_1 + p_{12}E_2,$$

or
$$\frac{E_2}{E_1} = -\frac{p_{11}}{p_{12}} = -\frac{V_1}{V_2}$$

by the preceding equation.

Hence if a charge be given to the first conductor, all the others being insulated, the ratio of the potential of the second conductor to that of the first will be equal in magnitude but opposite in sign to the charge induced on the first conductor, when connected to earth, by unit charge on the second conductor.

As an example of this result, suppose that the first conductor is a sphere with its centre at O, and that the second conductor is a small body at a point P outside the sphere; then if unit charge be given to the sphere, the potential of the body at P is a/OP times the potential of the sphere, where a is the radius of the sphere; hence, by the theorem of this article, when unit charge is placed at P, and the sphere is connected to the earth, there will be a negative charge on the sphere equal to a/OP.

Another example of this result is when the first

conductor completely surrounds the second; then since the potential inside the first conductor is constant when all the conductors inside are free from charge, the potential of the second conductor when a charge is given to the first conductor will be the same as that of the first. Hence from the above result it follows that when the first conductor is connected to earth, and a charge given to the second, the charge induced on the first conductor will be equal and opposite to that given to the second.

Another consequence of this result is that if S be an equipotential surface when the first conductor is charged, all the others being insulated, then if the first conductor be connected to earth the charge induced on it by a charge on a small body P remains the same however P may be moved about, provided that P always keeps on the surface S.

31. As an example in the calculation of coefficients of capacity and induction, we shall take the case when the conductors are two concentric spherical shells. Let a be the radius of the inner shell, which we shall call the first conductor, b the radius of the outer shell, which we shall call the second conductor. Let E_1, E_2 be the charges of electricity on the inner and outer shells respectively, V_1, V_2 the corresponding potentials of these shells.

Then if there were no charge on the outer shell the charge E_1 on the inner would produce a potential E_1/a on its own surface, and a potential E_1/b on the surface of the outer shell; hence, Art. 26,

$$p_{11} = \frac{1}{a}; \quad p_{12} = \frac{1}{b}.$$

The charge E_2 on the outer shell would, if there were no charge on the inner shell, make the potential inside the outer shell constant and equal to the potential at the surface of the outer shell. This potential is equal to E_2/b, so that the potential of the first conductor due to the charge E_2 on the second is E_2/b, which is also equal to the potential of the second conductor due to the charge E_2; hence, by Art. 26,

$$p_{21} = \frac{1}{b}, \ p_{22} = \frac{1}{b}.$$

We have therefore

$$V_1 = p_{11}E_1 + p_{21}E_2 = \frac{E_1}{a} + \frac{E_2}{b},$$

$$V_2 = p_{12}E_1 + p_{22}E_2 = \frac{E_1}{b} + \frac{E_2}{b}.$$

Solving these equations, we get

$$E_1 = \frac{ab}{b-a} V_1 - \frac{ab}{b-a} V_2,$$

$$E_2 = -\frac{ab}{b-a} V_1 + \frac{b^2}{b-a} V_2.$$

Hence

$$q_{11} = \frac{ab}{b-a}, \quad q_{12} = q_{21} = -\frac{ab}{b-a}, \quad q_{22} = \frac{b^2}{b-a}.$$

We notice that q_{12} is negative; this, as we shall prove later, is always true whatever the shape and position of the two conductors.

32. Another case we shall consider is that of two spheres the distance between whose centres is very large compared with the radius of either. Let a be the radius of the first sphere, b that of the second, R the distance between their centres, E_1, E_2 the charges, V_1, V_2 the potentials of the two spheres. Then if there were no charge on the second sphere, the potential at the surface of the first sphere would, if the distance between the spheres were very great, be approximately E_1/a, while the potential of the second sphere would be approximately E_1/R; hence

$$p_{11} = \frac{1}{a}, \quad p_{12} = \frac{1}{R},$$

approximately.

Similarly, if there were no charge on the first sphere, but a charge E_2 on the second, the potential of the first sphere would be E_2/R, that of the second E_2/b, approximately; hence we have approximately

$$p_{21} = \frac{1}{R}, \quad p_{22} = \frac{1}{b}.$$

So that approximately

$$V_1 = \frac{E_1}{a} + \frac{E_2}{R},$$

$$V_2 = \frac{E_1}{R} + \frac{E_2}{b}.$$

Solving these equations we get

$$E_1 = \frac{aR^2}{R^2 - ab} V_1 - \frac{abR}{R^2 - ab} V_2,$$

$$E_2 = -\frac{abR}{R^2 - ab} V_1 + \frac{bR^2}{R^2 - ab} V_2,$$

hence when R is large compared with either a or b

$$q_{11} = \frac{aR^2}{R^2 - ab}, \quad q_{12} = q_{21} = -\frac{abR}{R^2 - ab}, \quad q_{22} = \frac{bR^2}{R^2 - ab},$$

approximately.

We see that as before q_{12} is negative. We also notice that q_{11} and q_{12} become larger the nearer the spheres are together.

33. Electric Screens. As an example of the use of coefficients of capacity we shall consider the case of three conductors, A, B, C, and shall suppose that the first of these conductors A is, as in Fig. 14, inside the third

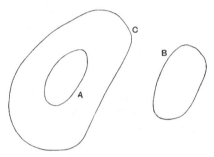

Fig. 14.

conductor C, which is supposed to be a closed surface, while the second conductor B is outside C. Then if E_1, V_1; E_2, V_2; E_3, V_3 denote the charges and potentials of the conductors A, B, C respectively, q_{11}, q_{22},... q_{12}... the coefficients of capacity and induction, we have

$$E_1 = q_{11}V_1 + q_{12}V_2 + q_{13}V_3 \dots\dots\dots\dots(1).$$
$$E_2 = q_{12}V_1 + q_{22}V_2 + q_{23}V_3 \dots\dots\dots\dots(2).$$
$$E_3 = q_{13}V_1 + q_{23}V_2 + q_{33}V_3 \dots\dots\dots\dots(3).$$

Now let us suppose that the conductor C is connected to earth so that V_3 is zero; then, since the potential inside a closed conductor is constant if it contains no charge, we see that if E_1 is zero, V_1 must vanish whatever may be the value of V_2. Hence it follows from equation (1) that q_{12} must vanish; putting q_{12} and V_3 both zero we see from (1) that

$$E_1 = q_{11} V_1,$$

and from (2) $$E_2 = q_{22} V_2.$$

Thus, in this case, the charge on A if its potential is given, or the potential if its charge is given, is entirely independent of E_2 and V_2, that is a charge on B produces no electrical effect on A, while a charge on A produces no electrical effect at B. Thus the interaction between A and B is entirely cut off by the interposition of the closed conductor at potential zero.

C is called an electric screen since it screens off from A all the effects that might be produced by B. This property of a closed metallic surface at zero potential has very important applications, as it enables us by surrounding our instruments by a metal covering connected with earth to get rid entirely of any electrical effects arising from charged bodies not under our control. Thus, in the experiment described in Art. 4, the gold leaves of the electroscope were protected from the action of external electrified bodies by enclosing them in a surface made of wire-gauze and connected with the earth.

34. Expression for the change in the energy of the system.   The energy of the system Q is, by Art. 23, equal to $\frac{1}{2}\Sigma EV$; hence we have, by Art. 27

$$Q = \tfrac{1}{2}p_{11}E_1^2 + \tfrac{1}{2}p_{22}E_2^2 + \dots p_{12}E_1E_2 + \dots.$$

If the charges are increased to E_1', E_2' &c. the energy Q' corresponding to these charges is given by the equation

$$Q' = \tfrac{1}{2}p_{11}E_1'^2 + \tfrac{1}{2}p_{22}E_2'^2 + \dots p_{12}E_1'E_2' + \dots.$$

The work done in increasing the charges is equal to $Q' - Q$. By the preceding equations

$$Q' - Q = (E_1' - E_1)\tfrac{1}{2}\{p_{11}(E_1 + E_1') + p_{12}(E_2 + E_2') + ...\}$$
$$+ (E_2' - E_2)\tfrac{1}{2}\{p_{12}(E_1 + E_1') + p_{22}(E_2 + E_2') + ...\}$$
$$+ ......$$
$$= (E_1' - E_1)\tfrac{1}{2}(V_1' + V_1) + (E_2' - E_2)\tfrac{1}{2}(V_2' + V_2) + ...$$

where V_1', V_2' ... are the potentials of the first, second, ... conductors when their charges are E_1', E_2'....

Thus the work required to increase the charges is equal to the sum of the products of the increase in the charge on each conductor into the mean of the potentials of the conductor before and after the charges are increased.

If we express Q and Q' by Art. 26 in terms of the potentials instead of the charge, we have

$$Q = \tfrac{1}{2}q_{11}V_1^2 + \tfrac{1}{2}q_{22}V_2^2 + q_{12}V_1V_2 + ...,$$
$$Q' = \tfrac{1}{2}q_{11}V_1'^2 + \tfrac{1}{2}q_{22}V_2'^2 + q_{12}V_1'V_2' + ...,$$

and we see that

$$Q' - Q = (V_1' - V_1)\tfrac{1}{2}(E_1 + E_1') + ...$$

So that the work required is equal to the sum of the products of the increase of potential of each conductor into the mean of the initial and final charges of that conductor.

35. Force tending to produce any displacement of the system. When the conductors are not connected with any external source of energy, i.e. when they are insulated, then by the principle of the Conservation of Energy, the work done by the system during any displacement will be equal to the electrical energy lost by the system in consequence of the displacement; and in this case the conductors will tend to move so as to make the electric energy diminish.

When, however, the potentials of the conductors are kept constant, as may be done by connecting them with galvanic batteries, we shall show that the system moves so that the electric energy increases. There is thus not merely work done by the system when it is displaced, but along with this expenditure of work there is an increase in the electric energy, and the batteries to which the conductors are attached are drained of a quantity of energy equal to the sum of the mechanical work done and the increase in the electric energy.

36. We shall now prove that if any small displacement of the system takes place the diminution in the electrical energy, when the charges are kept constant, is equal to the increase in the potential energy when the same displacement takes place and the potentials are kept constant.

Let E_1, V_1, E_2, V_2,... be the charges and potentials of the conductors before the displacement takes place,

E_1, V_1', E_2, V_2', ... the charges and potentials of the conductors after the displacement has taken place when the charges are constant,

E_1', V_1, E_2', V_2, ... the charges and potentials of the conductors after the displacement when the potentials are constant.

Then since the electric energy is one half the sum of the product of the charges and the potentials, the loss in electric energy by the displacement when the charges are constant is

$$\tfrac{1}{2}\left\{E_1(V_1 - V_1') + E_2(V_2 - V_2') + \ldots\right\}.$$

The gain in electric energy when the potentials are constant is

$$\tfrac{1}{2}\{V_1(E_1' - E_1) + V_2(E_2' - E_2) + ...\}.$$

The difference between the loss when the charges are constant and the gain when the potentials are constant is thus equal to

$$\tfrac{1}{2}\{(E_1 - E_1')(V_1 - V_1') + ...\} + \tfrac{1}{2}\{(E_1 V_1 - E_1' V_1') + ...\}.$$

Now for the displaced positions of the system E_1, V_1', E_2, V_2' ... are one set of corresponding values of the charges and the potentials, while E_1', V_1, E_2', V_2 ... are another set of corresponding values. Hence if p_{11}', p_{12}', ... denote the values of the coefficients of induction for the displaced position of the system

$$V_1 = p_{11}' E_1' + p_{12}' E_2' + ...$$
$$V_2 = p_{12}' E_1' + p_{22}' E_2' + ...$$
$$..............................$$

and
$$V_1' = p_{11}' E_1 + p_{12}' E_2 + ...$$
$$V_2' = p_{12}' E_1 + p_{22}' E_2 + ...$$
$$..............................$$

Thus

$$E_1 V_1 + E_2 V_2 + ... = p_{11}' E_1 E_1'$$
$$+ p_{22}' E_2 E_2' + ... p_{12}'(E_1 E_2' + E_1' E_2) + ...$$

and
$$E_1' V_1' + E_2' V_2' + ... = p_{11}' E_1 E_1'$$
$$+ p_{22}' E_2 E_2' + ... + p_{12}'(E_1 E_2' + E_1' E_2) + ...,$$

hence $\quad E_1 V_1 + E_2 V_2 + ... - (E_1' V_1' + ...) = 0.$

Thus the difference between the loss in electric energy when the charges are kept constant and the gain when the potentials are kept constant is equal to

$$\tfrac{1}{2}\{(E_1 - E_1')(V_1 - V_1') + ...\}.$$

Now when the displacements are very small $E - E'$ and $V - V'$ will each be proportional to the first power of the displacements, and hence the preceding expression is proportional to the square of the displacements, and may be neglected when the displacements are very small. Hence we see that the loss in electric energy for any small displacement when the charges are kept constant, is equal to the gain in potential energy for the same displacement, when the potentials are kept constant. When the potentials are kept constant, the batteries which maintain the potentials of the conductors at their constant value, will be called upon to furnish twice the amount of mechanical work done by the electric forces. For they will have to furnish energy equal to the sum of the mechanical work done and the increase in the electric energy of the system ; the latter is, as we have just seen, equal to the decrease in the electric energy of the system while the charges are kept constant, and this is equal by the principle of the Conservation of Energy to the mechanical work done.

37. Mechanical Force on each unit of area of a charged conductor. The electric intensity is at right angles to the surface of the conductor, so that the force on any small portion of the surface surrounding a point P will be along the normal to the surface at P.

To find the magnitude of this force let us consider a small electrified area round P. Then the electric intensity in the neighbourhood of P may conveniently be regarded as arising from two causes, (1) the electrification on the small area round P, and (2) the electrification on the rest of the surface of the conductor and on any other surfaces there may be in the electric field. To find the force on

the small area we must find the value of the second part
of the electric intensity, for the electric intensity due to
the electrification on the small area will evidently not
have any tendency to move this area one way or another.

Let R be the total electric intensity along the outward drawn normal just outside the surface at P, R_1 that
part of it due to the electrification on the small area round
P, R_2 the part due to the electrification of the rest of the
system.   Then $R = R_1 + R_2$.

Compare now the electric intensities at two points Q,
S (Fig. 15) close together and near to P, but so placed

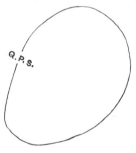

Fig. 15.

that Q is just outside and S just inside the surface of
which the small area forms a part.   Then the part of the
electric intensity at S in the direction of the outward
normal at P, which is due to the electrification on the
conductors other than the small area, will be equal to R_2
its value at Q since these points are close together.   The
part of the electric intensity due to the small area will
have at S the same magnitude as at Q, but will be in the
opposite direction, since Q is on one side of the small area,
while S is on the other.   Thus the electric intensity at S

due to this area in the direction of the outward drawn
normal will be $-R_1$, that due to the rest of the electrification R_2. The total intensity at S will therefore be
$-R_1 + R_2$. But this must be zero, since the intensity
inside a closed equipotential surface enclosing no charge
is zero. Thus $R_2 = R_1$, and therefore since

$$R = R_1 + R_2,$$
$$R_2 = \tfrac{1}{2}R.$$

Now the force on the area ω in the direction of the
outward normal is $R_2\omega\sigma$ if σ is the surface density at P;
thus if F is the mechanical force per unit area in the
direction of the outward normal

$$F\omega = R_2\omega\sigma = \tfrac{1}{2}R\omega\sigma,$$
or $$F = \tfrac{1}{2}R\sigma \quad \dots\dots\dots\dots\dots\dots(1).$$

Since by Coulomb's Law, Art. 22,

$$R = 4\pi\sigma,$$

we have the following expressions for the force per unit
area

$$F = \frac{R^2}{8\pi} \quad \dots\dots\dots\dots\dots\dots(2),$$
$$F = 2\pi\sigma^2 \quad \dots\dots\dots\dots\dots(3).$$

Since Coulomb's Law requires modification when the
medium surrounding the conductor is not air, the expressions (2) and (3) are only true for air: the equation (1) is
always true whatever be the insulator surrounding the
conductor.

When the electric intensity at the surface of a conductor exceeds a certain value the air ceases to insulate
and the electrification of the conductor is discharged.
The value of the electric intensity when the electrification

begins to escape from the conductor, depends upon a great number of circumstances, such as the pressure of the air and the proximity of other conductors. When the pressure of the air is about 760 mm. of mercury and the temperature about $15°$ C., the greatest value of R is about 100, unless the conductor is within a fraction of a millimetre of other conductors; hence the greatest value of F in dynes per square centimetre is

$$10^4/8\pi.$$

The pressure of the atmosphere is about 10^6 dynes per square centimetre, hence the greatest tension along the normal to an electrified surface in air is about $1/800\pi$ of the atmospheric pressure. That is, a pressure due to about ·3 of a millimetre of mercury would equal in magnitude the greatest tension on a conductor placed in air at ordinary pressure.

CHAPTER II

38. Expression of the properties of the Electric Field in terms of Faraday Tubes. The results we have hitherto obtained only depend upon the fact that two charged bodies are attracted towards or repelled from each other with a force varying inversely as the square of the distance between them; we have made no assumption as to how this force is produced, whether, for example, it is due to the action at a distance of the charged bodies upon each other or to some action taking place in the medium between the bodies.

Great advances have been made in our knowledge of electricity through the introduction by Faraday of the view that electrical effects are due to the medium between the charged bodies being in a special state, and do not arise from any action at a distance exerted by one charged body on another.

We shall now proceed to consider Faraday's method of regarding the electric field—a method which enables us to form a vivid mental picture of the processes going on in such a field, and to connect together with great ease many of the most important theorems in Electrostatics.

We have seen in Art. 15 that a line of force is a curve
such that its tangent at any point is in the direction of
the electric intensity at that point.   As these lines of
force are fundamental in the method employed in this
and subsequent chapters for considering the properties
of the electric field, we give below some carefully drawn
diagrams of the lines of force in some typical cases.

Figure 16 represents the lines of force due to two
equal and opposite charges.   In this case all the lines of
force start from the positive charge and end on the

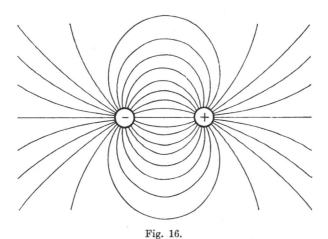

Fig. 16.

negative. Fig. 17 represents the lines of force due to
two equal positive charges; in this case the lines of
force do not pass between the charged bodies, but lines
start from each of the bodies and travel off to an infinite
distance.

Figure 18 represents the lines of force due to a
positive charge equal to 4 at A, and a negative charge

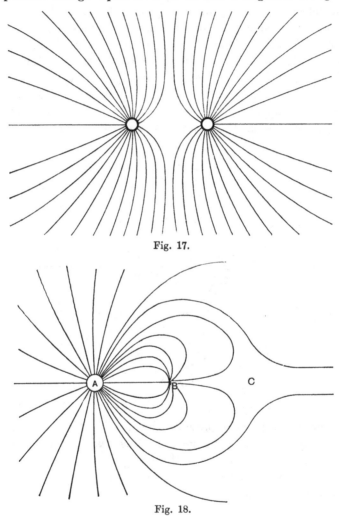

Fig. 17.

Fig. 18.

equal to −1 at B. In this case all the lines of force which fall on B start from A, but since the charge at A is numerically greater than that at B, lines of force will start from A which do not fall on B but travel off to an infinite distance.

The lines of force which pass between A and B are separated from those which proceed from A and go off to an infinite distance by the line of force which passes through C, the point of equilibrium, where

$$AC = 2AB.$$

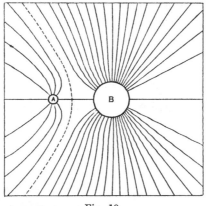

Fig. 19.

Figure 19 represents the lines of force due to a charge 1 at A and 4 at B.

Figure 20 represents the lines of force due to a charged conductor formed by two spheres intersecting at right angles. The electric intensity vanishes along the intersection of the spheres.

Figure 21 represents the lines of force between two finite parallel places; between the plates but away from

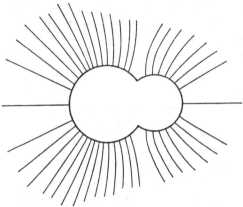

Fig. 20.

the edges of the plates the lines of force are straight lines at right angles to the planes, but nearer the

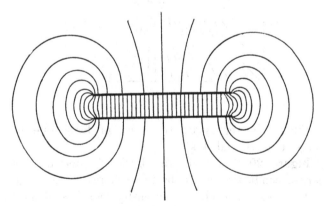

Fig. 21.

edges of the plates they curve out; some lines also pass from the back of one plate to the back of the other.

39. Tubes of Force. If we take any small closed curve in the electric field and draw the lines of force, which pass through each point of the curve, these lines will form a tubular surface which is called a tube of force. These tubes possess the property that the electric intensities at any two points on a tube are inversely proportional to the cross sections of the tube, made by planes cutting the tube at right angles at these points, provided that the cross sections are so small that the electric intensity may be regarded as constant over each section. For let Fig. 22 represent a closed surface formed

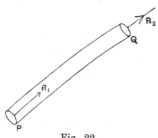

Fig. 22.

by the tube and its normal sections. Let ω_1 be the area of the cross section of the tube at P, ω_2 its cross section at Q; R_1, R_2 the electric intensities at P and Q respectively. Now consider the total normal electric induction over the surface. The only parts of the surface which contribute anything to this are the flat ends, as the sides of the tube are by hypothesis parallel to the electric intensity, so that this has no normal component over the sides. Thus the total normal induction over the closed surface PQ is equal to

$$R_2\omega_2 - R_1\omega_1,$$

the minus sign being given to the second term because, as drawn in the figure, the electric intensity at P is in the direction of the inward-drawn normal. Now, by Gauss's theorem, the total normal electric induction over any closed surface is equal to 4π times the charge inside the surface; hence if the surface does not include any charge, we have

$$R_2\omega_2 - R_1\omega_1 = 0,$$

or the electric intensity at P is to that·at Q inversely as the cross section of the tube of force at P is to that at Q.

The tubes of force will start from positive electrification and go on until they end on a negative electrified body. If the points P and Q are on the surfaces of positively and negatively electrified conductors, then if σ_P is the surface density at P, σ_Q that at Q,

$$R_1 = 4\pi\sigma_P, \quad R_2 = -4\pi\sigma_Q;$$

thus the equation

$$R_2\omega_2 - R_1\omega_1 = 0,$$

is equivalent to $\sigma_Q\omega_2 = -\sigma_P\omega_1.$

Now $\sigma_P\omega_1$ is the charge enclosed by the tube where it leaves the positively electrified conductor, and $-\sigma_Q\omega_2$ the charge enclosed by the tube where it arrives at the negatively electrified conductor, hence we see that the positive charge at the beginning of the tube is equal in magnitude to the negative charge at the end. We may draw these tubes so that they each enclose one unit of electricity at their origin, each of these tubes will therefore include unit negative charge at its end. Such tubes are sometimes called unit tubes of force, we shall for brevity, call them Faraday tubes. Each unit of positive charge will be the origin, each unit of negative charge the end of a Faraday tube. The total charge on a conductor will be the excess of the number of tubes which leave the conductor over the number which arrive at the conductor.

Since the Faraday tubes run in the direction of the electric intensity in air, they begin at places of high and travel to places of low potential. No Faraday tube can have its ends at the same potential, that is no Faraday tube can pass from one surface to another if the two surfaces are at the same potential.

40. The electric intensity at any point in the field is proportional to the number of Faraday tubes which pass through unit area of a plane drawn at right angles to the direction of the electric intensity at the point or, what is the same thing, through unit area of the equipotential surface passing through the point.

For let A be a small area drawn at right angles to the electric intensity, and let the tubes which pass through this area be prolonged until they arrive at the positively electrified surface from which they start; let B be the portion of this surface over which these tubes are spread, R the electric intensity at any point on B, ω' the area at B. Let F be the electric intensity, and ω the area enclosed by the tubes, at A. Then applying Gauss's theorem (Art. 10) to the tubular surface formed by the prolongations backwards of the tubes through A, we get

$$F\omega - R\omega' = 0.$$

But as σ is the surface density of the electrification at B, we have, by Coulomb's law (Art. 22), when the medium surrounding B is air,

$$R = 4\pi\sigma,$$

and hence $\qquad F\omega = 4\pi\sigma\omega'.$

But since $\sigma\omega'$ is the charge of electricity on B, it is equal to N the number of Faraday tubes which start from B, and which pass through A, hence

$$F\omega = 4\pi N,$$

or if ω is unity $\qquad F = 4\pi N.$

Thus the electric intensity at any point in air is 4π times the number of Faraday tubes passing through unit area of a plane drawn through the point at right angles to the electric intensity.

41. The properties of the Faraday tubes enable us to prove with ease many important theorems relating to the electric field.

Thus, for example, we see that on the conductor at the highest potential in the field the electrification must be entirely positive. Any negative electrification would imply that Faraday tubes arrived at the conductor; these tubes must however arrive at a place which is at a lower potential than the place from which they start. Thus, if the potential of the conductor we are considering is the highest in the field it is impossible for a Faraday tube to arrive at it, for this would imply that there was some other conductor at a still higher potential from which the tube could start.

Similar reasoning shews that the electrification on the conductor or conductors at the lowest potential in the field must be entirely negative.

When one conductor has a positive charge while all the other conductors are connected to earth, we see from the last result that the charges on the uninsulated conductors must be all negative, and since the potentials of these conductors are all equal and the same as that of the earth, no Faraday tubes can pass from one of these conductors to another, or from one of these to the earth. Hence all the tubes which fall on these conductors must have started from the conductor at highest potential. Thus the sum of the number of tubes which fall on the un-insulated conductors cannot exceed the number which leave the positively charged conductor, that is, the sum of the negative charges induced on the conductors connected to earth cannot exceed the positive charge on the insulated conductor.

42. These results give us important information as to the coefficients of capacity and of induction defined in Art. 26.

For let us take the first conductor as the insulated one with the positive charge; then since V_2, V_3 ... are all zero we have, using the notation of that Article,

$$E_1 = q_{11}V_1, \quad E_2 = q_{12}V_1, \quad E_3 = q_{13}V_1 \ldots \ldots$$

Since E_1 and V_1 are positive, while E_2, E_3, &c. are all negative, we see that q_{11} is positive, while q_{12}, q_{13}, &c. are all negative. Again, since the positive charge on the first conductor is numerically not less than the sum of the negative charges on the other conductors,

E_1 is numerically not less than $E_2 + E_3 + \ldots$,

i.e. q_{11} is numerically not less than $q_{12} + q_{13} + q_{14} + \ldots$.

If one of the conductors, say the second, completely surrounds the first, and if there is no conductor other than the first inside the second, and if all the conductors except the first are at zero potential, then all the tubes which start from the first must fall on the second. Thus the negative charge on the second must be numerically equal to the positive charge on the first (see Art. 30). There can be no charges on any of the other conductors, for all the tubes which might fall on these conductors must come from the first conductor, and all the tubes from this conductor are completely intercepted by the second surface. Thus if the second conductor encloses the first conductor, and if there are no other conductors between the first and the second, then $q_{11} = -q_{12}$, and q_{13}, q_{14}, q_{15}... are all zero.

43. Expression for the Energy in the Field. When we regard the Faraday tubes as the agents by which the phenomena in the electric field are produced we are naturally led to suppose that the energy in the electric field is in that part of the field through which the tubes pass, i.e. in the dielectric between the conductors. We shall now proceed to find how much energy there must be in each unit of volume if we regard the energy as distributed throughout the electric field. We have seen in Art. 23, that the electric energy is one half the sum of the products got by multiplying the charge on each conductor by the potential of that conductor. We may regard each unit charge as having associated with it a Faraday tube, which commences at the charge if that is positive and ends there if the charge is negative. Let us now see how the energy in the field can be expressed in terms of these tubes. Each tube will contribute twice to the expression for the electric energy $\frac{1}{2}\Sigma EV$, the first time corresponding to the positive charge at its origin, the second time corresponding to the negative charge at its end. Thus, since there is unit charge at each end of the tube, the contribution of each tube to the expression for the energy will be $\frac{1}{2}$ (the difference of potential between its beginning and end). The difference of potential between the beginning and end of the tube is equal to $\Sigma R \cdot PQ$, where PQ is a small portion of the length of the tube so small that along it R, the electric intensity, may be regarded as constant: the sign Σ denotes that the tube between A and B, A being a unit of positive and B a unit of negative charge, is to be divided up into small pieces similar to PQ, and that the sum of the products of the length of each piece into the electric intensity along it is to be taken. Thus the whole tube AB contributes $\frac{1}{2}\Sigma R \cdot PQ$ to the electric energy, so that we may suppose that each unit length of the tube contributes an amount of energy equal to one half the electric intensity. Any finite portion CD of the tube will therefore contribute an amount of energy numerically equal to one half the difference of potential between C and D. We may therefore regard the electrical energy as distributed throughout the field and that each of the Faraday tubes has associated with it an amount of energy per unit length numerically equal to one half the electric intensity.

Let us now consider the amount of energy per unit volume. Take a small cylinder surrounding any point P in the field with its axis parallel to the electric intensity at P, its ends being at right angles to the axis. Then if R is the electric intensity at P and l the length of the cylinder, the amount of energy due to each tube passing through the cylinder is $\frac{1}{2}Rl$. If ω is the area of the cross section of the cylinder, N the number of tubes passing through unit area, the number of tubes passing through the cylinder is $N\omega$. Thus the energy in the cylinder is

$$\tfrac{1}{2}RlN\omega$$

but in air, by Art. 40,

$$4\pi N = R,$$

so that the energy in the cylinder is

$$\frac{1}{8\pi} R^2 l\omega .$$

But $l\omega$ is the volume of the cylinder, hence the energy per unit volume is equal to

$$\frac{R^2}{8\pi} .$$

Thus we may regard the energy as distributed throughout the field in such a way that the energy per unit of volume is equal to $R^2/8\pi$.

44. If we divide the field up by a series of equi-

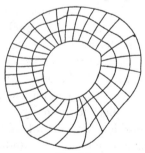

Fig. 23.

potential surfaces, the potentials of successive surfaces decreasing in arithmetical progression, and if we then

draw a series of tubular surfaces cutting these equipotential surfaces at right angles, such that the number
of Faraday tubes passing through the cross section of
each of the tubular surfaces is the same for all the
tubes, the electric field will be divided up into a
number of cells which will all contain the same amount
of energy. For the potential difference between the
places where a Faraday tube enters and leaves a cell is
the same for all the cells, and thus the energy of the
portion of each Faraday tube passing through a cell will
be constant for all the cells, and since the same number
of Faraday tubes pass through each cell, the energy in
each cell will be constant.

**45. Force on a conductor regarded as arising
from the Faraday Tubes being in a state of tension.** We have seen, Art. 37, that on each unit of area of
a charged conductor there is a pull equal to $\frac{1}{2}R\sigma$, where
σ is the surface density of the electricity, and R the electric
intensity. Now σ is equal to the number of Faraday
tubes which fall on unit area of the surface, and hence
the force on the surface is the same as if each of the
tubes exerted a pull equal to $\frac{1}{2}R$. Thus the mechanical
forces on the conductors in the electric field are the same
as they would be if the Faraday tubes were in a state
of tension, the tension at any point being equal to one
half the electric intensity at that point. Thus the tension
at any point of a Faraday tube is numerically equal to
the energy per unit length of the tube at that point.

If we have a small area ω, at right angles to the
electric intensity, the tension over this area is equal to

$$\tfrac{1}{2}NR\omega,$$

where N is the number of Faraday tubes passing through unit area, and R is the electric intensity.   By Art. 40

$$N = \frac{R}{4\pi}.$$

Hence the tension parallel to the electric intensity is

$$\frac{1}{8\pi} R^2 \omega.$$

The tension across unit area is therefore equal to

$$\frac{R^2}{8\pi}.$$

46.   This state of tension will not however leave the dielectric in equilibrium unless the electric field is uniform, that is unless the tubes are straight and parallel to each other. If however there is in addition to this tension along the lines of force a pressure acting at right angles to them and equal to $R^2/8\pi$ per unit area the dielectric will be in equilibrium, and since this pressure is at right angles to the electric intensity it will not affect the normal force acting on a conductor. To show that this pressure is in equilibrium with the tensions along the Faraday tubes, consider a small volume whose ends are portions of equipotential surfaces and whose sides are lines of force.

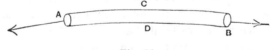

Fig. 24.

Let us now consider the forces acting on this small volume parallel to the electric intensity at A. The forces are the tensions in the Faraday tubes and the pressures at right angles to the sides. Resolve these parallel to the outward-drawn normal at A. The number n' of Faraday tubes which pass through A is the same as the number which pass through B. If R, R' are the electric intensities at A and B respectively, then the force exerted on the volume in the direction of the outward-drawn normal at A by the Faraday tubes at A will be $n'R/2$, while the force in the opposite direction exerted by the Faraday tubes at B is $n'R'\cos\epsilon/2$, where ϵ is the small angle between the directions of the Faraday tubes at A and B. Since ϵ is a very small angle we may replace $\cos\epsilon$ by unity; thus the resultant force on the volume in the direction of the outward-drawn normal at A due to the tension in the Faraday tubes is

$$n'(R - R')/2.$$

Let N be the number of tubes passing through unit area, ω, ω' the areas of the ends A and B respectively; then, Art. 40,

$$n' = N\omega = \frac{R}{4\pi}\,\omega = \frac{R'}{4\pi}\,\omega',$$

so that the resultant in the direction of the outward-drawn normal at A is

$$\frac{R}{4\pi}\,\omega\,(R - R')/2\,;$$

since $$R'\omega' - R\omega,$$

we may write this as

$$\frac{RR'}{8\pi}\,(\omega' - \omega),$$

or approximately, since R' is very nearly equal to R

$$\frac{R^2}{8\pi}\,(\omega' - \omega).$$

Let us now consider the effect of the pressure p at right angles to the lines of force; this has a component in the direction of the outward-drawn normal at A as in consequence of the curvature of the tube the normals to its surface are not everywhere at right angles to the outward-drawn normal at A; the angle between the pressure and the normal at A will always however be nearly a right angle. If this angle is $\frac{\pi}{2} - \theta$ at a point where the pressure is p', the component of the pressure along the normal at A will be proportional to $p' \sin \theta$. But since p' only differs from p, the value of the pressure at A, by a small quantity, and θ is small, the component of the pressure will be equal to $p \sin \theta$, if we neglect the squares of small quantities; that is, the effect along the normal at A of the pressure over the surface will be approximately the same as if that pressure were uniform. To find the effect of the pressure over the sides we remember that a uniform hydrostatic pressure over any closed surface is in equilibrium; hence the force due to the pressures over the sides C, D will be equal and opposite to the force due to the pressures over the ends A and B. But the force due to the pressure over these ends is $p\omega' - p\omega$; hence the resultant effect in the direction of the outward-drawn normal at A of the pressure over the sides is $p(\omega - \omega')$. Combining this with the effect due to the tension in the tubes we see that the total force on the element parallel to the outward-drawn normal at A is

$$\frac{R^2}{8\pi}(\omega' - \omega) + p(\omega - \omega');$$

this vanishes if $\qquad p = \frac{R^2}{8\pi} = \frac{NR}{2}.$

Thus the introduction of this pressure will maintain equilibrium as far as the forces parallel to the electric intensity are concerned.

Now consider the force at right angles to the electric intensity. Let $PQRS$, Fig. 25, be the section of the surface in Fig. 24 by the plane of the paper, PS, QR being sections

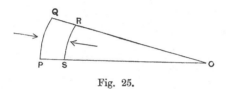

Fig. 25.

of equipotential surfaces, and PQ, SR lines of force. Let t be the depth of the volume at right angles to the plane of the paper. We shall assume that the section of the figure by the plane through PQ at right angles to the plane of the paper is a rectangle. Let R be the electric intensity along PQ, R' that along SR, s the length PQ, s' that of SR. Since the difference of potential between P and Q is the same as that between S and R,

$$Rs = R's'.$$

Consider the forces parallel to PS. First take the tensions along the Faraday tubes; the force due to those at PS will have no component along PS: in each tube at Q there is a tension $R/2$, the component of which along PS is $(R \sin \theta)/2$, where θ is the angle between PS and QR. Since θ is very small this component is equal to $R\theta/2$. Let PS and QR meet in O,

$$\theta = \frac{RS}{OR} = \frac{PQ}{OQ} = \frac{PQ - SR}{OQ - OR} = \frac{s - s'}{RQ}.$$

Thus the component along PS due to the tension at Q is

$$\frac{R}{2} \cdot \frac{s - s'}{RQ}.$$

The number of tubes which pass through the end of the figure through RQ at right angles to the plane of the paper is $N . QR . t$, where N is the number of tubes which pass through unit area.

The total component along PS due to the tensions in these tubes is thus

$$\frac{R}{2} \frac{(s - s')}{RQ} . N . QR . t$$

$$= \frac{R^2}{8\pi}(s - s') t.$$

Now the component along PS due to the pressures at right angles to the electric intensity is equal to

$$pst - p's't,$$

where p and p' are the pressures over PQ, RS respectively.

If $$p = \frac{R^2}{8\pi}, \quad p' = \frac{R'^2}{8\pi};$$

$$pst - p's't = \left(\frac{R^2}{8\pi} s - \frac{R'^2}{8\pi} s'\right) t$$

$$= \frac{RR'}{8\pi}(s' - s) t, \text{ (since } Rs = R's'),$$

or approximately, since R' is very nearly equal to R,

$$= \frac{R^2}{8\pi}(s' - s) t.$$

Thus the component in the direction of PS due to the tensions is equal and opposite to the component due to the pressures; thus the two are in equilibrium as far as the component in the plane of the diagram at right angles to the electric intensity is concerned; we easily see that the same is true for the component at right angles to the plane of the paper. We have already proved that the tensions and pressures balance as far as the component along the direction of the electric intensity is concerned; thus the system of pressures and tensions constitutes a system in equilibrium.

47. This system of tensions along the tubes of force and pressures at right angles to them is thus in equilibrium at any part of the dielectric where there is no charge, and gives rise to the forces which act on electrified bodies when placed in the electric field. Faraday introduced this method of regarding the forces in the electric field; he expressed the system of tensions and pressures which we have just found, by saying that the tubes tended to contract and that they repelled each other. This conception enabled him to follow the processes of the electric field without the aid of mathematical analysis.

Since
$$Rs = R's',$$

and
$$\frac{s}{s'} = \frac{OQ}{OR} = 1 + \frac{RQ}{OR},$$

we have
$$\frac{R - R'}{RQ} = -\frac{R}{OR}.$$

Now OR is the radius of curvature of the line of force; denoting this by ρ we have

$$\frac{1}{\rho} = -\frac{\dfrac{dR}{d\nu}}{R}$$

where $d\nu$ is an element of length at right angles to the electric force; we see from this equation that the lines of force are concave to the stronger parts of the field.

The lines of force arrange themselves as a system of
elastic strings would do if acted on by forces whose
potential for unit length of string was $R/2$.

48.   The student will find much light thrown on the
effects produced in the electric field by the careful study
from this point of view of the diagrams of the lines of
force given in Art. 38.   Thus, take as an example the
diagram given in Fig. 18, which represents the lines of
force due to two charges A and B of opposite signs, the
ratio of the charges being $4:1$. We see from the diagram
that though more tubes of force start from the larger
charge A, and the tension in each of these is greater than
in a tube near the smaller charge B, the tubes are much
more symmetrically distributed round A than round B.
The approximately symmetrical distribution of the tubes
round A makes the pulls exerted on A by the taut Faraday
tubes so nearly counterbalance each other that the resultant
pull of these tubes on A is only the same as that exerted
on B by the tubes starting from it; since these, though
few in number, are less symmetrically distributed, and
so do not tend to counterbalance each other to nearly
the same extent.   The tubes of force in the neighbourhood of the point of equilibrium are especially interesting.
Since the charge on A is four times that on B, only $\frac{1}{4}$ of
the tubes which start from A can end on B, the remaining
$\frac{3}{4}$ must go off to other bodies, which in the case given in
the diagram are supposed to be at an infinite distance.
The point of equilibrium corresponds as it were to the
'parting of the ways' between the tubes of force which
go from A to B and those which go off from A to an
infinite distance.

When the charges A and B are of the same sign, as in Fig. 19, we see how the repulsion between similar tubes causes the tubes to congregate on the side of A remote from B, and on the side of B remote from A.

We see again how much more symmetrically the tubes are distributed round A than round B; this more symmetrical distribution of the tubes round A makes the total pull on A the same as that on B.

We see too from this example that the repulsion between the charges of the same sign and the attraction between charges of opposite signs are both produced by the same mechanism, i.e. a system of pulls; the difference between the cases being that the pulls are so distributed that when the charges are of the same sign the pulls tend to pull the bodies apart, while when the charges are of opposite signs the pulls tend to pull the bodies together.

The diagram of the lines of force for the two finite plates (Fig. 21) shows how the Faraday tubes near the edges of the plates get pushed out from the strong parts of the field and are bent in consequence of the repulsion exerted on each other by the Faraday tubes.

49. As an additional example of the interpretation of the processes in the electric field in terms of the Faraday tubes, let us consider the effect of introducing an insulated conductor into an electric field.

Let us take the field due to a single positively charged body at A; before the introduction of the conductor the Faraday tubes were radial, but when the conductor is introduced the tubes, which previously existed in the region occupied by the conductor, are annulled; thus the repulsion previously exerted by these tubes on the surrounding ones ceases, and a tube such as AB, which was previously straight, is now, since the pressure below it is diminished, bent down towards the conductor; the tubes near the conductor are bent down so much that they strike against it, they then divide and form two tubes, with negative electrification at the end C, positive at the end D.

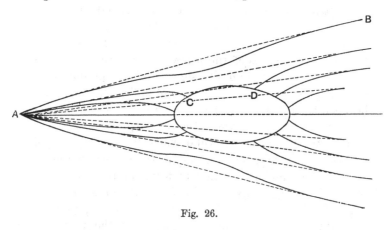

Fig. 26.

50. Force on an uncharged conductor placed in an electric field. If a small conductor is placed in the field at P, the Faraday tubes inside the conductor disappear, and, if the introduction of the conductor did not alter the tubes outside it, the diminution of energy due to the annihilation of the tubes in the conductor would be proportional to $R^2/8\pi$ per unit volume, where R is the electric intensity in the field at P before the conductor was introduced. If the conductor is moved to a place where the electric intensity is R', the diminution in the electric energy in the field is $R'^2/8\pi$ per unit volume. Now it is a general principle in mechanics that a system always tends to move from rest in such a way as to diminish the potential energy as much as possible, and the force tending to assist a displacement in any direction is equal to the rate of diminution of the potential energy in that direction. The conductor will thus tend to move so as to produce the greatest possible diminution in the electric energy, that is, it will tend to get into the parts of the field where the electric intensity is as large as possible ; it will thus move from the weak to the strong parts of the field.

The presence of the conductor will however disturb the electric field in its neighbourhood; thus **R**, the actual electric intensity, will differ from R, the electric intensity at the same point before the conductor was introduced. By differentiating $R^2/8\pi$ we shall get an inferior limit to the force acting on the conductor per unit volume. For suppose we introduce a conductor into the electric field, then $R^2/8\pi$ would be the diminution in electric energy per unit volume due to the disappearance of the Faraday tubes from the inside of the conductor, the tubes outside being supposed to retain their original position. In reality however the tubes outside will have to adjust themselves so as to be normal to the conductor, and this adjustment will involve a further diminution in the energy, thus the actual change in the energy is greater than that in $R^2/8\pi$ and the force acting per unit volume will therefore be greater than the rate of diminution of this quantity. If we take the case when the force is due to a charge e at a point, the rate of diminution of $R^2/8\pi$ is $e^2/2\pi r^5$, and thus the force on a small conducting sphere of radius a will be greater than $(4\pi a^3/3)\,(e^2/2\pi r^5)$, that is greater than $2e^2a^3/3r^5$. The actual value (see Art. 87) is $2e^2a^3/r^5$.

CHAPTER III

51. The capacity of a conductor is defined to be the numerical value of the charge on the conductor when its potential is unity, all the other conductors in the field being at zero potential.

Two conductors insulated from each other and placed near together form what is called a condenser; in this case the charge on either conductor may be large, though the difference between their potentials is small.

In many instruments the two conductors are so arranged that their charges are equal in magnitude and opposite in sign; in such cases the magnitude of the charge on either conductor when the potential difference between the conductors is unity is called the *capacity* of the *condenser*.

If the difference of potential between two conductors, produced by giving a charge $+q$ to one conductor and $-q$ to the other, is V, then q/V is defined to be the capacity between the conductors.

52. Capacity of a Sphere placed at an infinite distance from other conductors. Let a be the radius of the sphere, V its potential, e its charge, the corresponding charge of opposite sign being at an infinite distance. Then (Art. 17), the potential due to the charge on the sphere at a distance r from the centre is e/r; therefore the potential at the surface of the sphere is e/a.

Hence we have

$$V = \frac{e}{a}.$$

When V is unity, e is numerically equal to a: hence, Art. 51, the capacity of the sphere is numerically equal to its radius.

53. Capacity of two concentric spheres. Let us first take the case when the outer sphere and any conductors which may be outside it are connected to earth, while the inner sphere is maintained at potential V. Then, since the outer sphere and all the conductors outside are connected to earth, no Faraday tubes can start from or arrive at the outer surface of the outer sphere, for Faraday tubes only pass between places at different potentials, and the potentials of all places outside the sphere are the same, being all zero. Again, all tubes which start from the inner sphere will arrive at the internal surface of the outer shell, so that the charge on the inner surface of this shell will be equal and opposite to the charge on the inner sphere. Let a be the radius of the inner sphere, b the radius of the internal surface of the outer sphere, e the charge on the inner sphere, then $-e$ will be the charge on the interior of the outer sphere.

Consider the work done in moving a unit of electricity from the surface of the inner sphere to the inner surface of the outer sphere; the charge on the outer sphere produces no electric intensity at a point inside, so that the electric intensity, which produces the work done on the unit of electricity, arises entirely from the charge on the inner sphere. The electric intensity due to the charge on this sphere is, by Art. 11, the same as that which would be

due to the charge e collected at the centre O. The work
done on unit of electricity when it moves from the inner
sphere to the outer one is thus the same as the work done
on a unit charge when it moves from a distance a to a
distance b from a small charged body placed at the centre
of the spheres; this, by Art. 17, is equal to

$$\frac{e}{a} - \frac{e}{b},$$

and is by definition equal to V, the potential difference
between the two spheres; hence we have

$$V = \frac{e}{a} - \frac{e}{b},$$

or
$$e = \frac{ab}{b-a} V.$$

Thus, when $b-a$ is very small, that is, when the radii of
the two spheres are very nearly equal, the charge is very
large. When $V = 1$, the charge is

$$\frac{ab}{b-a};$$

so that this is, by Art. 51, the capacity of the two spheres.
The value of this quantity when the radii of the two spheres
are very nearly equal is worthy of notice. In this case,
writing t for $b-a$, the distance between the spheres, the
capacity is equal to

$$\frac{ab}{t} = \frac{a(a+t)}{t};$$

this, since t is very small compared with a, is approximately

$$\frac{a^2}{t} = \frac{4\pi a^2}{4\pi t}$$

$$= \frac{\text{surface of the sphere}}{4\pi t}.$$

Thus the capacity in this case is equal per unit area of surface to $1/4\pi$ times the distance between the conductors. The case of two spheres whose distance apart is very small compared with their radii is however approximately the case of two parallel planes; hence the capacity of such planes per unit area of surface is equal to $1/4\pi$ times the distance between the planes. This is proved directly in Art. 56.

If, after the spheres are charged, the inner one is insulated, and the outer one removed to an infinite distance (to enable this to be done we may suppose that the outer sphere consists of two hemispheres fitted together, and that these are separated and removed), the charge on the sphere will remain equal to e, i.e. $\dfrac{ab}{b-a} V$, but the potential of the sphere will rise; when it is alone in the field the potential will be e/a, i.e.

$$\frac{b}{b-a} V.$$

Thus by removing the outer sphere the potential difference between the sphere and the earth has been increased in the proportion of b to $b - a$. By making $b - a$ very small compared with b, we can in this way increase the potential difference enormously and make it capable of detection by means which would not have been sufficiently sensitive before the increase in the potential took place.

It was by the use of this principle that Volta succeeded in demonstrating by means of the gold-leaf electroscope and two metal plates, the difference of potential between the terminals of a galvanic cell; this difference is so small that the electroscope is not deflected when the cell is directly connected to it; by connecting the terminals of the cell to two plates placed very close together, and then removing one of the plates *after* severing the connections between the plates and the cells, Volta was able to increase the potential of the other plate to such an extent that it produced an appreciable deflection of an electroscope with which it was connected.

Work has to be done in separating the two conductors; this work appears as increased electric energy. Thus, to take the case of the two spheres, when both spheres were in position the electric energy, which, by Art. 23 is equal to $\frac{1}{2}\Sigma EV$, is

$$\frac{1}{2}\frac{ab}{b-a}V^2.$$

When the outer sphere which is at zero potential is removed the potential of the sphere is e/a, so that the electric energy is

$$\frac{1}{2}\frac{e^2}{a},$$

$$\frac{1}{2}\frac{ab^2}{(b-a)^2}V^2,$$

and has thus been increased in the proportion of b to $b-a$.

54. Let us now take the case when the inner sphere is connected to earth while the outer sphere is at the potential V. In this case we can prove exactly as before that the charge on the inner sphere is equal and opposite to the charge on the internal surface of the outer sphere, and that, if e is the charge on the inner sphere,

$$e = -\frac{ab}{b-a}V.$$

In this case, in addition to the positive charge on the internal surface of the outer sphere, there will be a positive charge on the external surface, since this surface is at a higher potential than the surrounding conductors. If c is the radius of the external surface of the outer sphere, the sum of the charges on the two spheres must be Vc. Since the charge on the inner surface of the outer sphere is equal and opposite to the charge on the inner sphere, the charge on the external surface of the outer sphere must be equal to Vc. Thus the total charge on the outer sphere is equal to

$$\frac{ab}{b-a}\,V + cV.$$

55. The charge on the outside of the outer sphere will be affected by the presence of other conductors. Let us suppose that outside the external sphere there is a small sphere connected to earth; let r be the radius of this sphere, R the distance of its centre from O the centre of the concentric spheres. Let e' be the total charge on the two concentric spheres, e'' the charge on the small sphere. The potential due to e' at a great distance R from O is e'/R, similarly the potential due to e'' is at a distance R equal to e''/R.

Since the surface of the outer sphere is at the potential V, we have

$$V = \frac{e'}{c} + \frac{e''}{R},$$

and, since the potential of the small sphere is zero, we have

$$0 = \frac{e'}{R} + \frac{e''}{r},$$

hence
$$V = \frac{e'}{c} \left\{ 1 - \frac{rc}{R^2} \right\},$$

$$e' = \frac{cV}{1 - \dfrac{rc}{R^2}},$$

that is, the presence of the small sphere increases the charge on the outer sphere in the proportion of

$$1 \text{ to } 1 - rc/R^2.$$

It is only the charge on the external surface of the outer sphere which is affected. The charges on the inner sphere and on the internal surface of the outer sphere are not altered by the presence of conductors outside the latter sphere.

56. Parallel Plate Condensers. Condensers are frequently constructed of two parallel metallic plates; the theory of the case, when the plates are so large in comparison with their distance apart that they may be regarded as infinite in area, is very simple.

In this case the Faraday tubes passing between the plates will be straight and at right angles to the plates, and the electric intensity between the plates is constant since in passing from one plate to the other each Faraday tube has a constant cross section; let R be its value, then if d is the distance between the plates, the work done on unit charge of electricity as it passes from the plate where the potential is high to the one where the potential is low is Rd, and this by definition is equal to V, the difference of potential between the plates. Hence

$$V = Rd.$$

If σ is the surface-density of the charge on the plate at high potential, that on the plate of low potential will be $-\sigma$, and by Coulomb's law, Art. 22,

$$R = 4\pi\sigma.$$

Hence
$$V = 4\pi\sigma d,$$

or
$$\sigma = \frac{V}{4\pi d} \quad \dots\dots\dots\dots\dots(1),$$

and if V is equal to unity, σ is equal to

$$\frac{1}{4\pi d}.$$

The charge on an area A of one of the plates when the potential difference is unity is thus $A/4\pi d$, this by definition is the capacity of the area A. We arrived at the same result in Art. 53 from the consideration of two concentric spheres. The electrical energy of the condenser is, by Art. 23, equal to

$$\tfrac{1}{2}\Sigma EV,$$

which in this case is equal to

$$\frac{V^2 A}{8\pi d},$$

or, if E is the charge on one of the plates, to

$$\frac{2\pi d E^2}{A}.$$

57. Guard Ring. In practice it is of course impossible to have infinite plates, and when the plates are finite, then, as the diagram, Fig. 21, Art. 38, shows, the Faraday tubes near the edges of the plates are no longer straight, and the electrification ceases to be uniform, and is no longer given by the expression (1), Art. 56. Thus to express the quantity of electricity on the finite plate, we should have to add to the expression a correction for the inequality of the distribution over the ends of the plates. This correction can be calculated, but the necessity for it may be avoided in practice by making use of a device due to Lord Kelvin, and called a guard ring.

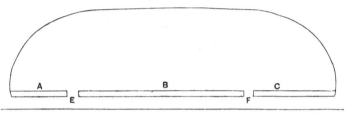

Fig. 27.

Suppose one of the plates, say the upper one, is divided into three portions flush with each other and separated by the narrow gaps E, F. Then if, in charging the condenser the portions A, B, C are connected metallically with each other, the places where the electrification is not uniform will be on A and C, so that apart from the effects of the narrow gaps E, F, the electrification on B will, if we neglect the effect of the gaps, be uniform and the total charge on B will be equal to $SV/4\pi d$, where S is the area of the plate B. The capacity of B is thus equal to $S/4\pi d$.

If, as ought to be the case, the widths of the gaps at E and F are very small compared with the distance between the plates, we can easily calculate the effect of the gaps. For if the gaps are very narrow the electrification of the lower plate will be approximately uniform. The Faraday tubes in the neighbourhood of the gaps will be distributed as in Fig. 28. We see

from this, if we consider the gap E, that all the Faraday
tubes which would have fallen on a plate whose breadth

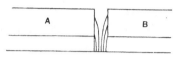

Fig. 28.

was E, if there had been no gap, will fall on one
or other of the plates A and B, Fig. 28, and from the
symmetry of the arrangement half of these tubes will
fall on B, the other half on A; thus the actual amount
of electricity on B will be the same as if we supposed B
to extend halfway across the gap, and to be uniformly
charged with electricity whose surface density is $V/4\pi d$.
We see then that, allowing for the effects of the gaps,
the capacity of B will be equal to $S'/4\pi d$, where

$\qquad S' = $ area of plate B

$\qquad\qquad + \frac{1}{2}$ (the sum of the areas of the gaps E and F).

If the plate B is not at zero potential, there will be
some electrification on the back of the plate arising from
Faraday tubes which go from the back of B to other
conductors in its neighbourhood and to earth. The electrification of the back of B may be obviated by covering
this side of A, B, C with a metal cover connected with
A and C. It can also be obviated by making B the low
potential plate (i.e. the one connected to earth), care being
taken that the other conductors in the neighbourhood are
also connected to earth.

58. Capacity of two coaxial cylinders. Let us
take the case of two coaxial cylinders, the inner one being

at potential V, the outer one being at potential zero. Then if E is the charge per unit length on the inner cylinder, $-E$ will be the charge per unit length on the inner surface of the outer one, since all the Faraday tubes which start from the inner cylinder end on the outer one.

The electric intensity at a distance r from the axis of the cylinders is, by Art. 13, equal to

$$\frac{2E}{r}.$$

Thus the work done on unit charge, when it goes from the outer surface of the inner cylinder to the inner surface of the outer cylinder, is equal to

$$\int_a^b \frac{2E}{r}\, dr,$$

where a is the radius of the inner cylinder, b the radius of the inner surface of the outer cylinder.

This work is, however, by definition equal to V, the difference of potential between the cylinders, and hence

$$V = \int_a^b \frac{2E}{r}\, dr$$

$$= 2E \log \frac{b}{a}.$$

When V is unity, E, the charge per unit length, is equal to

$$\frac{1}{2 \log \dfrac{b}{a}},$$

and this, by definition, is the capacity of the condenser per unit length.

If the radii of the cylinders are nearly equal, and if $b - a = t$, t will be small compared with a ; in this case the capacity per unit length

$$= \frac{1}{2 \log \dfrac{a+t}{a}}$$

$$= \frac{1}{2 \dfrac{t}{a}} \text{ approximately}$$

$$= \frac{1}{2} \frac{a}{t}$$

$$= \frac{2\pi a}{4\pi t} .$$

Since $2\pi a$ is the area of unit length of the inner cylinder, the capacity per unit area is $1/4\pi t$; we might have deduced this result from the case of two parallel planes.

When the two cylinders are coaxial, there is no force tending to move the inner cylinder; thus since the system is in equilibrium, the potential energy, if the charges are given, must be either a maximum or a minimum. The equilibrium is, however, evidently unstable, for, if the inner cylinder is displaced, the force due to the electric field tends to make the cylinders come into contact with each other and thus increase the displacement. Since the equilibrium is unstable the potential energy is a maximum when the cylinders are coaxial. The potential energy, however, is, by Art. 23, equal to

$$\frac{1}{2} EV = \frac{1}{2} \frac{E^2}{C} ,$$

where C is the capacity of the condenser. Thus if the potential energy is a maximum the capacity must be a minimum. Thus any displacement of the inner cylinder will produce an increase in the capacity, but since the capacity is a minimum when the cylinders are coaxial, the increase in the capacity will be proportional to square and higher powers of the distance between the axes of the cylinders.

59. Condensers whose capacities can be varied. For some experimental purposes it is convenient to use a condenser whose capacity can be altered continuously, and in such a way that the alteration in the capacity can be easily measured. For this purpose a condenser made of two parallel plates, one of which is fixed, while the other can be moved by means of a screw, through known distances, always remaining parallel to the fixed plate, is useful. In this case the capacity is inversely proportional to the distance between the plates, provided that this distance is never greater than a small fraction of the radius of the plates.

Another arrangement which has been used for this purpose is shown in Fig. 29. It consists of three

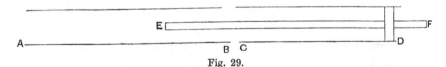

Fig. 29.

coaxial cylinders, two of which, AB, CD, are of the same radius and are insulated from each other, while the third, EF, is of smaller radius and can slide parallel to its axis. The cylinder EF is connected metallically with CD, so that these two are always at the same potential, and the cylinder AB is at a different potential, then when the cylinder EF is moved about so as to expose different amounts of surface to AB the capacity of the condenser formed by AB and EF will alter, and the increase in the capacity will be proportional to the increase in the area of the surface of EF brought within AB.

60. Electrometers.

Consider the case of two parallel conducting plates; let V be the potential difference between the plates, d their distance apart. The force on a conductor per unit area is, by Art. 37, equal to $\frac{1}{2} R\sigma$, where R is the electric intensity at the conductor and σ the surface density; but $R = \frac{V}{d}$, while $\sigma = \frac{1}{4\pi} R$ by Coulomb's law; we see therefore that the attraction of one plate on the other is per unit area equal to $\frac{1}{8\pi} \frac{V^2}{d^2}$. Hence the force on an area A of one of the plates is equal to

$$\frac{A}{8\pi} \frac{V^2}{d^2} \quad \cdots\cdots\cdots\cdots\cdots\cdots(1).$$

Thus, if we measure the mechanical force between the plates, we can deduce the value of V, the potential difference between them. This is the principle of Lord Kelvin's attracted disc electrometer. This instrument measures the force necessary to keep a moveable disc surrounded by a fixed guard ring in a definite position; when this force is known the value of the potential difference is given by the expression (1).

Quadrant Electrometer. The effect measured by the instrument just described varies as the square of the

potential difference; thus when the potential difference
is diminished the attraction between the plates diminishes
with great rapidity.   For this reason the instrument is
not suited for the measurement of very small potential
differences.   To measure these another electrometer, also
due to Lord Kelvin, called the quadrant electrometer, is
frequently employed.

This instrument is represented in Fig. 30: it consists
of a cage, made by the four quadrants A, B, C, D; each
quadrant is supported by an insulating stem, while the
opposite quadrants A and C are connected by a metal wire,
as are also B and D; thus A and C are always at the same
potential and so also are B and D.   Each pair of quadrants
is in connection with an electrode, E or F, by means of
which it can easily be put in metallic connection with any
body outside the case of the instrument.   Inside the quadrants and insulated from them is a flat piece of aluminium
shaped like a figure of eight.   This is suspended by a
silk fibre and can rotate, with its plane horizontal, about
a vertical axis.   A fine metal wire hangs from the lower
surface of this aluminium needle and dips into some
sulphuric acid contained in a glass vessel, the outside
of which is coated with tin-foil and connected with earth.
This vessel, with the conductors inside and outside, forms
a condenser of considerable capacity; it requires therefore
a large charge to alter appreciably the potential of this jar,
and therefore of the needle.   To use the instrument the jar
is charged to a high potential C; the needle will then also
be at the potential C.   Now if the two pairs of quadrants
are at the same potential, the needle is inside a conductor
symmetrical about the axis of rotation of the needle, and
at one potential.   There will evidently be no couple on

the needle arising from the electric field, and the needle
will take up a position in which the couple arising from
the torsion of the thread supporting the needle vanishes.
If, however, the two pairs of quadrants are not at the same
potential the needle will swing round until, if there is
nothing to stop it, the whole of its area will be inside the

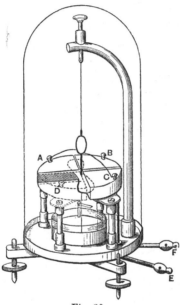

Fig. 30.

pair of quadrants whose potential differs most widely
from its own.   As it swings round, however, the torsion of
the thread produces a couple tending to bring the needle
back to the position from which it started.   The needle
finally takes up a position in which the couple due to the
torsion in the thread balances that due to the electric

field. The angle through which the needle is deflected gives us the means of estimating the potential difference between the quadrants.

The way in which the couple acting on the needle depends upon the potentials of the quadrants and the needle can be illustrated by considering a case in which the electric principles involved are the same as in the quadrant electrometer, but where the geometry is simpler.

Let E, F (Fig. 31) be two large co-planar surfaces insulated from each other by a small air gap. Let G be another plane surface, parallel to E and F, and free to move in its own plane. Let t be the distance between G and the planes E and F. Let A, B, C be the potentials of the planes F, E, G respectively. Let l be the width of

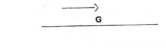

Fig. 31.

the planes at right angles to the plane of the paper. If Xl is the force tending to move the plane G in the direction of the arrow, then, if this plane be moved through a short distance x in this direction, the work done by the electric forces is Xlx. If the electric system is left to itself, i.e. if it is not connected to any batteries, etc., so that the charges remain constant, this work must have been gained at the expense of the electric energy; we have therefore, by the principle of the Conservation of Energy,

$Xlx = $ *decrease* in the electric energy of the system, the charges remaining constant, when the plane G is displaced through the distance x;

or by Art. 36,

$Xlx = increase$ in the electric energy of the system, the
potentials remaining constant, when the plane G is
displaced through the same distance x ....(1).

Consider the change in the electric energy when the
plane G is moved through a distance x. The area of G
opposite to F will be increased by lx, and in consequence
the energy will be increased by the energy in a parallel
plate condenser, whose area is lx, the potentials of whose
plates are A and C respectively, and the distance between the plates is t; this, by Art. 56, is equal to

$$\frac{lx}{8\pi t}(C - A)^2.$$

At the same time as the area of G opposite to F is increased by lx, that opposite to E is decreased by the same
amount, so that the electric energy will be decreased by
the energy in a parallel plate condenser whose area is lx,
the potentials of the plates B and C and their distance
apart t; this, by Art. 56, is equal to

$$\frac{lx}{8\pi t}(C - B)^2.$$

Thus the total increase in the electric energy when G
is displaced through x, the potentials being constant,
is equal to

$$\frac{1}{8\pi}\frac{lx}{t}\left\{\overline{C - A}^2 - \overline{C - B}^2\right\}$$

$$= \frac{1}{4\pi}\frac{lx}{t}(B - A)\left\{C - \frac{1}{2}(A + B)\right\}.$$

Thus, by equation (1),

$$Xlx = \frac{1}{4\pi}\frac{lx}{t}(B - A)\left\{C - \frac{1}{2}(A + B)\right\},$$

or

$$X = \frac{1}{4\pi t}(B - A)\left\{C - \frac{1}{2}(A + B)\right\}.$$

If $(C - A)^2$ is greater than $(C - B)^2$, X is positive, that is, the plate G tends to bring as much of its surface as it can over the plate from which it differs most in potential.

In the quadrant electrometer the electrical arrangements are similar to the simple case just discussed, and hence the force will vary with the potential differences in a similar way. Hence we conclude that if the needle in the quadrant electrometer be at potential C, the couple tending to twist it from the quadrant whose potential is B to that whose potential is A, will be proportional to

$$(B - A) \left\{ C - \frac{1}{2}(A + B) \right\};$$

we may put it equal to

$$n(B - A) \left\{ C - \frac{1}{2}(A + B) \right\},$$

where n is some constant.

When the needle is in equilibrium, this couple will be balanced by the couple due to the torsion in the suspension of the needle.

The torsional couple is proportional to the angle θ through which the needle is deflected. Let the couple equal $m\theta$. Hence we have when the needle is in equilibrium

$$m\theta = n(B - A) \left\{ C - \frac{1}{2}(A + B) \right\},$$

$$\theta = \frac{n}{m}(B - A) \left\{ C - \frac{1}{2}(A + B) \right\} \dots\dots\dots\dots(2).$$

If, as is generally the case when small differences of potential are measured, the jar containing the sulphuric acid is charged up so that its potential is very high compared with that of either pair of quadrants, C will be very large compared with A or B, and therefore with

$$\frac{1}{2}(A + B),$$

so that the expression (2) is very approximately

$$\theta = \frac{n}{m}(B - A)\,C.$$

Hence, in this case, the difference of potential is proportional to the deflection of the needle. This furnishes a very convenient method of comparing differences of potential, and though it does not give at once the absolute measure of the potential, this may be deduced by measuring the deflection produced by a standard potential difference of known absolute value such as that between the electrodes of a Clark's cell.

The quadrant electrometer may also be used to measure large differences of potential; to do this, instead of charging the jar independently, connect the jar and therefore the needle to one pair of quadrants, say the pair whose potential is A. Then, since $C = A$, the expression (2) becomes

$$\theta = -\frac{n}{2m}(A - B)^2;$$

thus the needle is deflected towards the pair of quadrants whose potential is B, and the deflection of the needle is, in this case, proportional to the *square* of the potential difference between the quadrants. Thus, if the quadrants are connected respectively to the inside and outside coatings of a condenser, the deflection of the electrometer will be proportional to the energy in the condenser.

61. Use of the Electrometer to measure a charge of electricity. Let α and β denote the two pairs of quadrants. If to begin with α and β are both connected with the earth, there will be a charge Q_0 on the quadrants α induced by the charge on the needle; let α now be disconnected from β and from the earth, insulated, and given a charge Q' of electricity, the needle will be deflected; let θ be the angle of deflection, A the potential of the quadrants α, then if C is the potential of the needle, we have, by Art. 26, since the charge on α is $Q_0 + Q'$

$$Q_0 + Q' = q_{11}A + q_{13}C \ldots\ldots\ldots\ldots\ldots(1),$$

where q_{11}, q_{13} are the coefficients of capacity and induction for the displaced position of the needle. Since Q_0 is the charge on α when A is zero

$$Q_0 = (q_{13})_0 C,$$

where $(q_{13})_0$ is the value of q_{13} when $\theta = 0$; hence by (1)

$$Q' = q_{11}A + (q_{13} - (q_{13})_0) C.$$

Let $\qquad q_{13} - (q_{13})_0 = -\mu\theta,$

θ being taken as positive when measured in the direction of deflection due to a positive value of A, then if the charge on the needle is negative Q_0 the positive charge on α induced by the needle will evidently increase with θ so that as C is negative μ is a positive quantity; we have also by equation (2), page 103, when C is large compared with A,

$$\theta = -\frac{n}{m}AC,$$

hence $\qquad Q' = -\theta\left\{\frac{q_{11}m}{nC} + \mu C\right\},$

or $\qquad \theta = -\frac{Q'\,nC}{q_{11}m + \mu nC^2}\ldots\ldots\ldots\ldots(2).$

It is interesting to notice that when the potential of the needle is increased beyond a certain point the deflection of the needle due to a given charge on the quadrants diminishes as the potential of the needle increases, hence to obtain the greatest sensitiveness when measuring electrical charges we must be careful not to charge the needle too highly. We see from (2) that the greatest deflection θ' due to the charge Q' is given by the equation

$$\theta' = \tfrac{1}{2} Q' \sqrt{\frac{n}{\mu m q_{11}}} \; ;$$

when the deflection is greatest the potential of the needle

$$= -(q_{11} m / \mu n)^{\frac{1}{2}}.$$

To get from the readings of the electrometer the value of the charge in absolute measure, connect one plate of a condenser whose capacity is Γ with the quadrants α, and connect the other plate with the earth; the coefficient q_{11} will now be increased by Γ and, if θ_1 is the deflection of the electrometer for the same charge, then by (2)

$$\theta_1 = -\frac{Q'nC}{(q_{11} + \Gamma)\, m + \mu n C^2} \quad \dots\dots\dots\dots(3).$$

Hence from (2) and (3)

$$\theta = -\frac{\theta - \theta_1}{\theta_1}\frac{Q'nC}{\Gamma m} \dots\dots\dots\dots\dots(4).$$

If the deflection of the electrometer when the potential of α is V is ϕ, then

$$\phi = -\frac{n}{m} CV,$$

hence, from (4),

$$Q' = \Gamma V \frac{\theta}{\phi} \cdot \frac{\theta_1}{\theta - \theta_1}.$$

61 a. A gold leaf electroscope is for some purposes preferable to an electrometer, on account of its much smaller capacity, its portability and the ease with which it can be shielded from external disturbances. With suitably designed electroscopes it is possible to obtain with ease a deflection of the gold leaf of 70 or 80 scale divisions for a change of 1 volt in the potential of the gold leaf, these divisions are those of a micrometer eye-piece in a reading microscope through which the gold leaf is observed. The behaviour of these sensitive electroscopes may be illustrated by the consideration of a very simple case. Suppose that we have two parallel plates D and E maintained at potentials A and $-A$ respectively, let us represent the gold leaf by another parallel plate C which can move backwards and forwards and is pulled to a position midway between D and E by a spring, which when C is displaced a distance x from the mid-position pulls it back again with a force equal per unit area of C to μx.

If V is the potential of C, $2d$ the distance between D and E, x the displacement of C towards the negative plate, then for the equilibrium of the plate we must have

$$\frac{1}{8\pi}\frac{(A+V)^2}{(d-x)^2} - \frac{1}{8\pi}\frac{(A-V)^2}{(d+x)^2} = \mu x,$$

or if $V = yA$, $x = \xi d$,

$$\frac{(1+y)^2}{(1-\xi)^2} - \frac{(1-y)^2}{(1+\xi)^2} = \frac{8\pi\mu d^3}{A^2}\cdot\xi = 4\mu'\xi$$

if

$$\mu' = \frac{2\pi\mu d^3}{A^2}.$$

If y and ξ are small, this equation becomes

$$(\mu'-1)\xi = y,$$

or

$$x = \frac{d\cdot\dfrac{V}{A}}{\dfrac{2\pi\mu d^3}{A^2} - 1}.$$

The equilibrium will be unstable unless $2\pi\mu d^3/A^2$ is greater than 1, when this quantity exceeds unity by a small fraction the denominator in the expression for x is small so that x itself tends to become large, i.e. a small potential difference V will produce a large displacement of the plate. In addition to the value of x given above there is a second value corresponding to another position of equilibrium, the equilibrium in this case is unstable and if C were in this position it would move up to D. When $V = 0$ the two positions of equilibrium are given by $x = 0$ and

$$\frac{x^2}{d^2} = 1 - \frac{1}{\sqrt{\mu'}}.$$

Thus when the instrument is very sensitive, i.e. when μ' is nearly unity, the second value of x is very small, thus the unstable position of equilibrium is close to the stable one, so that a slight deflection from the latter will make the gold leaf unstable, and it will fly up to one of the plates. As $\cdot V$ increases the value of x for the stable position increases while that for the unstable one diminishes, so that the two get nearer together, for a certain value of V they coincide, while for greater values there is no position of equilibrium.

In practice the office of the spring in the preceding example is performed by the weight of the gold leaf; the leaf is hung so as to be vertical when midway between the plates, when it is disturbed from this position gravity tends to bring it back. The successful use of instruments of this type depends upon having means to keep the potential of the fixed plates accurately constant. Except for very small values of V, the deflection is not directly proportional to V, so that it is necessary to calibrate the instrument by charging the gold leaf to known potentials and observing the deflection.

The sensitiveness of the instrument can be adjusted by altering V or d. In a type of instrument invented by Mr C. T. R. Wilson and called the tilted electroscope (Fig. 31 a), where the instrument can be tilted by means

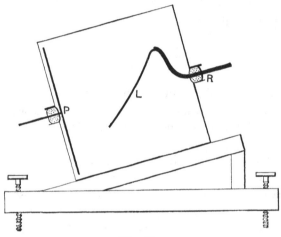

Fig. 31 a.

of foot-screws, the adjustment is effected by altering the tilt. The plate P is charged to a high potential, the case of the instrument to earth, and initially the gold leaf is to earth, it takes up a position of equilibrium from which it is displaced as soon as its potential is altered.

62. Test for the equality of the capacities of two condensers. The test can easily be made in the following way. Suppose A and B, Fig. 32, are the plates of one condenser, C and D those of the other. First connect A to C, and B to D, and charge the condensers by connecting A and B with the terminals of a battery or some other suitable means. Then disconnect A and B from the battery. Disconnect A from C and B from D. Then, if the capacities of the two condensers are equal, their charges will be equal since they have been charged to equal potentials. The charge on A will be equal and opposite to that on D, while that on B will be equal and opposite to that on C. Thus, if A be connected with D and C with B, the positive charge on the one plate will counterbalance the negative on the other, so that if after

Fig. 32.

this connection has been made A and B are connected with the electrodes of an electrometer, no deflection will occur.

63. Comparison of two condensers. If a condenser whose capacity can be varied is available, the capacity of a condenser can be compared with known capacities by the following method.

Let A and B (Fig. 33) be the plates of the condenser whose capacity is required, C and D, E and F, G and H, the plates of three condensers whose capacities are known. Connect the plates B and C together and to one electrode of an electrometer, also connect F and G together and to the other electrode of the electrometer. Connect D and

E together and to one pole of a battery, induction coil or other apparatus for producing a difference of potential, and connect A and H together and to the other pole of this battery. In general this will cause a deflection of the electrometer; if there is a deflection, then we must alter the capacity of the condenser whose capacity is variable until the vanishing of this deflection shows that the plates BC, FG are at the same potential. When this is the case a simple relation exists between the capacities.

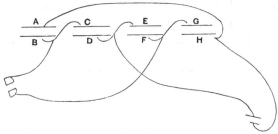

Fig. 33.

Let C_1, C_2, C_3, C_4 be the capacities of the condensers AB, CD, EF, GH respectively, let V_0 be the potential of A and H, x the potential of B and C and y that of F and G, V the potential of D and E. To fix our ideas, let us suppose that V is greater than V_0, then there will be a negative charge on A, a positive one on B, a negative charge on C, and a positive one on D; then since B and C form an insulated system which was initially without charge, the positive charge on B must be numerically equal to the negative charge on C.

The positive charge on B

$$= C_1 (x - V_0),$$

while the negative one on C is numerically equal to

$$C_2(V - x),$$

which is a positive quantity; hence, since these are equal, we have

$$C_1(x - V_0) = C_2(V - x)\dots\dots\dots(1).$$

Again, since F and G are insulated the positive charge on G must be numerically equal to the negative charge on F.

The positive charge on G is equal to

$$C_4(y - V_0),$$

while the negative charge on F is numerically equal to

$$C_3(V - y);$$

since these are equal

$$C_4(y - V_0) = C_3(V - y)\dots\dots\dots(2).$$

When there is no deflection of the electrometer the potential of F and G is equal to that of B and C, i.e. $y = x$. When this is the case we see by comparing equations (1) and (2), that

$$\frac{C_1}{C_4} = \frac{C_2}{C_3},$$

or

$$C_1 = \frac{C_2 C_4}{C_3}.$$

Hence, if we know the capacities of the other condensers, we know C_1.

Thus, if we have standard condensers whose capacities are known, we can measure the capacity of other condensers.

There is a close analogy between the methods of measuring capacity and those of measuring electrical resistance. It is convenient to indicate that analogy here, although the methods of measuring electrical resistance have not yet been discussed.

The arrangement of the condensers in the last method can also be represented by the diagram (Fig. 34). In this diagram C is the coil and G the electrometer. This arrangement is analogous to that of resistances in a Wheatstone's Bridge, see Art. 191, and the condition for

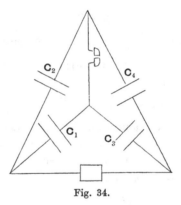

Fig. 34.

the balance of the condensers is the same as that of resistances in a Wheatstone bridge if each condenser were replaced by a resistance inversely proportional to its capacity.

63 a. De Sauty's method. If two of the condensers C_3 and C_4 in the last method are replaced by resistances R_3 and R_4, the electrometer by a galvanometer and the induction coil by a battery with a key for making and breaking the circuit, we get the arrangement known as De Sauty's method, Fig. 35. In this method the re-

sistances R_3 and R_4 are adjusted so that there is no kick
of the galvanometer on making the battery circuit. If i_3
and i_4 are the transient currents flowing through R_3 and
R_4 at some short interval after making the circuit, then
neglecting self-induction, the potential difference at this
time between the terminals of the galvanometer will, by
Ohm's law, be $R_3i_3 - R_4i_4$, and this will be proportional to
the current through the galvanometer at this time. The
quantity of electricity flowing through the galvanometer
during charging will thus be proportional to $\int (R_3i_3 - R_4i_4)dt$.
when the integration extends over the time of charging.

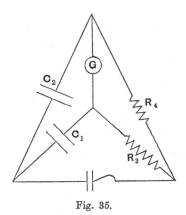

Fig. 35.

If no current flows through the galvanometer, the current
i_3 goes into the condenser (1) and i_4 into condenser (2), so
that

$$\int i_3 dt = Q_1, \quad \int i_4 dt = Q_2,$$

where Q_1 and Q_2 are the final charges in condensers (1)
and (2) respectively. Thus

$$\int (R_3i_3 - R_4i_4)\, dt = R_3Q_1 - R_4Q_2,$$

since there is no kick of the galvanometer this vanishes,
so that
$$R_3 Q_1 = R_4 Q_2.$$
But when the condensers are charged there is the same
potential difference between the plates of (1) as between
those of (2), hence
$$Q_1 : Q_2 = C_1 : C_2,$$
where C_1, C_2 are the capacities of the condensers, hence
when there is no kick of the galvanometer
$$R_3 C_1 = R_4 C_2,$$
thus the ratio C_1/C_2 is found as the ratio of two resistances.
We see that again the condition is the same as for the
balance in a Wheatstone bridge in which the condensers
have been replaced by resistances inversely proportional
to their capacity.

Other methods of determining capacity which require
for their explanation a knowledge of the principles of
electro-magnetism, will be described in the part of the
book dealing with that subject.

64. Leyden jar. A convenient form of condenser
called a Leyden jar is represented in Fig. 36. The

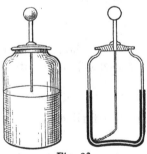

Fig. 36.

condenser consists of a vessel made of thin glass; the inside and outside surfaces of this vessel are coated with tin-foil. An electrode is connected to the inside of the jar in order that electrical connection can easily be made with it. If A is the area of each coat of tin-foil, t the thickness of the glass, i.e. the distance between the surfaces of tin-foil, then, if the interval between these surfaces was filled with air, the capacity would be approximately

$$\frac{A}{4\pi t},$$

since this case is approximately that of two parallel planes provided the thickness of the glass is very small compared with the dimensions of the vessel. The effect of having glass within the tin-foil surfaces will, as we shall see in the next chapter, have the effect of increasing the capacity so that the capacity of the Leyden jar will be

$$K\frac{A}{4\pi t},$$

where K is a quantity which depends on the kind of glass of which the vessel is made. K varies in value from 4 to 10 for different specimens of glass.

Systems of Condensers.

65. If we have a number of condensers we can connect them up so as to make a condenser whose capacity is either greater or less than that of the individual condensers.

Thus suppose we have a number of condensers which in the figures are represented as Leyden jars, and suppose we connect them up as in Fig. 37, that is, connect all the insides of the jars together and likewise all the outsides; this is called connecting the condensers in parallel. We thus get a new condenser, one plate of which consists of all the insides, and the other plate of all the outsides of the jars. If C is the capacity of the compound condenser, Q the total charge in this condenser, V the difference of potential between the plates, then by definition

$$Q = CV.$$

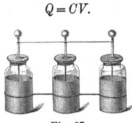

Fig. 37.

If Q_1, Q_2, Q_3, ... are the charges in the first, second, third, etc. condensers, C_1, C_2, C_3, ... the capacities of these condensers

$$Q_1 = C_1 V, \quad Q_2 = C_2 V, \quad Q_3 = C_3 V, \text{ etc.};$$

but $\quad Q = Q_1 + Q_2 + Q_3 + \ldots = (C_1 + C_2 + C_3 + \ldots) V,$

hence $\qquad C = C_1 + C_2 + C_3 + \ldots,$

or the capacity of a system of condensers connected in this way, is the sum of the capacities of its components. Thus the capacity of the compound system is greater than that of any of its components.

Next, let the condensers be connected up as in Fig. 38, where the condensers are insulated, and where the outside of the first is connected to the inside of the second, the outside of the second to the inside of the third, and so on.

This is called connecting the condensers up in cascade or in series. One plate of the compound system thus formed is the inside of the first condenser, the other plate is the outside of the last.

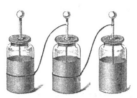

Fig. 38.

Let C be the capacity of the system, C_1, C_2, C_3, ... the capacities of the individual condensers; then, since the condensers are insulated, the charge on the outside of the first is equal in magnitude and opposite in sign to the charge on the inside of the second, the charge on the outside of the second is equal in magnitude and opposite in sign to the charge on the inside of the third, and so on. Since the charge on the inside of any jar is equal and opposite to the charge on the outside, we see that the charges of the jars are all equal. Let Q be the charge of any jar, V_1, V_2... the differences of potential between the inside and outside of the first, second, ... jars. Then

$$V_1 = \frac{Q}{C_1}, \quad V_2 = \frac{Q}{C_2}, \quad V_3 = \frac{Q}{C_3} \dots.$$

If V is the difference of potential between the outside of the last jar and the inside of the first, then

$$V = V_1 + V_2 + V_3 \dots$$

$$= Q \left(\frac{1}{C_1} + \frac{1}{C_2} + \frac{1}{C_3} + \dots \right),$$

so that
$$Q = \frac{V}{\dfrac{1}{C_1} + \dfrac{1}{C_2} + \dfrac{1}{C_3} + \dots} .$$

But, since C is the capacity of the compound condenser of which Q is the charge, and V the potential difference,

$$Q = CV,$$

hence
$$\frac{1}{C} = \frac{1}{C_1} + \frac{1}{C_2} + \frac{1}{C_3} + \dots .$$

Thus the reciprocal of the capacity of the system made by connecting up in cascade the series of condensers, is equal to the sum of the reciprocals of the capacities of the condensers so connected up.

We see that the capacity of the compound condenser is less than that of any of its constituents.

66. If we connect a condenser of small capacity in cascade with a condenser of large capacity, the capacity of the compound condenser will be slightly less than that of the small condenser; while if we connect them in parallel, the capacity of the compound condenser is slightly greater than that of the large condenser.

67. As another example on the theory of condensers, let us take the case when two condensers are connected in parallel, the first having before connection the charge Q_1, the second the charge Q_2. Let C_1 and C_2 be the capacities of these condensers respectively. When they are put in connection they form a condenser whose capacity is $C_1 + C_2$, and whose charge is $Q_1 + Q_2$.

Now the electric energy of a charged condenser is one half the product of the charge into the potential difference, while the potential difference is equal to the charge divided by the capacity. Thus if Q is the charge, C the capacity, the energy is

$$\frac{1}{2}\frac{Q^2}{C}.$$

Thus the total electric energy of the two jars before they are connected is

$$\frac{1}{2}\frac{Q_1^2}{C_1} + \frac{1}{2}\frac{Q_2^2}{C_2};$$

after they are connected it is

$$\frac{1}{2}\frac{(Q_1 + Q_2)^2}{C_1 + C_2};$$

Now

$$\frac{1}{2}\left(\frac{Q_1^2}{C_1} + \frac{Q_2^2}{C_2}\right) - \frac{1}{2}\frac{(Q_1 + Q_2)^2}{(C_1 + C_2)}$$

$$= \frac{1}{2C_1C_2(C_1 + C_2)}(C_2^2Q_1^2 + C_1^2Q_2^2 - 2C_1C_2Q_1Q_2)$$

$$= \frac{1}{2C_1C_2(C_1 + C_2)}(C_2Q_1 - C_1Q_2)^2,$$

an essentially positive quantity which only vanishes if

$$Q_1/C_1 = Q_2/C_2,$$

that is, when the potentials of the jars before connection are equal. In this case the energy after connection is the same as before the connections are made. If the potentials are equal before connection, connecting the jars will evidently make no difference, as all that connection does is to make the potentials equal. In every other case electric energy is lost when the connection is made; this energy is accounted for by the work done by the spark which passes when the jars are connected.

CHAPTER IV

68. Specific Inductive Capacity. Faraday found that the charge in a condenser between whose surfaces a constant difference of potential was maintained depended upon the nature of the dielectric between the surfaces, the charge being greater when the interval between the surfaces was filled with glass or sulphur than when it was filled with air.

Thus the 'capacity' of a condenser (see Art. 51) depends upon the dielectric between the plates. Faraday's original experiment by which this result was established was as follows: he took two equal and similar condensers, A and B, of the kind shown in Fig. 39, made of concentric spheres; in one of these, B, there was an opening by which melted wax or sulphur could be run into the interval between the spheres. The insides of these condensers were connected together, as were also the outsides, so that the potential difference between the plates of the condenser was the same for A as for B. When air was the dielectric between the spheres Faraday found, as might have been expected from the equality of the condensers, that any charge given to the condensers was equally distributed between A and B. When however the interval in B was filled with sulphur and the condensers again charged he found that the charge in *B* was three or four times that in *A*, proving that the capacity of *B* had been increased three or four times by the substitution of sulphur for air.

Fig. 39.

This property of the dielectric is called its specific inductive capacity. The measure of the specific inductive capacity of a dielectric is defined as the ratio of the capacity of a condenser when the region between its plates is entirely filled by this dielectric, to the capacity of the same condenser, when the region between its plates is entirely filled with air. As far as we know at present, the specific inductive capacity of a dielectric in a condenser does not depend upon the difference of potential established between the plates of that condenser, that is, upon the electric intensity acting on the dielectric. We may therefore conclude that, at any rate for a wide range of electric intensities, the specific inductive capacity is independent of the electric intensity.

The following table contains the values of the specific inductive capacities of some substances which are frequently used in a physical laboratory:

Solid paraffin	2·29.
Paraffin oil	1·92.
Ebonite	3·15.
Sulphur	3·97.
Mica	6·64.
Dense flint-glass	7·37.
Light flint-glass	6·72.
Turpentine	2·23.
Distilled water	76.
Alcohol	26.

The specific inductive capacity of gases depends upon the pressure, the difference between K, the specific inductive capacity, and unity being directly proportional to the pressure.

The specific inductive capacity of some gases at atmospheric pressure is given in the following table; the specific inductive capacity of air at atmospheric pressure is taken as unity:

Hydrogen	·999674.
Carbonic acid	1·000356.
Carbonic oxide	1·0001.
Olefiant gas	1·000722.

69. It was the discovery of this property of the dielectric which led Faraday to the view we have explained, in Art. 38, that the effects observed in the electric field are not due to the action at a distance of one electrified body on another, but are due to effects in the dielectric filling the space between the electrified bodies.

The results obtained in Chapters II and III were deduced on the supposition that there was only one dielectric, air, in the field; these require modification in the general case when we have any number of dielectrics in the field. We shall now go on to consider the theory of this general case.

We assume that each unit of positive electricity, whatever be the medium by which it is surrounded, is the origin of a Faraday tube, each unit of negative electricity the termination of one. Let us consider from this point of view the case of two parallel plate condensers A and B, the plates of A and of B being at the same distance apart, but while the plates of A are separated by air, those of B are separated by a medium whose specific inductive capacity is K. Let us suppose that the charge per unit area on the plates of the condensers A and B is the same. Then, since the capacity of the condenser B is K times that of A and since the charges are equal, the potential difference between the plates of B is only $1/K$ of that between the plates of A.

Now if V_P is the potential at P, V_Q that at Q, R the electric intensity along PQ, then, whatever be the nature of the dielectric, when PQ is small enough to allow of the intensity along it being regarded as constant,

$$R \cdot PQ = V_P - V_Q \dots\dots\dots\dots\dots(1),$$

for by definition R is the force on unit charge, hence the left-hand side of this expression is the work done on unit charge as it moves from P to Q, and is thus by definition, (Art. 16), equal to the right-hand side of (1).

The electric intensity between the plates both of A and of B is uniform, and is equal to the difference of potential between the plates divided by the distance between the plates; this distance is the same for the plates A and B, so that the electric intensity between the plates of A is to that between the plates of B as the potential difference between the plates of A is to that between the plates of B. That is, the electric intensity in A is K times that in B.

Consider now these two condensers. Since the charges on unit area of the plates are equal the number of Faraday tubes passing through the dielectric between the plates is the same, while the electric intensity in B is only $1/K$ that in air. Hence we conclude that when the number of Faraday tubes which pass through unit area of a dielectric whose specific inductive capacity is K is the same as the number which pass through unit area in air, the electric intensity in the dielectric is $1/K$ of the electric intensity in air.

By Art. 40, we see that if N is the number of Faraday tubes passing through unit area in air, and R is the electric intensity in air,

$$R = 4\pi N.$$

Hence, when N tubes pass through unit area in a medium whose specific inductive capacity is K, the electric intensity, R, in this dielectric is given by the equation

$$R = \frac{4\pi}{K} N.$$

70. Polarization in a dielectric. We define the polarization in the direction PQ where P and Q are two points close together as the excess of number of Faraday tubes which pass from the side P to the side Q over the number which pass from the side Q to the side P of a plane of unit area drawn between P and Q at right angles to PQ. We may express the result in Art. 68 in the form

(electric intensity in any direction at P)

$= \dfrac{4\pi}{K}$ (polarization in the dielectric in that direction at P).

The polarization in a dielectric is mathematically identical with the quantity called by Maxwell the electric displacement in the dielectric.

71. Thus the polarization along the outward-drawn normal at P to a surface is the excess of the number of Faraday tubes which leave the surface through unit area at P over the number entering it. If we divide any closed surface up as in Art. 9 into a number of small meshes, each of these meshes being so small that the polarization over the area of any mesh may be regarded as constant, then if we multiply the area of each of the meshes by the normal polarization at this mesh measured outwards, the sum of the products taken for all the meshes which cover the surface is defined to be the total normal polarization outwards from the surface. We see that it is equal to the excess of the number of Faraday tubes which leave the surface over the number which enter it.

Now consider any tube which does not begin or end inside the closed surface, then if it meets the surface at all it will do so at two places, P and Q; at one of these it will be going from the inside to the outside of the surface, at the other from the outside to the inside. Such a tube will not contribute anything to the total normal polarization outwards from the surface, for at the place where it leaves the surface it contributes $+1$ to this quantity, which is neutralized by the -1 which it contributes at the place where it enters the surface.

Now consider a tube starting inside the surface; this tube will leave the surface but not enter it, or if the surface is bent so that the tube cuts the surface more than once, it will leave the surface once oftener than it enters it. This tube will therefore contribute $+1$ to the total outward normal polarization: similarly we may show that each tube which ends inside the surface contributes -1 to the total outward normal polarization. Thus if there are N tubes which begin, and M tubes which end inside the surface, the total normal polarization is equal to $N - M$. But each tube which begins inside the surface corresponds to a unit positive charge, each tube which ends in the surface to a unit negative one, so that $N - M$ is the difference between the positive and negative charges inside the surface, that is, it is the total charge inside the surface.

Thus we see that the total normal polarization over a closed surface is equal to the charge inside the surface. Since the normal polarization is equal to $K/4\pi$ times the normal intensity where K is the specific inductive capacity, which is equal to unity for air, we see that when the dielectric is air the preceding theorem is identical with Gauss's theorem, Art. 10. In the form stated above it is applicable whatever dielectrics may be in the field, when in general Gauss's theorem as stated in Art. 10 ceases to be true.

72. Modification of Coulomb's equation. If σ is the surface density of the electricity on a conductor, then σ Faraday tubes pass through unit area of a plane drawn in the dielectric just above the conductor at right angles to the normal. Hence σ is the polarization in the dielectric in the direction of the normal to the conductor. Hence, by Art. 69, if R is the normal electric intensity

$$R = \frac{4\pi}{K}\sigma.$$

This is Coulomb's equation generalized so as to apply to the case when the conductor is in contact with any dielectric.

73. Expression for the Energy. The student will see that the process of Art. 23 by which the expression $\frac{1}{2}\Sigma(EV)$ was proved to represent the electric energy of the system will apply whatever the nature of the dielectric may be, as will also the immediate deduction from it in Art. 43 that the energy is the same as it would be if each Faraday tube possessed an amount of energy equal per unit length to one-half the electric intensity.

The expression for the energy per unit volume however requires modification. Consider, as in Art. 43, a cylinder whose axis is parallel to the electric intensity and whose flat ends are at right angles to it, let l be the length of the cylinder, ω the area of one of the ends, P the polarization, R the electric intensity. Then the portion of each Faraday tube inside the cylinder has an amount of energy equal to

$$\tfrac{1}{2}lR.$$

Now the number of such tubes inside the cylinder is equal to $P\omega$, hence the energy inside the cylinder is equal to

$$\tfrac{1}{2} l\omega PR.$$

Since $l\omega$ is the volume of the cylinder, the energy per unit volume is equal to

$$\tfrac{1}{2} PR;$$

but by Art. 69 $\qquad P = \dfrac{K}{4\pi} R,$

so that the energy per unit volume is equal to

$$\frac{K}{8\pi} R^2.$$

Thus, for the same electric intensity the energy per unit volume of the dielectric is K times as great as it is in air. Another expression for the energy per unit volume is

$$\frac{2\pi}{K} P^2,$$

so that for same polarization the energy per unit volume in the dielectric is only $1/K$th part of what it is in air.

We see, as in Art. 45, that the pull along each Faraday tube will still equal one-half the electric intensity R; the tension across unit area in the dielectric will therefore be $\dfrac{KR^2}{8\pi}$, the lateral pressure will also be equal to $KR^2/8\pi$.

74. Conditions to be satisfied at the boundary between two media of different specific inductive capacities. Suppose that the line AB represents the section by the plane of the paper of the plane of separation between two different dielectrics; let the specific inductive capacities of the upper and lower media respectively be K_1, K_2.

Let us consider the conditions which must hold at the surface. In the first place we see that the electric intensities parallel to the surface must be equal in the two media; for if they were not equal, and that in the medium K_1 were the greater, we could get an infinite amount of work by making unit charge travel round the closed circuit $PQRS$, PQ being just above, and RS just below the surface of separation. For, if PQ is the direction of T_1 the tangential component of the electric intensity in the upper medium, the work done on unit

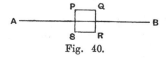

Fig. 40.

charge as its goes from P to Q is $T_1 \cdot PQ$; as QR is exceedingly small compared with PQ the work done on or by the charge as it goes from Q to R may be neglected if the normal intensity is not infinite; the work required to take the unit charge back from R to S is $T_2 \cdot RS$, if T_2 is the tangential component of the electric intensity in the lower dielectric, and the work done or spent in going from S to P will be equal to that spent or done in going from Q to R and may be neglected. Thus since the system is brought back to the state from which it started, the work done must vanish, and hence $T_1 \cdot PQ - T_2 \cdot RS$ must be zero. But since $PQ = RS$ this requires that $T_1 = T_2$ or the tangential components of the electric intensity must be the same in the two media.

Next suppose that σ is the surface density of the free electricity on the surface separating the two media. Draw a very flat circular cylinder shown in section at $PQRS$, the axis of this cylinder being parallel to the normal to the surface of separation, the top face of this cylinder being just above, the lower face just below this surface. As the length of this cylinder is very small compared with its breadth, the area of the curved surface of the cylinder will be very small compared with the area of its ends, and by making the cylinder sufficiently short we can make the ratio of the area of the curved surface to that of the ends as small as we please. Hence in considering the total outward normal polarization over the very short cylinder, we may leave out the effect of the curved surface and consider only the flat ends of the cylinder. But since the cylinder encloses the charge $\sigma\omega$, if ω is the area of one end of the cylinder, the total normal polarization over its surface must be equal to $\sigma\omega$. If N_1 is the normal polarization in the first medium measured upwards the total normal polarization over the top of the cylinder is $N_1\omega$; if N_2 is the normal polarization measured upwards in the second medium, the total normal polarization over the lower face of the cylinder is $-N_2\omega$; hence the total outward normal polarization over the cylinder is

$$N_1\omega - N_2\omega.$$

Since, by Art. 71, this is equal to $\sigma\omega$, we have

$$N_1 - N_2 = \sigma.$$

When there is no charge on the surface separating the two dielectrics, these conditions become (1) that the tangential electric intensities, and (2) the normal polarizations, must be equal in the two media.

75. Refraction of the lines of force. Suppose that R_2 is the resultant electric intensity in the upper medium, R_2 that in the lower; and θ_1, θ_2 the angles these make with the normal to the surface of separation. The tangential intensity in the first medium is $R_1 \sin \theta_1$, that in the second is $R_2 \sin \theta_2$, and since these are equal

$$R_1 \sin \theta_1 = R_2 \sin \theta_2 \ldots\ldots\ldots\ldots\ldots(1).$$

The normal intensity in the upper medium is $R_1 \cos \theta_1$, hence the normal polarization in the upper medium is

$$K_1 R_1 \cos \theta_1/4\pi,$$

that in the second is $K_2 R_2 \cos \theta_2/4\pi$, and since, if there is no charge on the surface, these are equal, we have

$$\frac{K_1}{4\pi} R_1 \cos \theta_1 = \frac{K_2}{4\pi} R_2 \cos \theta_2 \ \ldots\ldots\ldots\ldots(2);$$

dividing (1) by (2), we get

$$\frac{1}{K_1} \tan \theta_1 = \frac{1}{K_2} \tan \theta_2.$$

Hence, if $K_1 > K_2$, θ_1 is $> \theta_2$, and thus when a Faraday tube enters a medium of greater specific inductive capacity from one of less, it is bent away from the normal.

This is shown in the diagram Fig. 41 (from Lord Kelvin's Reprint of Papers on Electrostatics and Magnetism), which represents the Faraday tubes when a sphere, made of paraffin or some material whose specific inductive capacity is greater than unity, is placed in a field of uniform force such as that between two infinite parallel plates.

An inspection of the diagram shows the tendency of the tubes to run as much as possible through the sphere; this is an example of the principle that when a system is in stable equilibrium the potential energy is as small as possible. We saw, Art. 73, that when the polarization is P the energy per unit volume is $2\pi P^2/K$, thus for the same value of P, this quantity is less in paraffin than it is in air. Hence when the same number of tubes pass through the paraffin they have less energy in unit volume than when they pass through air, and there is therefore a tendency for the tubes to flock into the paraffin. The reason why all the tubes do not run into the sphere is that those which are some distance away from it would have to bend considerably in order to reach the paraffin,

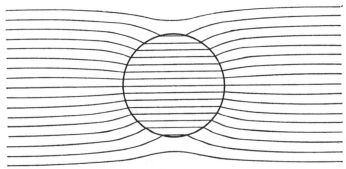

Fig. 41.

they would therefore have to greatly lengthen their path in the air, and the increase in the energy consequent upon this would not be compensated for in the case of the tubes some distance originally from the sphere by the diminution in the energy when they got in the sphere.

In Fig. 42 (from Lord Kelvin's Reprint of Papers on Electrostatics and Magnetism) the effect produced on a field of uniform force by a conducting sphere is given for comparison with the effects produced by the paraffin sphere. It will be noticed that the paraffin sphere produces effects similar in kind though not so great in degree as those due to the conducting sphere. This observation is true for all electrostatic phenomena, for we find that bodies having a greater specific inductive capacity than the surrounding dielectric behave in a similar way to conductors. Thus, they deflect the Faraday tubes in the same way though not to the same extent; again, as a conductor tends to move from the weak to the strong parts of

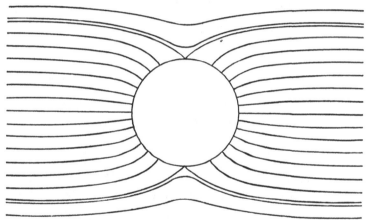

Fig. 42.

the field, so likewise does a dielectric surrounded by one of smaller specific inductive capacity. Again, the electric intensity inside a conductor vanishes, and just inside a dielectric of greater specific inductive capacity than the surrounding medium the electric intensity is less than that just outside. As far as electrostatic phenomena are concerned an insulated conductor behaves like a dielectric of infinitely great specific inductive capacity.

76. Force between two small charged bodies immersed in any dielectric. If we have a small body with a charge e immersed in a medium whose specific inductive capacity is K, then the polarization at a distance r from the body is $e/4\pi r^2$. To prove this, describe a sphere radius r, with its centre at the small body, then the polarization P will be uniform over the surface of the sphere and radial; hence the total normal polarization over the surface of the sphere will equal $P \times$ (surface of the sphere), i.e. $P \times 4\pi r^2$; but this, by Art. 71, is equal to e, hence

$$P \times 4\pi r^2 = e,$$

or
$$P = \frac{e}{4\pi r^2} \dots \dots \dots (1).$$

But, if R is the electric intensity, then, by Art. 70

$$R = \frac{4\pi}{K} \cdot P.$$

Hence, by (1), $R = \dfrac{e}{Kr^2}$;

the repulsion on a charge e' is Re', or ee'/Kr^2; hence the repulsion between the charges, when separated by a distance r in a dielectric whose specific inductive capacity is K, is only $1/K$th part of the repulsion between the charges when they are separated by the same distance in air. Thus, when the charges are given, the mechanical forces on the bodies in the field are diminished when the charges are imbedded in a medium with a large specific inductive capacity. We can easily show that the interposition between the charges of a spherical shell of the dielectric with its centre at either of the charges would not affect the force between these charges.

77. Two parallel plates separated by a dielectric. Let us first take the case of two parallel plates completely immersed in an insulating medium whose specific inductive capacity is K. Let V be the potential difference between the plates, σ the surface density of the electrification on the positive plate, and $-\sigma$ that on the negative. Let R be the electric intensity between the plates, and d the distance by which they are separated; then, by Art. 72,

$$4\pi\sigma = KR$$
$$= \frac{KV}{d}.$$

The force on one of the plates per unit area is, by Art. 37,

$$\tfrac{1}{2}R\sigma$$
$$= \frac{2\pi\sigma^2}{K}.$$

Hence if the charges are given the force between the plates is *inversely* proportional to the specific inductive capacity of the medium in which they are immersed.

Again, since

$$\frac{1}{2}R\sigma = \frac{1}{8\pi}KR^2 = \frac{K}{8\pi}\frac{V^2}{d^2},$$

we see that, if the potentials of the plates be given, the attraction between them is *directly* proportional to the specific inductive capacity. This result is an example of the following more general one which we leave to the reader to work out; if in a system of conductors maintained at given potentials and originally separated from each other by air we replace the air by a dielectric whose specific inductive capacity is K, keeping the positions of the conductors and their potentials the same as before, the forces between the conductors will be increased K times.

Thus, for example, if we fill the space between the needles and the quadrants of an electrometer with a fluid whose specific inductive capacity is K, keeping the potentials of the needles and quadrants constant, the couple on the needle will be increased K times by the introduction of the fluid. Thus, if we measure the couples before and after the introduction of the fluid, the ratio of the two will give us the specific inductive capacity of the fluid. This method has been applied to measure the specific inductive capacity of those liquids, such as water or alcohol, which are not sufficiently good insulators to allow the method described in Art. 82 to be applied.

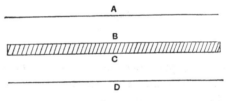

Fig. 43.

78. We shall next consider the case in which a slab of dielectric is placed between two infinite parallel conducting planes, the faces of the slab being parallel to the planes.

Let d be the distance between the planes, t the thickness of the slab, h the distance between the upper face of the slab and the upper plane. The Faraday tubes will go straight across from plane to plane, so that the polarization will be everywhere normal to the conducting planes and to the planes separating the slab of dielectric from the air.

We saw in Art. 74 that the normal polarization does not change as we pass from one medium to another, and as the tubes are straight the polarization will not change as long as we remain in one medium. Thus the polarization which we shall denote by P is constant between the planes. In air the electric intensity is $4\pi P$; in the dielectric of specific inductive capacity K, the electric intensity is equal to $4\pi P/K$.

Thus between A and B the electric intensity is $4\pi P$,

.... B and C .... $\dfrac{4\pi P}{K}$,

.... C and D .... $4\pi P$.

The difference of potential between the plates is the work done on unit charge when it is taken from one plate to the other. Now, when unit charge is taken across the space AB, the work done on it is

$$4\pi P \times h;$$

when it is taken across the plate of dielectric the work done is

$$\frac{4\pi P}{K} \times t;$$

when it is taken across CD the work done is

$$4\pi P \left\{d - (h + t)\right\}.$$

Hence V, the excess of the potential of the plate A above that of D, is equal to

$$4\pi Ph + \frac{4\pi P}{K} t + 4\pi P \left\{d - (h + t)\right\}$$
$$= 4\pi P \left\{d - t \left(1 - \frac{1}{K}\right)\right\}.$$

If σ is the surface density of the electricity on the positive plate, $\sigma = P$, so that

$$V = 4\pi\sigma \left(d - t + \frac{t}{K} \right) \quad \ldots\ldots\ldots\ldots(1).$$

Hence the capacity per unit area of the plate, i.e. the value of σ when $V = 1$, is

$$\frac{1}{4\pi \left(d - t + \dfrac{t}{K} \right)},$$

i.e. it is the same as if the plate of dielectric were replaced by a plate of air whose thickness was t/K. The presence of the dielectric increases the capacity of the condenser. The alteration in the capacity does not depend upon the position of the slab of dielectric between the parallel plates.

Let us now consider the force between the plates; the force per unit area

$$= \tfrac{1}{2} R\sigma,$$

where R is the electric intensity at the surface of the plate; but, since the surface of the plate is in contact with air, $R = 4\pi\sigma$, and thus the force per unit area on either plate

$$= 2\pi\sigma^2.$$

Hence if the charges on the plates are given, the attraction between them is not affected by the interposition of the plate of dielectric.

Next, let the potentials be given; we see from equation (1) that

$$\sigma = \frac{V}{4\pi \left(d - t + \dfrac{t}{K} \right)};$$

hence $2\pi\sigma^2$, the force per unit area, is equal to

$$\frac{V^2}{8\pi\left(d-t+\dfrac{t}{K}\right)^2}.$$

The force between the plates when there is nothing but air between them, is

$$\frac{V^2}{8\pi d^2}.$$

Now since K is greater than 1, $d-t+t/K$ is less than d, so that $1/(d-t+t/K)^2$ is greater than $1/d^2$. Thus, when the potentials are given, the force between the plates is increased by the interposition of the dielectric.

If K be very great, t/K is very small, thus $d-t+t/K$ is very nearly equal to $d-t$, and the effect of the interposition of the slab of dielectric both on the capacity and on the force between the plates is approximately the same as if the plates had been pushed towards each other through a distance equal to the thickness of the slab, the dielectric between the plates being now supposed to be air. This result, which is approximately true whenever the specific inductive capacity of the slab is very large, is rigorously true when the slab is made of a conducting material.

Effect of the slab of dielectric on the potential energy for given charges. The potential energy is, by Art. 23, equal to

$$\tfrac{1}{2}\Sigma(EV),$$

and thus the energy corresponding to the charge on each unit of area of the plates is equal to

$$\tfrac{1}{2}\sigma V;$$

by equation (1) this is equal to

$$2\pi\sigma^2 \left\{ d - t \left(1 - \frac{1}{K} \right) \right\},$$

and it is thus when $K > 1$ less than $2\pi\sigma^2 d$, the value of the energy for the same charges when no slab of dielectric is interposed. The interposition of the slab thus lowers the potential energy. We can easily see why this is the case. When the charges are given the number of Faraday tubes is given: and, when the plate of dielectric is interposed, the Faraday tubes in part of their journey between the plates are in the dielectric instead of in air, and we know from Art. 73 that when the Faraday tubes are in the dielectric their energy is less than when they are in air. Since the potential energy of a system always tends to become as small as possible, there will be a tendency to drag as much as possible of the slab of dielectric between the plates of the condenser. Thus, if the slab of dielectric projected on one side beyond the plates it would be drawn between the plates until as much of its area as possible was within the region between the plates.

Effect of the slab on the potential energy for a given difference of potential. The energy per unit area of the plates is as we have seen equal to

$$\tfrac{1}{2}\sigma V;$$

this by equation (1) is equal to

$$\frac{1}{8\pi} \frac{V^2}{\left\{ d - t \left(1 - \frac{1}{K} \right) \right\}}.$$

If the potential difference is given the energy when no slab is interposed is

$$\frac{V^2}{8\pi d},$$

so that when the potential difference is kept constant the electric energy is increased by the interposition of the slab.

79. Capacity of two concentric spheres with a shell of dielectric interposed between them. If we have two concentric conducting spheres with a concentric shell of dielectric between them, and if e be the charge on the inner sphere, a the radius of this sphere and b, c the radii of the inner and outer surfaces of the dielectric shell, and d the inner radius of the outer conducting sphere, then if V be the difference of potential between the conducting spheres, and K the specific inductive capacity of the shell, we may easily prove that

$$V = e \left\{ \frac{1}{a} - \frac{1}{b} + \frac{1}{K}\left(\frac{1}{b} - \frac{1}{c}\right) + \frac{1}{c} - \frac{1}{d} \right\}.$$

Thus the capacity of the system is equal to

$$\frac{1}{\dfrac{1}{a} - \dfrac{1}{d} - \left(1 - \dfrac{1}{K}\right)\left(\dfrac{1}{b} - \dfrac{1}{c}\right)}.$$

80. Two coaxial cylinders. As another example, we shall take the case of two coaxial cylinders with a coaxial cylindric shell of a dielectric, specific inductive capacity K, placed between them. If V be the difference of potential between the two conducting cylinders, E the charge per unit length on the inner cylinder, a the radius of this inner cylinder, b and c the radii of the inner and outer surfaces of the dielectric shell and d the inner radius of the outer cylinder, we easily find by the aid of Art. 58 that

$$V = 2E \log \left\{ \frac{b}{a} + \frac{1}{K} \log \frac{c}{b} + \log \frac{d}{c} \right\},$$

so that the capacity per unit length of this system is

$$\frac{1}{2 \left\{ \log \dfrac{b}{a} + \dfrac{1}{K} \log \dfrac{c}{b} + \log \dfrac{d}{c} \right\}}.$$

81. Force on a piece of dielectric placed in an electric field. If a piece of a dielectric such as sulphur or glass is placed in the electric field, then, when the Faraday tubes traverse the dielectric there is, Art. 73, less energy per unit volume than when the same number of Faraday tubes pass through air. Thus, as we see in Fig. 39, the Faraday tubes tend to run through the dielectric, because by so doing the potential energy is decreased. If the dielectric is free to move, it can still further decrease the energy by moving from its original position to one where the tubes are more thickly congregated, because the more tubes which get through the dielectric the greater the decrease in the potential energy. The body will tend to move so as to make the decrease in the energy as great as possible, thus it will tend to move so as to be traversed by as great a number of Faraday tubes as possible. It will therefore be urged towards the part of the field where the Faraday tubes are densest, i.e. to the strongest parts of the field. There will thus be a force on a piece of dielectric tending to make it move from the weak to the strong parts of the field. The dielectric will not move except in a variable field where it can get more Faraday tubes by its change of position. In a uniform field such as that between two parallel infinite plates the dielectric would have no tendency to move.

The force acting upon the dielectric differs in another respect from that acting on a charged body, inasmuch as it would not be altered if the direction of the electric intensity at each point in the field were reversed without altering its magnitude.

82. Measurement of specific inductive capacity. The specific inductive capacity of a slab of dielectric can be measured in the following way, provided we have a parallel plate condenser one plate of which can be moved by means of a screw through a distance which can be accurately measured. To avoid the disturbance due to the irregular distribution of the charge near the edges of the plates (see Art. 57) care must be taken that the distance between the plates never exceeds a small fraction of the diameter of the plates. Let us call this parallel plate condenser A; to use the method described in Art. 63, first take the condenser A and before inserting the slab of dielectric adjust the other variable condenser used in that method until there is no deflection of the electrometer. If the slab of dielectric be now inserted between the plates of A the capacity will be increased, A will no longer be balanced by the other condensers and the electrometer will be deflected. The capacity of A can be diminished by screwing the plates further apart, and when the plates have been moved through a certain distance, the diminution in the capacity due to the increase in the distance between the plates will balance the increase due to the insertion of the slab of dielectric; the stage when this occurs will be indicated by there being again no deflection of the electrometer.

Suppose that when the deflection of the electrometer is zero before the slab is inserted, the distance between the plates of the condenser is d, while the distance after the slab is inserted, when the electrometer is again in equilibrium, is d'. Then the capacity of A in these two cases is the same. But if A is the area of the plate of A the capacity before the slab is inserted is

$$\frac{A}{4\pi d}.$$

If t is the thickness of the slab and K its specific inductive capacity, the capacity after the insertion of the slab is (see Art. 78) equal to

$$\frac{A}{4\pi \left(d' - t + \dfrac{t}{K} \right)},$$

but since the capacities are equal

$$d = d' - t + \frac{t}{K},$$

so that

$$d' - d = t \left\{ 1 - \frac{1}{K} \right\}.$$

But $d' - d$ is the distance through which the plate has been moved, so that if we know this distance and t we can determine K the specific inductive capacity of the slab. It should be noticed that this method does not require a knowledge of the initial or final distances between the plates, but only the difference of these quantities, and this can be measured with great accuracy by the screw attached to the moveable plate.

CHAPTER V.

ELECTRICAL IMAGES AND INVERSION.

83. We shall now proceed to discuss some geometrical methods by which we can find the distribution of electricity in several very important cases. We shall illustrate the first method by considering a very simple example; that of a very small charged body placed in front of an infinite conducting plane maintained at potential zero. Let P, Fig. 44, be the charged body, AB the conducting plane.

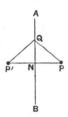

Fig. 44.

Any solution of the problem must satisfy the following conditions in the region to the right of the plane AB; (α) it must make the potential zero over the plane AB, and (β) it must make the total outward normal induction taken over any closed surface enclosing P equal to $4\pi e$, where e is the charge at P, while if the closed surface does not enclose P the total normal induction over it must vanish. We shall now prove that there is only one solution which satisfies these conditions. Suppose there were two different solutions, which we shall call (1) and (2). Take the solution corresponding to (2) and reverse the sign of all the charges of electricity in the field, including that at P; this new solution, which we shall denote by (-2), will correspond to a field in which the electric intensity at any point is equal and opposite to that due to the solution (2) at the same point. The solution (-2) corresponds to a field in which the electric potential is zero over AB and at any point at an infinite distance from P; it also makes the total normal induction over any closed surface enclosing P equal to $-4\pi e$, that is equal and opposite to the total induction over the same surface due to the solution (1); and the total induction over any other closed surface in the region to the right of AB zero. Now consider the field got by superposing the solutions (1) and (-2): it will have the following properties; the potential over AB will be zero and the total normal induction over any closed surface in the region to the right of AB will vanish. Since the normal induction vanishes over all closed surfaces in this region, there will in the field corresponding to this solution be no charge of electricity. We may regard the region as the inside of a closed surface at zero potential (bounded by the plane AB and an equipotential surface at an infinite distance): by Art. 18, however, the electric intensity must vanish throughout this region as there is no charge inside it. Thus, the electric intensity in the field corresponding to the superposition of the solutions (1) and (-2) is zero: that is, the electric intensity in the solution (1) is equal and opposite to that in (-2). But the electric intensity in (-2) is equal and opposite to that in (2). Hence the electric intensity in (1) is at all points the same as (2), in other words, the solutions give identical electric fields. Hence, if we get in any way a solution satisfying the conditions (α) and (β), it must be the only solution of the problem.

84. Let P' be a point on the prolongation of the perpendicular PN let fall from P on the plane, such that $P'N = PN$, and let a charge equal to $-e$ be placed at P'. Consider the properties, in the region to the right of AB, of the field due to the charge e at P and the charge $-e$ at P'.

The potential due to $-e$ at P' and $+e$ at P at a point Q on the plane AB is equal to

$$\frac{e}{PQ} - \frac{e}{P'Q}.$$

But since AB bisects PP' at right angles $PQ = P'Q$, thus the potential at Q vanishes. Again, any closed surface drawn in the region to the right of the plane AB does not enclose P', and thus the charge at P' is without effect upon the total induction over any such surface. The total induction over such a surface is zero or $4\pi e$ according as the closed surface does not or does include P. In the region to the right of AB the electric field due to e at P and $-e$ at P' thus satisfies the conditions (α) and (β) and therefore represents the state of the electric field. Thus the electrical effect of the electricity induced on the conducting plane AB will be the same as that of the charge $-e$ at P' at all points to the right of AB. This charge at P' is called the electrical image of the charge P in the plane.

The attraction on P towards the plane will be the same as the attraction between the charges e at P, and $-e$ at P', that is

$$\frac{e^2}{(2PN)^2} = \frac{1}{4}\frac{e^2}{PN^2}.$$

Thus the attraction on the charged body varies inversely as the square of its distance from the plane.

To find the surface density of the electricity induced on the plane AB we require the electric intensity at right angles to the plane. The electric intensity at right angles to the plane AB at a point Q on the plane due to the charge e at P is equal to

$$\frac{e}{PQ^2}\cdot\frac{PN}{PQ},$$

and acts from right to left. The electric intensity at Q due to $-e$ at P' in the same direction is

$$\frac{e}{P'Q^2}\cdot\frac{P'N}{P'Q}.$$

Hence since $PQ = P'Q$ and $PN = P'N$ the resultant normal electric intensity at Q is

$$\frac{2ePN}{PQ^3}.$$

This, by Coulomb's law, is equal to $4\pi\sigma$, if σ is the surface density of the electricity at Q, and hence

$$\sigma = \frac{e}{2\pi}\frac{PN}{PQ^3},$$

or the surface density varies inversely as the cube of the distance from P.

The total charge of electricity on the plane is $-e$, as all the tubes which start from P end on the plane.

The electrical energy is equal to $\frac{1}{2}\Sigma EV$, so that if the small body at P is a sphere of radius a, the energy in the field is equal to

$$\frac{1}{2}\frac{e^2}{a} - \frac{1}{4}\frac{e^2}{PN}.$$

The dielectric in this case is supposed to be air. The electric intensity vanishes in the region to the left of AB

85. Electrical images for spherical conductors
In applying the method of images to spherical conductors we make great use of the following theorem due to Apollonius. If S, Fig. 45, is a point on a sphere whose centre is O and radius a, and P and Q are two fixed points on a straight line passing through O, such that $OP \cdot OQ = a^2$, then QS/PS is constant wherever S may be on the sphere.

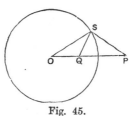

Fig. 45.

Consider the triangles QOS, POS. Since

$$OQ \cdot OP = OS^2, \quad \frac{OQ}{OS} = \frac{OS}{OP},$$

hence these triangles have the angle at O common and the sides about this angle proportional. They are therefore similar triangles, so that

$$\frac{QS}{OQ} = \frac{PS}{OS},$$

or
$$\frac{QS}{PS} = \frac{OQ}{OS} = \frac{OS}{OP}.$$

Hence QS/PS is constant whatever may be the position of S on the sphere.

86. Now suppose that we have a spherical shell (Fig. 45) at potential zero whose centre is at O and that a small body with a charge e of electricity is placed at P and that we wish to find the electric field outside the sphere. There is no field inside the sphere, as the sphere is an equipotential surface with no charge inside it.

Let $OP = f, OS = a$. Consider the field due to a charge e at P, and e' at Q where $OQ . OP = a^2$. The potential at a point S on the sphere due to the two charges is

$$\frac{e}{PS} + \frac{e'}{QS}.$$

But by Art. 85,

$$QS = PS . \frac{a}{f}.$$

Thus the potential at $S = \left\{ e + e' \frac{f}{a} \right\} \frac{1}{PS}.$

Hence, if $e' = -ea/f$, the potential is zero over the surface. Thus, under these circumstances the field satisfies condition (α) of Art. 83, and it obviously satisfies the condition that the total normal induction over any closed surface not enclosing the sphere is zero or $4\pi e$ according as the surface does not or does enclose P, so that, by Art. 83 this is the actual field due to the sphere and the charged body. Hence, at a point outside the sphere, the effect of the electricity induced on the sphere by the

charge at P is the same as that of a charge $-ea/f$ at Q.
This charge at Q is called the *electrical image of P in the
sphere*. Since this charge produces the same effect as
the electrification on the sphere, the total charge on the
sphere must equal the charge at Q, i.e. it must be equal to
$-ea/f$ (compare Art. 30). Thus of the Faraday tubes
which start from P the fraction a/f fall on the sphere.

The force on P is an attraction towards the sphere and
is equal to

$$\frac{a}{f}\frac{e^2}{PQ^2} = \frac{a}{f}\frac{e^2}{(OP-OQ)^2} = \frac{a}{f}\frac{e^2}{\left(f-\dfrac{a^2}{f}\right)^2} = \frac{e^2 fa}{(f^2-a^2)^2}.$$

We see from this result that, when the distance of P
from the centre of the sphere is large compared with the
radius, the force varies inversely as the *cube* of the
distance from the centre of the sphere: while when P
is close to the surface of the sphere the force varies
inversely as the *square* of the distance from the nearest
point on the surface of the sphere. When P is very
near to the surface of the sphere, the problem becomes
practically identical with that of a charge placed in front
of a plane at potential zero. We shall leave it as an
exercise for the student to deduce the solution for the
plane as the limit of that of the sphere.

If the body at P is a small sphere of radius b, then
since the electric energy is equal to $\frac{1}{2}\Sigma EV$, it is in this case

$$\frac{1}{2}e\left\{\frac{e}{b} - \frac{ea}{f}\frac{1}{PQ}\right\}$$

or

$$\frac{1}{2}e^2\left\{\frac{1}{b} - \frac{a}{f^2-a^2}\right\}.$$

87. To find the surface density at a point S on the surface of the sphere, we must find the electric intensity along the normal.

The electric intensity at S due to the charge e at P can by the triangle of forces be resolved into the two components

$$(\alpha) \quad \frac{e}{PS^2}\frac{OS}{PS} \text{ along } OS,$$

$$(\beta) \quad \frac{e}{PS^2}\frac{OP}{PS} \text{ parallel to } PO,$$

while the electric intensity at S due to the charge $-ea/f$ at Q can be resolved into the components

$$(\gamma) \quad -\frac{ea}{f}\frac{1}{QS^2}\frac{OS}{QS} \text{ along } OS,$$

$$(\delta) \quad -\frac{ea}{f}\frac{1}{QS^2}\frac{OQ}{QS} \text{ parallel to } PO.$$

Hence the components of the resultant intensity are $\alpha+\gamma$ along the normal OS, and $\beta+\delta$ parallel to PO.

Now the resultant intensity is along the normal, so that the component $\beta+\delta$ must vanish, and the resultant intensity along the normal is equal to $\alpha+\gamma$, i.e. to

$$e.OS\left\{\frac{1}{PS^3}-\frac{a}{f}\frac{1}{QS^3}\right\}$$

or to
$$\frac{e.OS}{PS^3}\left\{1-\frac{a}{f}\left(\frac{PS}{QS}\right)^3\right\}.$$

Since PS/QS is constant, the quantity inside the brackets is constant.

If σ is the surface density of the electrification at S, then, by Coulomb's law,

$$4\pi\sigma = \frac{eOS}{PS^3}\left\{1 - \frac{a}{f}\left(\frac{PS}{QS}\right)^3\right\} = \frac{ea}{PS^3}\left\{1 - \frac{f^2}{a^2}\right\},$$

so that the surface density of the electrification varies inversely as the cube of the distance from P, and is, since f is greater than a, everywhere negative.

88.   If the sphere is insulated instead of being at zero potential, the conditions are that the potential over the sphere should be constant and that the charge on the sphere should be zero. The charge on the sphere in the last case was $-ea/f$. Hence if we superpose on the last solution the field due to a quantity of electricity equal to ea/f placed at the centre of the sphere, which will give rise to a uniform potential over the sphere, the resulting field at points outside the sphere will have the following properties; (1) the potential over the sphere is constant, (2) the total charge on the sphere is zero, (3) the total normal induction over any closed surface is equal to $4\pi e$ if the surface encloses P and is zero if it does not.   Hence it is the solution in the region outside the sphere when a charge e is placed at P in front of an insulated conducting sphere.   Thus, outside the insulated sphere the electric field is the same as that due to the three charges, e at P, $-ea/f$ at Q, ea/f at O.   Let us consider the potential of the sphere: the charges at P and Q together produce zero potential over the sphere, so that the potential will be that due to the charge ea/f, at O; this charge produces at any point on the sphere a potential equal to e/f, so that by the presence of e at

P the potential of the sphere is raised by e/f. This result was proved by a different method in Art. 29.

The force on P in this case is an attraction equal to

$$\frac{e^2}{PQ^2}\frac{a}{f} - \frac{e^2 a}{f \cdot f^2}$$

$$= \frac{e^2 a}{f}\left\{\frac{f^2}{(f^2 - a^2)^2} - \frac{1}{f^2}\right\}$$

$$= \frac{e^2 a^3}{f^3} \cdot \frac{2f^2 - a^2}{(f^2 - a^2)^2},$$

so that in this case, when f is very large compared with a the force varies inversely as the fifth power of the distance. When the point is very close to the surface of the sphere the force is the same as if the sphere were at zero potential.

The potential energy, $\frac{1}{2}\Sigma EV$ is, if the body at P is a small sphere of radius b, equal to

$$\frac{1}{2}e\left\{\frac{e}{b} - \frac{ea}{f \cdot PQ} + \frac{ea}{f^2}\right\}$$

$$= \frac{1}{2}e\left\{\frac{e}{b} - \frac{ea^3}{f^2(f^2 - a^2)}\right\}.$$

To find the surface density at S, we must superpose on the value given in Art. 87, the uniform density

$$\frac{ea}{f \cdot 4\pi a^2}.$$

Thus

$$4\pi\sigma = -\frac{ea}{PS^3}\left(\frac{f^2}{a^2} - 1\right) + \frac{e}{af} \quad\ldots\ldots\ldots(1).$$

At R the point on the sphere nearest to P,

$$PR = f - a,$$

so that the surface density at R is equal to

$$- \frac{1}{4\pi} \frac{e}{a} \left\{ \frac{f+a}{(f-a)^2} - \frac{1}{f} \right\}$$

$$= - \frac{e}{4\pi} \frac{(3f-a)}{f(f-a)^2}.$$

At R' the point on the sphere most remote from P,

$$PR' = f+a,$$

and the surface density at R' is equal to

$$\frac{e}{4\pi} \frac{(3f+a)}{f(f+a)^2}.$$

Since the total charge on the sphere is zero, the surface density of the electricity must be negative on one part of the sphere, positive on another part. The two parts will be separated by a line on the sphere along which there is no electrification. To find the position of this line put σ equal to zero in equation (1), we get if S is a point on this line

$$PS^3 = (f^2 - a^2)f = f^2 \left(f - \frac{a^2}{f} \right)$$

$$= OP^2 \times PQ,$$

hence the points at which the electrification vanishes will be at a distance $(OP^2 \times PQ)^{\frac{1}{3}}$ from P.

The parts of the surface of the sphere whose distances from P are less than this value are charged with electricity of the opposite sign to that at P, the other parts of the sphere are charged with electricity of the same sign as that at P.

89. If the sphere instead of being insulated and without charge is insulated and has a charge E, we can deduce the solution by superposing on the field discussed in Art. 88 that due to a charge E uniformly distributed over the surface of the sphere; this at a point outside the sphere is the same as that due to a charge E at O. Thus the field outside the sphere is in this case the same as that due to charges

$$E + \frac{ea}{f} \text{ at } O, \quad -\frac{ea}{f} \text{ at } Q, \quad e \text{ at } P.$$

The repulsive force acting on P is equal to

$$\left(E + \frac{ea}{f}\right)\frac{e}{f^2} - \frac{e^2 a}{f \cdot PQ^2}$$

$$= \frac{Ee}{f^2} - \frac{e^2 a^3 (2f^2 - a^2)}{f^3 (f^2 - a^2)^2}.$$

When the point is very near the sphere we may put $f = a + x$, where x is small, and then the repulsion is approximately equal to

$$\frac{Ee}{a^2} - \frac{e^2}{4x^2},$$

and this is negative, i.e. the force is attractive unless

$$E > e\frac{a^2}{4x^2}.$$

Thus, when the charges are given, and when P gets within a certain distance of the sphere, P will be attracted towards the sphere even though the sphere is charged with electricity of the same sign as that on P. When we recede from the sphere we reach a place where the attraction changes to repulsion, and at this point there is no force on P. Thus if P is placed at this point, it will be in equilibrium. The equilibrium will, however, be unstable, for if we displace P towards the sphere the force on it becomes attractive and so tends to bring P still nearer to the sphere, that is to increase its displacement, while if we displace P away from the sphere the force on it becomes repulsive and tends to push P still further away from the sphere, thus again increasing the displacement. This is an example of a more general theorem due to Earnshaw that no charged body (whether charged by induction or otherwise) can be in stable equilibrium in the electrostatic field under the influence of electric forces alone.

90. If the potential of the sphere is given instead of the charge, we can still use a similar method to find the field round the sphere. Thus if the potential of the sphere is V, then the field outside the sphere is the same as that due to a charge Va at O, $-ea/f$ at Q, and e at P.

91. Sphere placed in a uniform field. As the point P moves further and further away from O the Faraday tubes due to the charge at P get to be in the neighbourhood of the sphere more and more nearly parallel to OP, thus when P is at a very great distance from the sphere the problems we have just considered become in the limit problems relating to the distribution of electricity on a sphere placed in a uniform electric field.

Suppose that, as the charged body P travels away from the sphere, the charge e increases in such a way that the electric intensity at the centre of the sphere due to this charge remains finite and equal to F, we have thus

$$\frac{e}{f^2} = F.$$

Now consider the problem of an insulated sphere without charge placed in this uniform field. We see by Art. 88 that the electrification on the sphere produces the same effect at points outside the sphere as would be produced by two charges, one equal to ea/f placed at the centre O, the other equal to $-ea/f$ at Q the image of P. If we express these charges in terms of F we see that they are equal respectively to $\pm Faf$; when f is infinite they are also infinite. Since $OQ = a^2/f$ the distance between these charges diminishes indefinitely as f increases, and we see that the product of either of the charges into the distance between them is equal to Fa^3 and is finite. The electrification over the surface of the sphere when placed in a uniform field produces the same effect therefore as an electrical system consisting of two oppositely charged bodies, placed at a very short distance apart, the charges on the bodies being equal in magnitude and so large that the product of either of the charges into the distance between them is finite. Such a system is called an electrical doublet and the product of either of the charges into the distance between them is called the moment of the doublet.

92. Electric field due to a doublet. Let A, B be the two charged bodies, let e be the charge at A, $-e$

Fig. 46.

that at B; let O be the middle point of AB, M the moment of the doublet. Let C be a point at which the electric intensity is required, and let the angle $AOC = \theta$. The intensity at right angles to OC is equal to

$$\frac{e}{AC^2} \sin ACO + \frac{e}{BC^2} \sin BCO$$

$$= \frac{e}{AC^3} AO \sin \theta + \frac{e}{BC^3} BO \sin \theta$$

$$= \frac{e}{OC^3} AB \sin \theta$$

$$= \frac{M \sin \theta}{OC^3},$$

approximately, since AO is very small compared with OC.

The intensity in the direction OC is equal to

$$\frac{e}{AC^2} \cos ACO - \frac{e}{BC^2} \cos BCO,$$

but we have approximately

$$AC = OC - AO \cos \theta,$$

$$BC = OC + BO \cos \theta.$$

Hence putting $\cos ACO = 1$, $\cos BCO = 1$ and using the Binomial Theorem we find that the electric intensity along OC is approximately

$$\frac{e}{OC^2}\left(1 + \frac{2AO}{OC} \cos \theta\right) - \frac{e}{OC^2}\left(1 - \frac{2BO \cos \theta}{OC}\right)$$

$$= \frac{2eAB \cos \theta}{OC^3}$$

$$= \frac{2M \cos \theta}{OC^3}.$$

93. Let us now return to the case of the sphere placed in the uniform field: the moment of the doublet which represents the effect of the electrification over the sphere is Fa^3. Hence, when the sphere is placed in a uniform field F parallel to PO, the intensity at a point C is the resultant of electric intensities, F parallel to PO, $Fa^3 \sin \theta / OC^3$ at right angles to OC, and $2Fa^3 \cos \theta / OC^3$ along CO; θ denotes the angle POC.

At the surface of the sphere where $OC = a$, the resultant intensity along the outward drawn normal is

$$- F \cos \theta - 2F \cos \theta,$$

or $$- 3F \cos \theta;$$

but by Coulomb's law, if σ is the surface density of the electrification on the sphere,

$$4\pi\sigma = - 3F \cos \theta,$$

or $$\sigma = - \frac{3}{4\pi} F \cos \theta.$$

Hence we see, that when an insulated conducting sphere is placed in a uniform field, the surface density at any point on the sphere is proportional to the distance of that point from a plane through the centre of the sphere at right angles to the electric intensity in the uniform field.

On account of the concentration of the Faraday tubes on the sphere the maximum intensity in the field is three times the intensity in the uniform field.

94. We have hitherto supposed the electrified body to be outside the sphere, but we can apply the same method when it is inside. Thus, if we have a charge e

at a point Q inside a spherical surface maintained at zero potential, then the effect, inside the sphere, of the electricity induced on the sphere will be the same as that due to a charge $-e \cdot a/OQ$ at P where $OP \cdot OQ = a^2$. The charge on the sphere is $-e$, since all the tubes which start from Q end on the sphere.

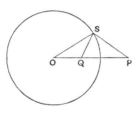

Fig. 47.

If the sphere is insulated, then the charge on the inside of the sphere and the force inside are the same as when it is at potential zero; the only difference is that on the outside of the sphere there is a charge equal to e uniformly distributed over the sphere, and the field outside is the same as that due to a charge e at the centre.

Again, if there is a charge E on the sphere, the effect inside is the same as in the two previous cases, only now there is a charge $E + e$ uniformly distributed over the surface of the sphere raising its potential to $(E + e)/a$.

In all these cases the surface density of the electrification at any point on the inner surface of the sphere varies inversely as the cube of the distance of that point from P.

95. Case of two spheres intersecting at right angles and maintained at unit potential. Let the figure represent the section of the spheres, A and B being their centres, and C a point on the circle in which they

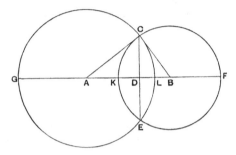

Fig. 48.

intersect, CD a part of the chord common to the two circles; then, since the spheres intersect at right angles ACB is a right angle and CD is the perpendicular let fall from C on AB.

Then we have by Geometry

$$AD \cdot AB = AC^2,$$

$$DB \cdot AB = BC^2.$$

Thus D and B are inverse points with regard to the sphere with centre A, and A and D are inverse points with regard to the sphere whose centre is B.

Let $AC = a$, $BC = b$, then $CD \cdot AB = AC \cdot BC$, so that

$$CD = \frac{ab}{\sqrt{a^2 + b^2}}.$$

Consider the effect of putting a positive charge at A numerically equal to the radius AC, a positive charge at B equal to BC, and a negative charge at D equal to CD.

The charges at A and D will together, by Art. 86, produce zero potential over the sphere with centre B. For A and D are inverse points with respect to this sphere, and the charge at D is to the charge at A as $-CD$ is to AC, i.e. as $-BC$ is to AB, so that the ratio of the charges is the same as that of those on a point and its image, which together produce zero potential at the sphere. Thus the value of the potential over the surface of this sphere is that due to the charge at B, but the charge is equal to the radius of the sphere, so that the potential at the surface, being equal to the charge divided by the radius, is equal to unity. Thus these three charges produce unit potential over the sphere with centre B; we can in a similar way show that they give unit potential over the sphere with centre A. The two spheres then are an equipotential surface for the three charges, and the electric effect of the conductor formed by the two spheres, when maintained at unit potential, is at a point outside the sphere the same as that due to the three charges.

Capacity of the system. The charge on the system is equal to the sum of the charges on the points inside it which produce the same effect. Thus the capacity of the system which, since the potential is unity, is equal to the charge is equal to

$$a + b - \frac{ab}{\sqrt{a^2 + b^2}}.$$

96. If b is very small compared with a, the system becomes a small hemispherical boss on a large sphere as shown in Fig. 49. The capacity is equal to

$$a + b - \frac{ab}{\sqrt{a^2 + b^2}},$$

or to

$$a \left\{ 1 + \frac{b}{a} - \frac{b}{a} \left(1 + \frac{b^2}{a^2} \right)^{-\frac{1}{2}} \right\};$$

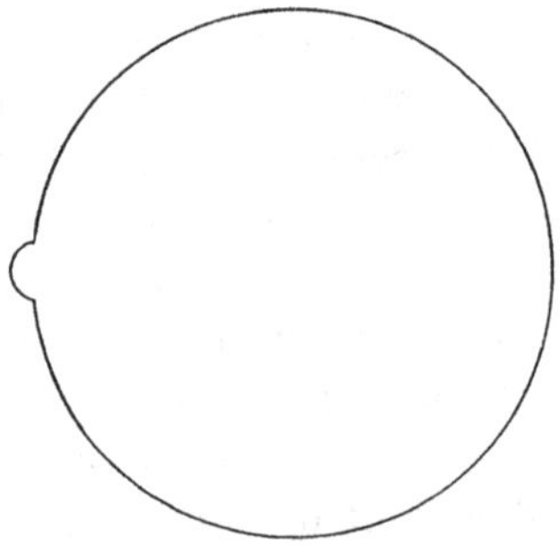

Fig. 49.

and, as in this case b/a is very small, the capacity is approximately equal to

$$a \left\{ 1 + \frac{b}{a} - \frac{b}{a} \left(1 - \frac{1}{2} \frac{b^2}{a^2} \right) \right\}$$

$$= a \left(1 + \frac{1}{2} \frac{b^3}{a^3} \right).$$

But

$$\frac{1}{2} \frac{b^3}{a^3} = \frac{\text{volume of boss}}{\text{volume of big sphere}}.$$

Thus we have, since a is the capacity of the large sphere without the boss,

$$\frac{\text{increase in capacity due to boss}}{\text{capacity of sphere}} = \frac{\text{volume of boss}}{\text{volume of sphere}}.$$

97. To compare the charges on the surface of the two spheres. The charge on the spherical cap EFC (Fig. 48) is, by Coulomb's law, equal to $1/4\pi$ of the total normal induction over EFC. Now the total normal induction is the sum of the total normal inductions due to the charges at A, B, D. Since B is the centre of the cap CFE the total normal induction due to B over CFE bears the same ratio to $4\pi b$ (the total normal intensity over the whole sphere) as the area of the cap CFE does to the area of the sphere. But the area of the surface of a sphere included between two parallel planes is proportional to the distance between the planes, thus

$$\frac{\text{area of } EFC}{\text{area of sphere}} = \frac{b+BD}{2b}.$$

Hence the total normal induction over CFE due to the charge at B

$$= 2\pi\,(b + BD).$$

The total normal induction due to the charge A over the closed surface $CFEL$ is zero, therefore the total normal induction due to A over CFE is equal in magnitude and opposite in sign to the total normal induction over CLE, that is, it is equal to the total normal induction over CLE reckoned outwards from the side A. But CLE is a portion of a sphere of which A is the centre, therefore the induction over CLE is to $4\pi a$ (the induction over the whole sphere with centre A) as the area of CLE is to the area of the sphere, that is as $DL : 2a$. Thus the induction due to A over CFE is equal to

$$2\pi DL.$$

Next consider the total normal induction over CFE due to the charge at D. Now of the tubes starting from D as many would go to the right as to the left if it were alone in the field, so that the induction over CFE will be half that due to D over a closed surface entirely surrounding it; the latter induction is equal to 4π times the charge at D, i.e. to $-4\pi \cdot CD$, hence the induction due to D over the surface CFE is

$$- 2\pi \cdot CD.$$

Thus the total induction over CFE due to the three charges is

$$2\pi \, (b + BD + DL - CD),$$

and the charge on CFE is therefore equal to

$$\frac{1}{2} \left(b + \frac{b^2}{\sqrt{a^2 + b^2}} + a - \frac{a^2}{\sqrt{a^2 + b^2}} - \frac{ab}{\sqrt{a^2 + b^2}} \right) \ldots (1).$$

The charge on CGE can be got by interchanging a and b in this expression, and is thus equal to

$$\frac{1}{2} \left(a + \frac{a^2}{\sqrt{a^2 + b^2}} + b - \frac{b^2}{\sqrt{a^2 + b^2}} - \frac{ab}{\sqrt{a^2 + b^2}} \right) \ldots (2).$$

98. In the case of a hemispherical boss on a large sphere, b is very small compared with a; in this case the expression (1) becomes approximately

$$\frac{1}{2} \left\{ b + \frac{b^2}{a} + a - a \left(1 - \frac{1}{2} \frac{b^2}{a^2} \right) - b \right\}$$

$$= \frac{3}{4} \frac{b^2}{a}.$$

This is equal to the charge on the boss. The mean density on the boss is this expression divided by $2\pi b^2$, the area of the surface of the boss, and is therefore

$$\frac{3}{8\pi a}.$$

When b/a is very small the expression (2) is approximately equal to a, thus the charge on the sphere is a and the mean density is got by dividing a by $4\pi a^2$ the area of the sphere. Thus the mean density on the sphere is

$$\frac{1}{4\pi a}.$$

Hence the mean density on the boss is to the mean density on the sphere as $3 : 2$.

99. Since a plane may be regarded as a sphere of infinite radius, this applies to a hemispherical boss of any radius on a plane surface. It thus applies to the case

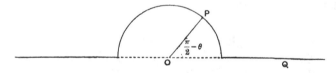

Fig. 50.

shown in Fig. 50. Since the mean density over the boss is $3/2$ of that over the plane, and since the area of the boss is twice the area of its base; there is three times as much electricity on the surface occupied by the boss as there is, on the average, on an area of the plane equal to the base of the boss.

100. When b is very small compared with a, the points B and D, Fig. 48, are close together, the distance between them being approximately b^2/a, which is small compared with b; the charge at B is b, that at D is

$$- \frac{ab}{\sqrt{a^2 + b^2}},$$

and, when b is very small compared with a, this is approximately equal to $-b$. Thus the charges at B and D form a doublet whose moment is b^3/a. The point A is very far away and the force at B or D due to its charge is $1/a$. Thus the moment of the doublet is b^3 times this force. This as far as the sphere is concerned is exactly the case considered in Art. 93. Hence if F is the force at the boss due to the charge A alone, the surface density at a point P, Fig. 50, on the boss is $\dfrac{3F}{4\pi} \cos \theta$, where θ is the angle OP makes with the axis of the doublet. Now if σ_0 is the surface density on the plane at some distance from the boss $F = 4\pi\sigma_0$. Hence, the surface density at P, a point on the boss, is equal to

$$3\sigma_0 \cos \theta,$$

where θ is the angle OP makes with the normal to the plane.

The electric intensity due to the doublet at Q, a point on the plane, is (Art. 92) equal to the moment of the doublet divided by OQ^3 and is at right angles to the plane, thus the normal electric intensity at Q is

$$F\left(1 - \frac{b^3}{OQ^3}\right)$$

and σ, the surface density at Q, is given by the equation

$$\sigma = \sigma_0 \left(1 - \frac{b^3}{OQ^3}\right).$$

We have thus found the distribution of electricity on a charged infinite plane with a hemispherical boss on it.

101.   In the general case when the two spheres are of any sizes the surface density on the conductor can be got by calculating the normal electric intensity due to the three charges.   We shall leave this as an example for the student, remarking that, since the potential of the conductor is the highest in the field, there can be no negative electrification over the surface and that the electrification vanishes along the intersection of the two spheres.

102. Effect of dielectrics.   We have hitherto only considered the case when the field due to the charge at P was disturbed by the presence of conductors, but by applying the principle that a solution which satisfies the electric conditions is the only solution, we can find the electric field in some simple cases when dielectrics are present.

103.   The first case we shall consider is that of a small charged body placed in front of an infinite mass of uniform

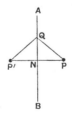

Fig. 51.

dielectric bounded by a plane face.   Let P be the charged body, AB the plane separating the dielectric from air, the medium to the right of AB being air, that to the left a dielectric whose specific inductive capacity is K. From P draw PN perpendicular to AB; produce PN to P', so that $PN = P'N$. Then we shall show that the field to the right of AB can be regarded as due to e at P and a charge e' at P', and that to the left of AB as due to e'' at P; these charges being supposed to produce the same field as if there was nothing but air in the field.

In the first place this field satisfies the conditions that the potential at an infinite distance is zero, also that the induction over any closed surface surrounding P is $4\pi e$, while the induction over any closed surface not enclosing P is zero. This is obvious if the surface is drawn entirely to the left or entirely to the right of AB. If it crosses this plane it can be regarded as two surfaces, one entirely to the left bounded by the portion of the surface to the left and the portion of the plane AB intersected by the surface, the other entirely to the right bounded by the same portion of the plane and the part of the surface to the right.

The only other conditions we have to satisfy are that along the plane AB the electric intensity parallel to the surface is the same in the air as in the dielectric, and that over this plane the normal polarization is the same in the air as in the dielectric.

At a point Q in AB the electric intensity parallel to AB is in the air

$$\frac{e}{PQ^2}\frac{QN}{PQ} + \frac{e'}{P'Q^2}\frac{QN}{P'Q}.$$

This, since $PQ = P'Q$, is equal to

$$(e+e')\frac{QN}{PQ^3}.$$

The electric intensity at Q parallel to AB in the dielectric is

$$e'' \frac{QN}{PQ^3};$$

this is equal to that in air if

$$e + e' = e'' \quad \dots\dots\dots\dots\dots(1).$$

Again, the polarization at Q at right angles to AB reckoned from right to left is in air

$$\frac{1}{4\pi}(e - e')\frac{PN}{PQ^3},$$

and that in the dielectric is

$$\frac{K}{4\pi} e'' \frac{PN}{PQ^3};$$

these are equal if

$$e - e' = Ke'' \quad \dots\dots\dots\dots\dots(2).$$

Hence both the boundary conditions are satisfied if e' and e'' satisfy (1) and (2), i.e. if

$$e'' = \frac{2}{1 + K} e,$$

$$e' = -\frac{(K - 1)}{K + 1} e.$$

The attraction of P towards the plane is equal to that between e and e' and is thus

$$-\frac{ee'}{(2PN)^2} = \frac{K - 1}{K + 1}\frac{e^2}{4PN^2}.$$

If K is infinite this equals

$$\frac{e^2}{4PN^2};$$

which is the same as when the dielectric to the left of AB is replaced by a conductor.

Thus if $K = 10$, as is the case for some kinds of heavy glass, the force on P when placed in front of the glass would be about 9/11 of the attraction when P is placed in front of a conducting plate. Inside the mass of dielectric the tubes are straight and pass through P; the effect of the dielectric is, while not affecting the direction of the electric intensity, to reduce its magnitude to $2/(1+K)$ of its value in air when the dielectric is removed. The lines of force when $K = 1\cdot7$ are shown in Fig. 52.

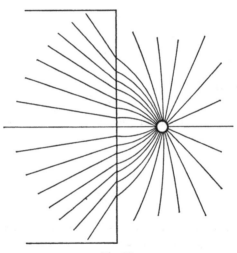

Fig. 52.

104. Case of a dielectric sphere placed in a uniform field. We have seen that, when a conducting sphere is placed in a uniform field, the effect of the electricity induced on the surface of the sphere can be represented at points outside the sphere by a doublet (see Art. 92) placed at the centre of the sphere. Since we have seen that the effects of a dielectric are similar in kind though different in degree to those due to a conductor, we are led to try if the disturbance produced by the presence of the sphere cannot be represented at a point outside the sphere by a doublet placed at its centre. With regard to the field inside the sphere we have as a guide the result obtained in the last article, that in the case when the radius of the sphere is infinitely large the field inside the dielectric is not altered in direction but only in magnitude by the dielectric.

We therefore try if we can satisfy the conditions which must hold when a sphere is placed in a uniform electric field by supposing the field inside the sphere to be uniform.

Let the uniform field before the insertion of the sphere be one where the electric intensity is horizontal and equal to H.

After the insertion of the sphere let the field outside consist of this uniform field plus the field due to a doublet whose moment is M placed at the centre of the sphere, the dielectric being removed.

Inside the sphere let the intensity be horizontal and equal to H'.

We shall see that it is possible to satisfy the conditions of the problem by a proper choice of M and H'.

The field at P due to the doublet is, by Art. 92, equivalent to an intensity $\dfrac{2M}{OP^3} \cos \theta$ along OP, and an intensity $\dfrac{M}{OP^3} \sin \theta$ at right angles to it, where θ is the angle OP

makes with the direction of the uniform electric intensity. Thus at a point Q just outside the sphere the intensity tangential to the sphere is equal to

$$H \sin \theta - \frac{M}{a^3} \sin \theta,$$

where a is the radius of the sphere.

The intensity in the same direction at a point close to Q but just inside the sphere is

$$H' \sin \theta.$$

The normal intensity at Q outside the sphere is

$$H \cos \theta + \frac{2M}{a^3} \cos \theta,$$

and at a point just inside the sphere it is $H' \cos \theta$.

The first boundary condition is that the tangential intensity at the surface of the sphere must be the same in the air as in the dielectric; this will be true if

$$H \sin \theta - \frac{M}{a^3} \sin \theta = H' \sin \theta,$$

or
$$H - \frac{M}{a^3} = H' \quad \dots\dots\dots\dots\dots(1).$$

The second boundary condition is that the normal polarization at the surface of the sphere must be the same in the air as in the dielectric, thus

$$\frac{1}{4\pi} \left\{ H \cos \theta + \frac{2M}{a^3} \cos \theta \right\} = \frac{K}{4\pi} H' \cos \theta,$$

or
$$H + \frac{2M}{a^3} = KH' \dots\dots\dots\dots\dots(2).$$

Equations (1) and (2) will be satisfied, if

$$H' = \frac{3H}{K + 2},$$

and if

$$M = \frac{H(K - 1)}{K + 2} a^3.$$

Thus, since, if H' and M have these values the conditions are satisfied, this will be the solution of the problem. We see that the intensity inside the sphere is $3/(K + 2)$ of that in the original field, so that the intensity of the field is less inside the sphere than outside; on the other hand the number of Faraday tubes which pass through unit area inside the sphere is $3K/(K + 2)$ times the number passing through unit area in the original uniform field. When K is very great $3K/(K + 2)$ is approximately equal to 3, so that the Faraday tubes in this case will be 3 times as dense inside the sphere as they are at a great distance away from it. This illustrates the crowding of the Faraday tubes to the sphere.

The diagram of the lines of force for this case was given in Fig. 41.

Method of Inversion.

105. This is a method by which, when we have obtained the solution of any problem in electrostatics, we can by a geometrical process obtain the solution of another.

Definition of inverse points. If O is a fixed point, P a variable one, and if we take P' on OP, so that

$$OP \cdot OP' = k^2,$$

where k is a constant, then P' is defined to be the inverse point of P with regard to O, while O is called the centre of inversion, and k the radius of inversion.

If the point P moves about so as to trace out a surface, then P' will trace out another surface which is called the surface inverse to that traced out by P.

We shall now proceed to prove some geometrical propositions about inversion.

106. The inverse surface of a sphere is another sphere. Let O be the centre of inversion, P a point on the sphere to be inverted, C the centre of this sphere.

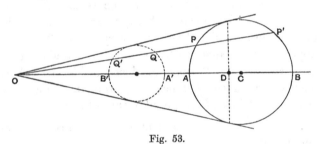

Fig. 53.

Let the chord OP cut the sphere again in P', let Q be the point inverse to P, Q' the point inverse to P', R the radius of the sphere to be inverted, then

$$OP . OQ = k^2.$$

But
$$OP . OP' = OC^2 - R^2,$$

and thus
$$OQ = \frac{k^2}{OC^2 - R^2} OP' ;$$

similarly
$$OQ' = \frac{k^2}{OC^2 - R^2} OP.$$

Thus
$$OQ \cdot OQ' = \frac{k^4}{(OC^2 - R^2)^2} OP \cdot OP'$$

$$= \frac{k^4}{OC^2 - R^2}.$$

Thus OQ bears a constant ratio to OP'; hence the locus of Q is similar to the locus of P', and is therefore a sphere. Thus a sphere inverts into a sphere. If

$$k^2 = OC^2 - R^2$$

the sphere inverts into itself.

To find the centre of the inverse sphere, let the diameter OC cut the sphere to be inverted in A and B. Let A', B' be the points inverse to A and B respectively and O' the centre of the inverted sphere; then

$$OO' = \frac{1}{2}(OA' + OB')$$

$$= \frac{1}{2}\left(\frac{k^2}{OC - R} + \frac{k^2}{OC + R}\right)$$

$$= k^2 \cdot \frac{OC}{OC^2 - R^2}.$$

If D is the point where the chord of contact of tangents from O to the sphere cuts OC, then

$$OD = \frac{OC^2 - R^2}{OC}.$$

Hence D inverts into the centre of the sphere.

The radius of the inverse sphere

$$= \tfrac{1}{2}(OA' - OB')$$

$$= k^2 \cdot \frac{R}{OC^2 - R^2}.$$

107. Since a plane is a particular case of a sphere a plane will invert into a sphere; this can be proved independently in the following way:

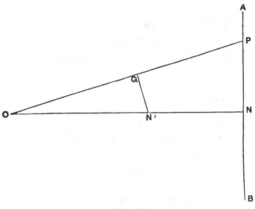

Fig. 54.

Let AB be the plane to be inverted, P a point on that plane, N the foot of the perpendicular let fall from O on the plane and Q and N' the points inverse to P and N respectively. Then since

$$OQ \cdot OP = ON' \cdot ON$$

$$\frac{OQ}{ON'} = \frac{ON}{OP};$$

thus the two triangles QON', PON have the angle at O common and the sides about this angle proportional, they are therefore similar, and the angle OQN' is equal to the angle ONP. Hence OQN' is a right angle and therefore the locus of Q is a sphere on ON' as diameter.

108.   Let O be the centre of inversion, PQ two points and $P'Q'$ the corresponding inverse points.

Then
$$\frac{OP'}{OQ'} = \frac{OQ}{OP};$$

thus the triangles POQ, $Q'OP'$ are similar, so that
$$\frac{PQ}{OP} = \frac{P'Q'}{OQ'}.$$

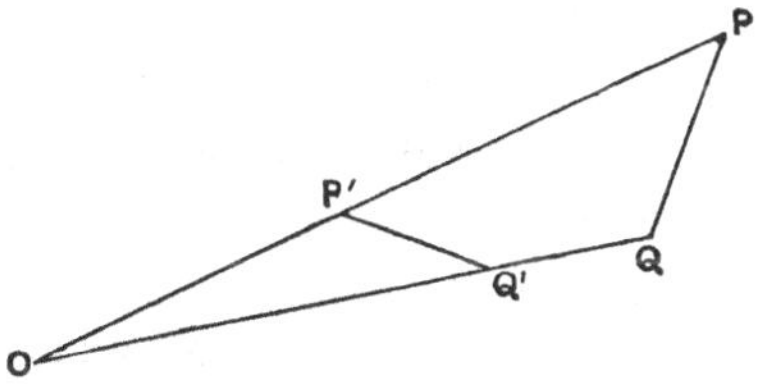

Fig. 55.

If we have a charge e at Q, and a charge e' at Q', then if V_P is the potential at P due to the charge at Q, and $V'_{P'}$ the potential at P' due to the charge at Q',

$$V_P : V'_{P'} = \frac{e}{PQ} : \frac{e'}{P'Q'} = \frac{e}{OP} : \frac{e'}{OQ'}.$$

Take
$$e : e' = OQ : k \dots\dots\dots\dots\dots\dots(1),$$

then
$$V'_{P'} = V_P \frac{k}{OP'}.$$

If we have any number of charges at different points and take the inverse of these points and place there charges given by the expression (1), then, if $\overline{V}_P$ be the potential at a point P due to the original assemblage of charges, $\overline{V}_{P'}$ the potential at P' (the point inverse to P) due to the charges on the inverted system,

$$\overline{V}_{P'} = \overline{V}_P \frac{k}{OP'}.$$

12—2

Thus, if the original assemblage of charges produces a constant potential V over a surface S, the inverted system will produce a potential $\dfrac{Vk}{OP'}$ at a point P' on the inverse of S. Hence, if we add to the inverted system a charge $-kV$ at the centre of inversion, the potential over the inverse of S will be zero.

If the charges on the original system are distributed over a surface instead of being concentrated at points the charges on the inverted system will also be distributed over a surface. Let σ be the surface density at Q, a place on the original system, σ' the surface density at Q', the corresponding point on the inverted system, α a small area at Q, α' the area into which it inverts; then by (1)

$$\sigma\alpha : \sigma'\alpha' = OQ : k$$

and, since α and α' are similar figures,

$$\alpha : \alpha' = OQ^2 : OQ'^2.$$

Hence

$$\sigma : \sigma' = OQ'^2 : kOQ$$

and thus

$$\sigma' = \sigma\,\frac{kOQ}{OQ'^2} = \sigma\,\frac{k^3}{OQ'^3}\ldots\ldots\ldots\ldots(2).$$

This expression gives the surface density of the inverted figure in terms of that at the corresponding point of the original figure.

109. As an example of the use of the method of inversion let us invert the system consisting of a sphere with a uniform distribution of electricity over it, the surface density being $V/4\pi a$; where a is the radius of the sphere. We know in this case that the potential is constant over the sphere and equal to V. Take the centre of inversion outside the sphere and choose the radius of inversion so that the sphere inverts into itself.

Then, if to the inverted system we add a charge $-kV$ at the centre of inversion the inverse sphere will be at potential zero. By equation (2) σ' the surface density in the inverted system at Q' is given by the equation

$$\sigma' = \frac{V}{4\pi a} \cdot \frac{k^3}{OQ'^3}.$$

If we put $e = -kV$, this equals

$$\frac{-e}{4\pi a} \cdot \frac{k^2}{OQ'^3} = \frac{-e \cdot (OC^2 - a^2)}{4\pi a \cdot OQ'^3},$$

where C is the centre of the sphere.

Thus a charge e at O induces on the sphere at zero potential a distribution of electricity such that the surface density varies inversely as the cube of the distance from O. In this way we get by inversion the solution of the problem which we solved in Art. 87 by the method of images.

110. As an example illustrating the uses of the method of inversion as well as that of images, let us consider the solution, by the method of images, of a charged body placed between two infinite conducting planes maintained at potential zero.

Let P be the charged point, AB and CD the two planes at potential zero, e the charge at P. Then if we place a charge $-e$ at P' where P' is the image of P in AB the potential over AB will be zero, it will not however be zero over CD; to make the potential over CD zero we must place a charge $-e$ at Q, the image of P in CD, and a charge e at Q_1, the image of P' in CD. These two charges will however disturb the potential of AB; to restore zero potential to AB we must introduce a charge $+e$ at P_1, the image of Q in AB, and a charge $-e$ at P'', the image of Q_1 in AB. The charges at P_1 and P'' will disturb the potential over the plane CD; to restore it to zero we must place a charge $-e$ at Q', the image of P_1 in CD, and a charge $+e$ at Q_2, the image of P'' in CD, and so on; we get in this way two infinite series of images to the right of AB and to the left of CD.

The images to the right of AB are (1) charges $-e$, at P', P'', $P'''\ldots$; and (2) charges $+e$, at P_1, P_2, $P_3\ldots$.

Now P'' is the image of Q_1 in AB, which is the image of P' in CD and hence

$$FP'' = FQ_1 = FE + EP' = 2FE + FP';$$

thus $FP'' - FP' = P'P'' = 2FE = 2c$, if c is the distance between the plates.

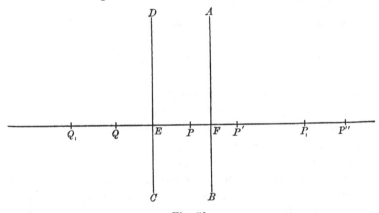

Fig. 56.

Similarly $P'P'' = P''P''' = \ldots = 2c$ and we can show in a similar way that $PP_1 = P_1P_2 = P_2P_3 = \ldots = 2c$. Thus on the right of AB we have an infinite series of charges equal to $-e$ at the distance $2c$ apart, beginning at P' the image of

P in AB, and a series of positive images at the same distance $2c$ apart, beginning at P_1, a point distant $2c$ from P.

Similarly to the left of CD we have an infinite series of images with the charge $-e$ at the distance $2c$ apart, beginning at Q, the image of P in CD, and an infinite series of images each with the charge $+e$, at points at a distance $2c$ apart, beginning at Q_1, a point distant $2c$ from P.

Now invert this system with respect to P. The two planes invert into two spheres touching each other at P, and maintained at a potential $-e/k$, the images to the right of AB invert into a series of charged points inside the sphere to the right of P and the images to the left of CD invert into a system of charged points inside the sphere to the left of P.

The system of charged points inside the spheres will produce a constant potential $-e/k$ over the surface of the spheres, and therefore at a point outside the spheres the electric field due to the two spheres in contact will be the same as that due to the system of the electrified points.

If a, b are the radii of the spheres into which the planes AB, CD invert, and if $PF = d$, then

$$2a = \frac{k^2}{d} \qquad 2b = \frac{k^2}{c-d} \qquad c = \frac{1}{2}\left\{\frac{k^2}{a} + \frac{k^2}{b}\right\}.$$

Consider now the series of images to the right of AB. The series of positive charges at the distance $2c$ apart invert into a series of charges inside the sphere, whose radius is a, of magnitudes

$$\frac{ek}{2c}, \quad \frac{ek}{4c}, \quad \frac{ek}{6c}, \quad \cdots,$$

since

$$\frac{\text{charge at inverted point}}{\text{charge at original point}}$$

$$= \frac{k}{\text{distance of original point from centre of inversion}}.$$

The series of negative images at the distance $2c$ apart invert into a series of negative charges

$$-\frac{ek}{2d}, \quad -\frac{ek}{2c+2d}, \quad -\frac{ek}{4c+2d}, \quad \dots$$

Similarly, inside the sphere into which the plane CD inverts, we have a series of positive charges

$$\frac{ek}{2c}, \quad \frac{ek}{4c}, \quad \frac{ek}{6c}, \dots,$$

and a series of negative ones

$$-\frac{ek}{2(c-d)}, \quad -\frac{ek}{4c-2d}, \quad -\frac{ek}{6c-2d}, \dots$$

Thus E_1, the sum of the charges on the points inside the first sphere, is given by the equation

$$E_1 = ek \left\{ \left(\frac{1}{2c} + \frac{1}{4c} + \frac{1}{6c} + \dots \right) \right.$$

$$\left. - \left(\frac{1}{2d} + \frac{1}{2c+2d} + \frac{1}{4c+2d} + \dots \right) \right\} \dots(1),$$

while E_2, the sum of the charges inside the second sphere, is given by the equation

$$E_2 = ek \left\{ \left(\frac{1}{2c} + \frac{1}{4c} + \frac{1}{6c} + \dots \right) \right.$$

$$\left. - \left(\frac{1}{2c-2d} + \frac{1}{4c-2d} + \dots \right) \right\} \dots(2).$$

Rearranging the terms, we may write

$$E_1 = -\frac{1}{2}ek\left\{\frac{1}{d} - \frac{d}{c\,(c+d)} - \frac{d}{2c\,(2c+d)} - \frac{d}{3c\,(3c+d)} - \cdots\right\},$$

$$E_2 = -\frac{1}{2}ek\frac{d}{c}\left\{\frac{1}{c-d} + \frac{1}{2\,(2c-d)} + \frac{1}{3\,(3c-d)} + \cdots\right\}.$$

Expanding the expressions for E_1 and E_2 in powers of d/c we get

$$E_1 = -\frac{1}{2}ek\left(\frac{1}{d} - \frac{d}{c^2}S_2 + \frac{d^2}{c^3}S_3 - \frac{d^3}{c^4}S_4\cdots\right)\quad\ldots\ldots(3),$$

$$E_2 = -\frac{1}{2}ek\frac{d}{c^2}\left(S_2 + \frac{d}{c}S_3 + \frac{d^2}{c^2}S_4 + \frac{d^3}{c^3}S_5 + \cdots\right),$$

where

$$S_n = \frac{1}{1^n} + \frac{1}{2^n} + \frac{1}{3^n} + \frac{1}{4^n} + \cdots.$$

The values of S_n are given in De Morgan's *Differential and Integral Calculus*, p. 554,

$$S_2 = \frac{\pi^2}{6} = 1\cdot645, \quad S_5 = 1\cdot037,$$

$$S_3 = 1\cdot202, \qquad S_6 = \frac{\pi^6}{945} = 1\cdot017,$$

$$S_4 = \frac{\pi^4}{90} = 1\cdot082, \quad S_7 = 1\cdot008.$$

Since E_1 can be got from E_2 by writing $c-d$ for d, we get

$$E_1 = -\frac{1}{2}ek\frac{(c-d)}{c^2}\left(S_2 + \frac{c-d}{c}S_3 + \frac{(c-d)^2}{c^2}S_4\cdots\right)\;(4).$$

Now, the total charge spread over the surface of the first sphere is equal to the sum of the charges at the points inside the sphere as these produce the same effect at external points as the electrification over the surface of the sphere: thus, E_1 will be the charge on the first sphere, E_2 that on the second. If V is the potential of the spheres

$$V = -\frac{e}{k},$$

$$E_1 = Va\left(1 - \frac{b}{a+b}\frac{b}{a+2b} - \frac{b}{2(a+b)}\frac{b}{2a+3b}\right.$$
$$\left. - \frac{b}{3(a+b)}\frac{b}{3a+4b} - \ldots\right) \quad (5),$$

$$E_2 = Vb\left(1 - \frac{a}{a+b}\frac{a}{2a+b} - \frac{a}{2(a+b)}\frac{a}{3a+2b}\right.$$
$$\left. - \frac{a}{3(a+b)}\frac{a}{4a+3b} - \ldots\right) \quad (6),$$

$$E_1 = Va\left\{1 - \left(\frac{b}{a+b}\right)^2 S_2 + \left(\frac{b}{a+b}\right)^3 S_3 - \ldots\right\} \quad \ldots\ldots\ldots(7),$$

and also

$$E_1 = V\frac{ba^2}{(a+b)^2}\left\{S_2 + \left(\frac{a}{a+b}\right)S_3 + \left(\frac{a}{a+b}\right)^2 S_4 + \ldots\right\} \quad (8).$$

The value of E_2 can be got by interchanging a and b in the expressions (7) and (8).

Let us now consider some special cases. Take first the case when $a = b$, then from equation (5) we have

$$E_1 = Va\left\{1 - \frac{1}{2}\frac{1}{3} - \frac{1}{4}\frac{1}{5} - \frac{1}{6}\frac{1}{7} - \ldots\right\}$$

$$= Va\left\{1 - \frac{1}{2} + \frac{1}{3} - \frac{1}{4} + \frac{1}{5} - \frac{1}{6} + \frac{1}{7} - \ldots\right\}$$

$$= Va\log 2,$$

the logarithm being the Napierian logarithm.

Since $\log 2 = \cdot 693$

$$E_1 = \cdot 693\,Va.$$

The charge on the second sphere is also E_1; thus the total charge on the two spheres is

$$1\cdot386\ Va.$$

When $V = 1$ the charge on the two spheres is equal to the capacity of the system; hence the capacity of two equal spheres in contact is $2a \log 2$ or $1\cdot386a$.

If the spheres had been an infinite distance apart, the capacity of the two would have been $2a$; if there had only been one sphere the capacity would have been a.

We can find from this the work done on an uncharged sphere when it moves under the attraction of a charged sphere of equal radius from an infinite distance into contact with the charged sphere. Let a be the radius of each sphere and e the charge on the charged sphere; then, when the spheres are at an infinite distance apart, the potential energy is $e^2/2a$ and when the spheres are in contact the potential energy is $e^2/2 \times 1\cdot386a$. Hence the work done by the electric field while the uncharged sphere falls from an infinite distance into contact with the charged sphere is

$$\frac{1}{2}\frac{e^2}{a}\left\{1 - \frac{1}{1\cdot386}\right\} = \cdot14\frac{e^2}{a}.$$

If one sphere has a charge E, the other the charge e, then, when they are at an infinite distance apart, the potential energy is $\dfrac{1}{2a}\{E^2 + e^2\}$.

When the spheres are in contact the potential energy is $\dfrac{1}{2 \times 1\cdot386a}\{E + e\}^2$.

Hence the potential energy is greater in the second case than in the first by

$$-\frac{1}{a}\left\{\cdot 14E^{2}-\frac{1}{1\cdot 386}\ Ee+\cdot 14e^{2}\right\}\dots\dots(9).$$

If $E=e$, this is equal to

$$+\frac{\cdot 44e^{2}}{a}.$$

This is the work required to push the spheres together against the repulsions exerted by their like charges.

The expression (9) vanishes when E/e is approximately 5 or 1/5; in this case the potential energy is the same when the spheres are in contact as when they are an infinite distance apart; thus no work is spent or gained in bringing them together. The attraction due to the induced electrification on the average balances the repulsion due to the like charges.

The next case we shall consider is where one sphere is very large compared with the other. Let b be very large compared with a. Now by (8) we have

$$E_{1}=\frac{Vba^{2}}{(a+b)^{2}}\left(S_{2}+\frac{a}{a+b}\ S_{3}+\dots\right),$$

or approximately, when b/a is large,

$$E_{1}=\frac{Va^{2}}{b}S_{2}$$

$$=V\frac{a^{2}}{b}\frac{\pi^{2}}{6}$$

$$=1\cdot 645\ \frac{Va^{2}}{b}.$$

Interchanging a and b in (7) we get

$$E_2 = Vb \left\{ 1 - \left(\frac{a}{a+b}\right)^2 S_2 + \left(\frac{a}{a+b}\right)^3 S_3 - \dots \right\}$$

or approximately, when b/a is large,

$$E_2 = V\left(b - \frac{a^2}{b} S_2\right)$$
$$= V\left(b - \frac{\pi^2}{6}\frac{a^2}{b}\right) = V\left(b - 1\text{·}645\,\frac{a^2}{b}\right).$$

The mean surface density over the small sphere is

$$\frac{E_1}{4\pi a^2} = \frac{V}{4\pi b}\cdot\frac{\pi^2}{6} = \frac{V}{4\pi b}\,1\text{·}645.$$

The mean surface density over the large sphere is approximately

$$\frac{E_2}{4\pi b^2} = \frac{V}{4\pi b}$$

and hence the mean surface density on the small sphere is $\pi^2/6$ or $1\text{·}645$ times that on the large sphere. We saw in Art. 98 that, when a small hemisphere was placed on a large sphere, the mean density on the hemisphere was $1\text{·}5$ times that on the sphere.

Since a plane may be regarded as a sphere of infinite radius, we see that if a sphere of any size is placed on a conducting plane the mean surface density of the electricity on the sphere is $\pi^2/6$ of that on the plane.

We have

$$E_1 + E_2 = Vb\left\{1 + \frac{2a^3}{(a+b)^3} S_3 + \dots\right\}$$
$$= Vb\left\{1 + 2\text{·}404\,\frac{a^3}{b^3}\right\} \text{ approximately.}$$

Thus, the capacity of the system of two spheres is approximately

$$b \left\{ 1 + 2 \cdot 404 \frac{a^3}{b^3} \right\}.$$

We have thus

$$\frac{\text{Increase of capacity due to small sphere}}{\text{Capacity of large sphere}}$$

$$= 2 \cdot 404 \frac{\text{volume of small sphere}}{\text{volume of large sphere}}.$$

Thus in this case, as in that discussed in Art. 96, the increase of capacity due to the small body is proportional to the volume of the small body.

From this result we can deduce the work done on a small uncharged sphere of radius a when it moves from an infinite distance up to a large sphere of radius b with a charge E.

For, when they are at an infinite distance apart, the potential energy is equal to

$$\frac{1}{2} \frac{E^2}{b};$$

when the spheres are in contact the potential energy is

$$\frac{1}{2} \frac{E^2}{b \left\{ 1 + 2 \cdot 404 \frac{a^3}{b^3} \right\}}.$$

The work done on the small sphere by the electrical forces is the difference between these expressions, or approximately,

$$E^2 \, 1 \cdot 202 \, \frac{a^3}{b^4}.$$

CHAPTER VI

MAGNETISM

111. A mineral called 'lodestone' or magnetic oxide of iron, which is a compound of iron and oxygen, is often found in a state in which it possesses the power of attracting small pieces of iron such as iron filings; if the lodestone is dipped into a mass of iron filings and then withdrawn, some of the iron filings will cling to the lodestone, collecting in tufts over its surface. The behaviour of the lodestone is thus in some respects analogous to that of the rubbed sealing-wax in the experiment described in Art. 1. There are however many well-marked differences between the two cases; thus the rubbed sealing-wax attracts all light bodies indifferently, while the lodestone does not show any appreciable attraction for anything except iron and, to a much smaller extent, nickel and cobalt.

If a long steel needle is stroked with a piece of lodestone, it will acquire the power possessed by the lodestone of attracting iron filings; in this case the iron filings will congregate chiefly at two places, one at each end of the needle, which are called the poles of the needle.

The piece of lodestone and the needle are said to be magnetized; the attraction of the iron filings is an example of a large class of phenomena known as magnetic. Bodies which exhibit the properties of the lodestone or the needle are called *magnets*, and the region around them is called the *magnetic field*.

The property of the lodestone was known to the ancients, and is frequently referred to by Pliny and Lucretius. The science of Magnetism is indeed one of the oldest of the sciences and attained considerable development long before the closely allied science of Electricity; this was chiefly due to Gilbert of Colchester, who in his work *De Magnete* published in 1600 laid down in an admirable manner the cardinal principles of the science.

112. Forces between Magnets. If we take a needle which has been stroked by a lodestone and suspend it by a thread attached to its centre it will set itself so as to point in a direction which is not very far from north and south. Let us call the end of the needle which points to the north, the north end, that which points to the south, the south end, and let us when the needle is suspended mark the end which is to the north; let us take another needle, rub it with the lodestone, suspend it by its centre and again mark the end which goes to the north. Now bring the needles together; they will be found to exert forces on each other, and the two ends of a needle will be found to possess sharply contrasted properties. Thus if we place the magnets so that the two marked ends are close together while their unmarked ends are at a much greater distance apart, the marked ends will be repelled from each other; again, if we place the magnets so that the two unmarked ends are close together while the marked ends are at a much greater distance apart, the unmarked ends will be found to be repelled from each other; while if we place the two magnets so that the marked end of one is close to the unmarked end of the other, while the other ends are much further apart, the two ends which are near each other will be found to be attracted towards each other. We see then that poles of the same kind are repelled from each other, while poles of opposite kinds are attracted towards each other. Thus the two ends of a magnet possess properties analogous to those shown by the two kinds of electricity.

113. We shall find it conduces to brevity in the statement of the laws of magnetism to introduce the term *charge of magnetism,* and to express the property possessed by the ends of the needles in the preceding experiment by saying that they are charged with magnetism, one end of the needle being charged with positive magnetism, the other end with negative. We regard the end of the needle which points to the north as having a charge of positive magnetism, the end which points to the south as having a charge of negative magnetism. It will be seen from the preceding experiment that two charges of magnetism are repelled from or attracted towards each other according as the two charges are of the same or opposite signs. It must be distinctly understood that this method of regarding the magnets and the magnetic field is only introduced as affording a convenient method of *describing* briefly the phenomena in that field and not as having any significance with respect to the constitution of magnets or the mechanism by which the forces are produced: we saw for example that the same terminology afforded a convenient method of describing the electric field, though we ascribe the action in that field to effects taking place in the dielectric between the charged bodies rather than in the charged bodies themselves.

114. Unit Charge of Magnetism, often called *pole of unit strength*. Take two very long, thin, uniformly magnetized needles, equal to each other in every respect (we can test the equality of their magnetic properties by observing the forces they exert on a third magnet), let A be one end of one of the magnets, B the like end of the other magnet, place A and B at unit distance apart in air, the other ends of the magnets being so far away that they exert no appreciable effect in the region about A and B: then each of the ends A and B is said to have a unit charge of magnetism or to be a pole of unit strength when A is repelled from B with the unit force. If the units of length, mass and time are respectively the centimetre, gramme and second the force between the unit poles is one dyne.

A charge of magnetism equal to 2, or a pole of strength 2, is one which would be repelled with the force of two dynes from unit charge placed at unit distance in air.

If m and m' are the charges on two ends of two magnets (or the strengths of the two poles), the distance between the charges being the unit distance, the repulsion between the charges is mm' dynes. If the charges are of opposite signs mm' is negative: we interpret a negative repulsion to mean an attraction.

115. Coulomb by means of the torsion balance succeeded in proving that the repulsion between like charges of magnetism varies inversely as the square of the distance between them. We shall discuss in Art. 132 a more delicate and convenient method of proving this result.

Since the forces between charges of magnetism obey

the same laws as those between electric charges we can apply to the magnetic field the theorems which we proved in Chap. II. for the electric field.

116. The Magnetic Force at any point is the force which would act on unit charge if placed at this point, the introduction of this charge being supposed not to influence the magnets in the field.

117. Magnetic Potential. The magnetic potential at a point P is the work which would be done on unit charge by the magnetic forces if it were taken from P to an infinite distance. We can prove as in Art. 17 that the magnetic potential due to a charge m at a distance r from the charge is equal to m/r.

118. The total charge of Magnetism on any magnet is zero. This is proved by the fact that if a magnet is placed in a uniform field the resultant force upon it vanishes. The earth itself is a magnet and produces a magnetic field which may be regarded as uniform over a space enclosed by the room in which the experiments are made. To show the absence of any horizontal resultant force on a magnet, we may mount the magnet on a piece of wood and let this float on a basin of water, then though the magnet will set so as to point in a definite direction, there will be no tendency for the magnet to move towards one side of the basin. There is a couple acting on the magnet tending to twist it so that the magnet sets in the direction of the magnetic force in the field, but there is no resultant horizontal force on the magnet. The absence of any vertical force is shown by the fact that the process of magnetization has no influence upon the

13—2

weight of a body. Either of these results shows that the total charge on the body is zero. For let m_1, m_2, m_3, &c. be the magnetic charges on the body, F the external magnetic force, then the total force acting on the body in the direction of F is

$$\Sigma Fm.$$

This, since the field is uniform, is equal to $F\Sigma m$.

As this vanishes $\Sigma m = 0$, i.e. the total charge on the body is zero. Hence on any magnet the positive charge is always equal to the negative one.

When considering electric phenomena we saw that it was impossible to get a charge of positive electricity without at the same time getting an equal charge of negative electricity. It is also impossible to get a charge of positive magnetism without at the same time getting an equal charge of negative magnetism; but whereas in the electrical case all the positive electricity might be on one body and all the negative on another, in the magnetic case if a charge of positive magnetism appears on a body an equal charge of negative magnetism must appear on the *same* body. This difference between the two cases would disappear if we regarded the *dielectric* in the electrical case as analogous to the magnets; the various charged bodies in the electrical field being regarded as portions of the surface of the dielectric.

119. Poles of a Magnet. In the case of very long and thin uniformly magnetized pieces of iron and steel we approximate to a state of things in which the magnetic charges can be regarded as concentrated at the ends of the magnet, which are then called its poles; the positive magnetism being concentrated at the end which points to the north, which is called the positive pole, the negative charge at the other end, called the negative pole.

In general however the magnetic charges are not localized to such an extent as in the previous case, they exist more or less over the whole surface of the magnet; to meet these cases we require a more extended definition of 'the pole of a magnet.'

Suppose the magnet placed in a uniform field, then the forces acting on the positive charges will be a series of parallel forces all acting in the same direction, these by statics may be replaced by a single force acting at a point P called the centre of parallel forces for this system of forces. This point P is called the positive pole of the magnet. Similarly the forces acting on the negative charges may be replaced by a single force acting at a point Q. This point Q is then called the negative pole of the magnet. The resultant force acting at P is by statics the same as if the whole positive charge were concentrated at P; this resultant is equal and opposite to that acting at Q.

120. Axis of a Magnet. The axis of a magnet is the line joining its poles, the line being drawn from the negative to the positive pole.

121. Magnetic Moment of a Magnet is the product of the charge of positive magnetism multiplied by the distance between the poles. It is thus equal to the couple acting on the magnet when placed in a uniform magnetic field where the intensity of the magnetic force is unity, the axis of the magnet being at right angles to the direction of the magnetic force in the uniform field.

122. The Intensity of Magnetization is the magnetic moment of a magnet per unit volume. It is to be regarded as having direction as well as magnitude, its direction being that of the axis of the magnet.

123. Magnetic Potential due to a Small Magnet. Let A and B, Fig. 57, represent the poles of a small magnet, m the charge of magnetism at B, $-m$ that

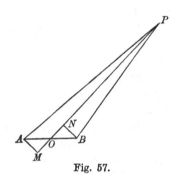

Fig. 57.

at A. Let O be the middle point of AB. Consider the magnetic potential at P due to the magnet AB. The magnetic potential at P due to m at B is $\dfrac{m}{BP}$, that due to $-m$ at A is $-\dfrac{m}{AP}$, hence the magnetic potential at P due to the magnet is

$$\frac{m}{BP} - \frac{m}{AP}.$$

From A and B let fall perpendiculars AM and BN on OP: since the angles BPO, APO are very small and the angles at M and N are right angles, the angles

PBN and PAM will be very nearly right angles, so that approximately

$$BP = PN = PO - ON,$$
$$AP = PM = PO + OM = PO + ON.$$

Then $\quad \dfrac{m}{BP} - \dfrac{m}{AP} = \dfrac{m}{PO - ON} - \dfrac{m}{PO + ON}$

$$= \frac{2m \cdot ON}{OP^2 - ON^2},$$

and this, since ON is very small compared with OP, is approximately equal to

$$\frac{2m \cdot ON}{OP^2}$$
$$= \frac{mAB \cos \theta}{OP^2},$$

where θ is the angle POB.

If M is the magnetic moment of the magnet

$$M = mAB,$$

hence the potential due to the magnet is equal to

$$\frac{M \cos \theta}{OP^2}.$$

124. Resolution of Small Magnets.

We shall first prove that the moment of a small magnet may be resolved like a force, i.e. if the moment of the magnet is M, and if a force M acting along the axis of the magnet be resolved into forces M_1, M_2, M_3, &c. acting in directions OL_1, OL_2, OL_3, &c., where O is the point midway between the poles, then the magnetic action of the original magnet at a distant point is the same as the combined effects of the magnets whose moments are M_1, M_2, M_3, &c., and whose axes are along OL_1, OL_2, OL_3, &c.

Now suppose a force M in the direction AB, Fig. 57, is the resultant of the forces M_1, M_2, M_3 in the directions OB_1, OB_2, OB_3, &c., let OB_1, OB_2, OB_3 make angles θ_1, θ_2, θ_3 with OP, then

$$M \cos \theta = M_1 \cos \theta_1 + M_2 \cos \theta_2 + \ldots,$$

and
$$\frac{M \cos \theta}{OP^2} = \frac{M_1 \cos \theta_1}{OP^2} + \frac{M_2 \cos \theta_2}{OP^2} + \ldots.$$

Now $M_1 \cos \theta_1/OP^2$ is the magnetic potential at P due to the magnet whose moment is M_1 and whose axis is along OB_1, $M_2 \cos \theta_2/OP^2$ is the potential due to the magnet whose moment is M_2 and whose axis is OB_2, and so on; hence we see that the original magnet may be replaced by a series of magnets, the original moment being the resultant of the moments of the magnets by which the magnet is replaced. In other words, the moment of a small magnet may be resolved like a force.

By the aid of this theorem the problem of finding the force due to a small magnet at any point may be reduced to that of finding the force due to a magnet at a point on its axis produced, and at a point on a line through its centre at right angles to its axis.

125. To find the magnetic force at a point on the axis produced. Let AB, Fig. 58, be the magnet, P the point at which the force is required. The magnetic force at P due to the charge m at B is equal to

$$\frac{m}{(OP - OB)^2}.$$

The magnetic force due to $-m$ at A is equal to

$$-\frac{m}{(OP + OB)^2}.$$

The resultant magnetic force at P is equal to

$$\frac{m}{(OP - OB)^2} - \frac{m}{(OP + OB)^2} = \frac{4m \cdot OB \cdot OP}{(OP^2 - OB^2)^2}$$

$$= \frac{4mOB \cdot OP}{OP^4}$$

approximately, since OB is small compared with OP.

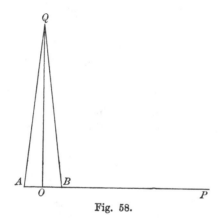

Fig. 58.

If M is the moment of the magnet $M = 2mOB$, thus the magnetic force at P is equal to

$$\frac{2M}{OP^3}.$$

The direction of this force is along OP.

126. To find the magnetic force at a point Q on the line through O at right angles to AB. Since Q is equidistant from A and B, Fig. 58, the forces due to A and B are equal in magnitude; the one being a repulsion, the other an attraction. The resultant of these forces is equal to

$$\frac{2m}{BQ^2}\frac{OB}{BQ}=\frac{M}{BQ^3}$$

$$=\frac{M}{OQ^3},$$

since BQ is approximately equal to OQ.

The direction of this force is parallel to BA and at right angles to OQ.

If Q, a point on the line through O at right angles to AB, is the same distance from O as P, a point on AB produced, we see from these results that the force at P is twice that at Q. This is the foundation of Gauss's method (see Art. 132) of proving that the force between two poles varies inversely as the square of the distance between them.

127. Magnetic force due to a small magnet at any point. Let AB, Fig. 59, represent the small magnet,

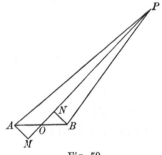

Fig. 59.

let M be its moment, O its centre, P the point at which the force is required, let OP make an angle θ with AB, the axis of the magnet. By Art. 124 the effect of M is

equivalent to that of two magnets, one having its axis
along OP and its moment equal to $M\cos\theta$, the other
having its axis at right angles to OP and its moment
equal to $M\sin\theta$.   Let $OP = r$.

The force at P due to the first is, by Art. 125, along
OP and equal to $2M\cos\theta/r^3$, the force at P due to the
second magnet is at right angles to OP and equal to
$M\sin\theta/r^3$, hence the force due to the magnet AB at
P is equivalent to the forces

$$\frac{2M\cos\theta}{r^3} \text{ along } OP,$$

and $\qquad \dfrac{M\sin\theta}{r^3}$ at right angles to OP.

Let the resultant magnetic force at P make an angle
ϕ with OP, then

$$\tan\phi = \frac{\dfrac{M\sin\theta}{r^3}}{\dfrac{2M\cos\theta}{r^3}} = \tfrac{1}{2}\tan\theta.$$

Let the direction of the resultant force at P cut AB
produced in T, draw TN at right angles to OP, then

$$\tan\phi = \frac{TN}{PN},$$

$$\tan\theta = \frac{TN}{ON},$$

and since $\tan\phi = \tfrac{1}{2}\tan\theta$, $PN = 2ON$.   Thus $ON = \tfrac{1}{3}OP$.
Thus, to find the direction of the magnetic force at P,
trisect OP at N, draw NT at right angles to OP to cut
AB produced in T, then PT will be the direction of the
force at P.

The magnitude of the resulting force is

$$\frac{M}{r^3}\sqrt{4\cos^2\theta + \sin^2\theta} = \frac{M}{r^3}\sqrt{1 + 3\cos^2\theta}\,;$$

for a given value of r it is greatest when $\theta = 0$ or π, i.e. at a point along the axis, and least when $\theta = \pi/2$ or $3\pi/2$, i.e. at a point on the line at right angles to the axis. The maximum value is twice the minimum one.

The curves of constant magnetic potential are represented by equations of the form

$$\frac{\cos\theta}{r^2} = C;$$

the lines of force which cut the equipotential curves at right angles are given by the equations

$$r = C\sin^2\theta,$$

where C is a variable parameter.

The radius of curvature of the line of force at a point P can easily be proved to equal

$$\frac{2r}{3\sin\phi\,(1 + \sin^2\phi)},$$

where ϕ is the angle the line of force makes with OP. Thus the radius of curvature at points on the line bisecting the magnet at right angles is one-third of the distance of the point from the magnet.

128. Couple on a Magnet in a Uniform Magnetic Field. If a magnet is placed in a uniform field the couple acting on the magnet, and tending to twist it about a line at right angles both to the axis of the magnet and the force in the external field, is

$$MH\sin\theta,$$

where M is the moment of the magnet, H the force in

the uniform field, and θ the angle between the axis of the magnet and the direction of the force.

Let AB be the magnet, the negative pole being at A, the positive one at B. Then if m is the strength of the pole at B, the forces on the magnet are a force mH at B in the direction of the external field and an equal and opposite force at A. These two forces are equivalent to a couple whose moment is $HmNM$, where NM is the distance between the lines of action of the two forces. But $NM = AB \sin \theta$, if θ is the angle between AB and H; hence the couple on the magnet is

$$HmAB \sin \theta = HM \sin \theta.$$

129. Couples between two Small Magnets.

Let AB, CD, Fig. 60, represent the two magnets; M, M' their moments; r the distance between their centres O, O'. Let AB, CD make respectively the angles θ, θ' with OO'.

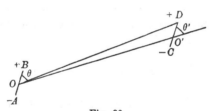

Fig. 60.

Consider first the couple on the magnet CD. The magnetic forces due to AB are

$$\frac{2M \cos \theta}{r^3} \text{ along } OO',$$

$$\frac{M \sin \theta}{r^3} \text{ at right angles to } OO'.$$

These may be regarded as constant over the space occupied by the small magnet CD.

The couple on CD tending to produce rotation in the direction of the hands of a watch, due to the first component, is

$$\frac{2M \cos \theta}{r^3} M' \sin \theta',$$

that due to the second is

$$\frac{M \sin \theta}{r^3} M' \cos \theta';$$

hence the total couple on CD is

$$\frac{MM'}{r^3} (2 \cos \theta \sin \theta' + \sin \theta \cos \theta').$$

This vanishes if $\tan \theta' = -\frac{1}{2} \tan \theta$, i.e. if CD is along the line of force due to AB, see Art. 127.

We may show in a similar way that the couple on AB due to CD tending to produce rotation in the direction of the hands of a watch is

$$\frac{MM'}{r^3} (2 \cos \theta' \sin \theta + \sin \theta' \cos \theta).$$

For both these couples to vanish, $\theta = 0$ or π, $\theta' = 0$ or π, or $\theta = \pm \frac{\pi}{2}$, $\theta' = \pm \frac{\pi}{2}$, so that the axes of the magnets must be parallel to each other, and either parallel or perpendicular to the line joining the centres of the two magnets.

We shall find it convenient to consider four special positions of the two magnets as standard cases.

Case I.

Fig. 61.

$\theta = 0$, $\theta' = 0$, couples vanish, equilibrium stable.

Case II.

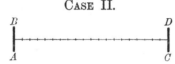

Fig. 62.

$\theta = \dfrac{\pi}{2}$, $\theta' = \dfrac{\pi}{2}$, couples vanish, equilibrium unstable.

Case III.

Fig. 63.

$\theta = 0$, $\theta' = \dfrac{\pi}{2}$, couple on $CD = \dfrac{2MM'}{r^3}$, couple on $AB = \dfrac{MM'}{r^3}$.

When the magnets are arranged as in this case, AB is said to be 'end on' to CD, while CD is 'broadside on' to AB.

Case IV.

Fig. 64.

$\theta = \dfrac{\pi}{2}$, $\theta' = 0$, couple on $CD = \dfrac{MM'}{r^3}$, couple on $AB = \dfrac{2MM'}{r^3}$.

In this case AB is broadside on to CD. We see that the couple exerted on CD by AB is twice as great when the latter is end on as when it is broadside on.

It will be noticed that the couples on AB and CD are not in general equal and opposite; at first sight it might appear that this result would lead to the absurd conclusion that if two magnets were firmly fastened to a board, and the board floated on a vessel of water, the board would be set in rotation and would spin round with gradually increasing velocity. The paradox will however be explained if we consider the *forces* exerted by one magnet on the other.

130. Forces between two Small Magnets. Let AB, CD (Fig. 60) represent the two magnets, O, O' the middle points of AB, CD respectively, θ, θ' the angles which AB, CD respectively make with OO'. Let ϕ be the angle DOO', $r = OO'$; m, m' the strengths of the poles of AB and CD.

The force due to the magnet AB on the pole at D consists of the component

$$\frac{2Mm'}{OD^3}\cos(\theta-\phi),$$

along OD, and

$$\frac{Mm'}{OD^3}\sin(\theta-\phi),$$

at right angles to OD.

These are equivalent to a force equal to

$$\frac{2Mm'\cos(\theta-\phi)\cos\phi}{OD^3}+\frac{Mm'\sin(\theta-\phi)\sin\phi}{OD^3},$$

along OO', and a force equal to

$$\frac{2Mm'\cos(\theta-\phi)\sin\phi}{OD^3} - \frac{Mm'\sin(\theta-\phi)\cos\phi}{OD^3},$$

acting upwards at right angles to OO'.

Neglecting squares and higher powers of CD/OO' we have

$$\cos\phi = 1, \quad \sin\phi = \frac{CD}{2r}\sin\theta',$$

$$OD = r + \frac{1}{2}CD\cos\theta', \quad \frac{1}{OD^3} = \frac{1}{r^3} - \frac{3}{2}\frac{CD}{r^4}\cos\theta'.$$

Substituting these values we see that the force exerted by AB on D is approximately equivalent to a component

$$\frac{2Mm'\cos\theta}{r^3} - \frac{3Mm'\,CD\cos\theta\cos\theta'}{r^4} + \frac{3}{2}\frac{Mm'\,CD\sin\theta\sin\theta'}{r^4},$$

along OO', and a component

$$-\frac{Mm'\sin\theta}{r^3} + \frac{3}{2}\frac{Mm'\,CD\sin\theta\cos\theta'}{r^4} + \frac{3}{2}\frac{Mm'\,CD\cos\theta\sin\theta'}{r^4},$$

acting upwards at right angles to OO'.

We may show in a similar way that the force exerted by AB on C is equivalent to a component

$$-\frac{2Mm'\cos\theta}{r^3} - \frac{3Mm'\,CD\cos\theta\cos\theta'}{r^4} + \frac{3}{2}\frac{Mm'\,CD\sin\theta\sin\theta'}{r^4},$$

along OO', and a component

$$\frac{Mm'\sin\theta}{r^3} + \frac{3}{2}\frac{Mm'\,CD\sin\theta\cos\theta'}{r^4} + \frac{3}{2}\frac{Mm'\,CD\cos\theta\sin\theta'}{r^4},$$

acting upwards at right angles to OO'.

Hence the force on the magnet CD, which is the

resultant of the forces acting on the poles C, D, is equivalent to a component

$$-\frac{3MM'}{r^4}(2\cos\theta\cos\theta' - \sin\theta\sin\theta'),$$

along OO', and a component

$$\frac{3MM'}{r^4}(\sin\theta\cos\theta' + \cos\theta\sin\theta'),$$

acting upwards at right angles to OO'.

The force on the magnet AB is equal in magnitude and opposite in direction to that on CD.

If we consider the two magnets as forming one system, the two forces at right angles to OO' are equivalent to a couple whose moment is

$$\frac{3MM'}{r^3}(\cos\theta\sin\theta' + \sin\theta\cos\theta'),$$

this couple is equal in magnitude and opposite in direction to the algebraical sum of the couples on the magnets AB, CD found in Art. 129: this result explains the paradox alluded to at the end of that article.

131. Force between the Magnets in the four standard positions. In the positions described in Art. 129, the forces between the magnets have the following values.

CASE I.   Fig. 61.

$\theta = 0$, $\theta' = 0$.   Force between magnets is an attraction along the line joining their centres equal to

$$\frac{6MM'}{r^4}.$$

Case II.　Fig. 62.

$\theta = \dfrac{\pi}{2}$, $\theta' = \dfrac{\pi}{2}$.　Force is a repulsion along the line joining the centres equal to

$$\frac{3MM'}{r^4}.$$

Case III.　Fig. 63.

$\theta = 0$, $\theta' = \dfrac{\pi}{2}$.　Force is at right angles to the line joining the centres and equal to

$$\frac{3MM'}{r^4}.$$

Case IV.　Fig. 64.

$\theta = \dfrac{\pi}{2}$, $\theta' = 0$.　Force is at right angles to the line joining the centres and equal to

$$\frac{3MM'}{r^4}.$$

The forces between the magnets vary inversely as the fourth power of the distance between their centres, while the couples vary inversely as only the cube of this distance. The directive influence which the magnets exert on each other thus diminishes less quickly with the distance than the translatory forces, so that when the magnets are far apart the directive influence is much the more important of the two.

132.　Gauss's proof that the force between two magnetic poles varies inversely as the square of the distance between them. We saw, Art. 129, that, the distance between the magnets remaining the same, the couple exerted by the first magnet on the second was twice as great when the first magnet was 'end on' to the second as when it was 'broadside on.' This is equivalent to the result proved in Art. 127, that when P and Q are two points at the same distance from the centre of the magnet, P being on the axis of the magnet and Q on the line through the centre at right angles to the axis, the magnetic force at P is twice that at Q. This result only holds when the force varies inversely as the square of the distance; we shall proceed to show that if the force varied inversely as the pth power of the distance the magnetic force at P would be p times that at Q.

If the magnetic force varies inversely as the pth power of the distance, then if m is the strength of one of the poles of the magnet, the magnetic force at P, Fig. 58, due to the magnet AB is equal to

$$\frac{m}{\overline{BP}^p} - \frac{m}{\overline{AP}^p}$$

$$= \frac{m}{(OP - OB)^p} - \frac{m}{(OP + OB)^p}$$

$$= \frac{2mp \cdot OB}{OP^{p+1}},$$

approximately, if OB is very small compared with OP; if M is the moment of AB this is equal to

$$\frac{pM}{OP^{p+1}}.$$

The force at $Q = \dfrac{m}{BQ^p}\dfrac{OB}{BQ} + \dfrac{m}{AQ^p}\dfrac{OA}{AQ}$

$$= \frac{M}{OP^{p+1}},$$

approximately.

Thus the magnetic force at P is p times that at Q. We see from this that if we have two small magnets the couple on the second when the first magnet is 'end on' to it is p times the couple when the first magnet is 'broadside on.' Hence by comparing the value of the couples in these positions we can determine the value of p.

This can be done by an arrangement of the following kind. Suspend the small magnet which is to be deflected so that it can turn freely about a vertical axis : a convenient way of doing this and one which enables the angular motion of the magnet to be accurately determined, is to place the magnet at the back of a very light mirror and suspend the mirror by a silk fibre. When the deflecting magnet is far away the suspended magnet will under the influence of the earth's magnetic field point magnetic north and south. When this magnet is at rest bring the deflecting magnet into the field and place it so that its centre is due east or west of the centre of the deflected magnet, the axis of the deflecting magnet passing through the centre of this magnet. The couple due to the deflecting magnet will make the suspended magnet swing from the north and south position until the couple with which the earth's magnetic force tends to bring the magnet back to its original position just balances the deflecting couple.

Let H be the magnetic force in the horizontal plane due to the earth's magnetic field. Then when the deflected magnet has twisted through an angle θ the couple due to the earth's magnetic field is, see Art. 128, equal to

$$HM' \sin \theta,$$

where M' is the moment of the deflected magnet.

The other magnet may be regarded as producing a field such that the magnetic force at the centre of the deflected magnet is east and west and equal to

$$\frac{Mp}{r^{p+1}},$$

where M is the moment of the deflecting magnet, r the distance between the centres of the deflected and deflecting magnets. Thus the couple on the deflected magnet due to this magnet is

$$\frac{MM'\, p \cos \theta}{r^{p+1}}.$$

The suspended magnet will take up the position in which the two couples balance: when this is the case

$$HM' \sin \theta = \frac{MM'p \cos \theta}{r^{p+1}},$$

or $\qquad \tan \theta = \dfrac{Mp}{Hr^{p+1}}$ ....(1).

Now place the deflecting magnet so that its centre is north or south of that of the suspended magnet, and at the same distance from it as in the last experiment, the axis of the deflecting magnet being again east and west. Let the suspended magnet be in equilibrium when it has twisted through an angle θ'. The couple due to the earth's magnetic field is

$$HM' \sin \theta'.$$

The couple due to the deflecting magnet is

$$\frac{MM' \cos \theta'}{r^{p+1}}.$$

Since the suspended magnet is in equilibrium these couples must be equal, hence

$$HM' \sin \theta' = \frac{MM' \cos \theta'}{r^{p+1}},$$

hence
$$\tan \theta' = \frac{M}{Hr^{p+1}} \dots\dots\dots\dots(2).$$

Thus
$$\frac{\tan \theta}{\tan \theta'} = p.$$

Hence if we measure θ and θ' we can determine p. By experiments of this kind Gauss showed that $p = 2$, i.e. that the force between two poles varies inversely as the square of the distance between them.

If we place the deflecting magnet at different distances from the deflected we find that $\tan \theta$ and $\tan \theta'$ vary as $1/r^3$, and thus obtain another proof that $p = 2$.

133. Determination of the Moment of a Small Magnet and of the horizontal component of the Earth's Magnetic Force. Suspend a small auxiliary magnet in the same way as the deflected magnet in the experiment just described, and place the magnet A whose moment is to be determined, so that its centre is due east or west of the centre of the auxiliary magnet, and its axis passes through the centre of the suspended magnet. Let θ be the deflection of the suspended magnet, H the horizontal component of the earth's magnetic force, M the moment of A: we have, by equation (1), Art. 132, putting $p = 2$

$$\frac{M}{H} = \tfrac{1}{2} r^3 \tan \theta;$$

hence if we measure r and θ we can determine M/H.

To determine MH suspend the magnet A so that it can rotate freely about a vertical axis, passing through its centre, taking care that the magnetic axis of A is horizontal. When the magnet makes an angle θ with the direction in which H acts, i.e. with the north and south line, the couple tending to bring it back to its position of equilibrium is equal to

$$MH \sin \theta.$$

Hence if K is the moment of inertia of the magnet about the vertical axis the equation of motion of the magnet is

$$K\frac{d^2\theta}{dt^2} + MH \sin \theta = 0,$$

or if θ is small

$$K\frac{d^2\theta}{dt^2} + MH\theta = 0.$$

Hence T, the time of a small oscillation, is given by the equation

$$T = 2\pi \sqrt{\frac{K}{MH}},$$

or $$MH = \frac{4\pi^2 K}{T^2},$$

hence if we know K and T we can determine MH; and knowing M/H from the preceding experiment we can find both M and H. The value of H at Cambridge is about ·18 c.g.s. units.

134. Magnetic Shell of Uniform Strength. A magnetic shell is a thin sheet of magnetizable substance magnetized at each point in the direction of the normal to the sheet at that point.

The **strength** of the shell at any point is the product of the intensity of magnetization into the thickness of the shell measured along the normal at that point, it is thus equal to the magnetic moment of unit area of the shell at the point.

To find the potential of a shell of uniform strength. Consider a small area α of the shell round the point Q, Fig. 65, let I be the intensity of magnetization of the shell

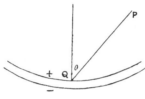

Fig. 65.

at Q, t the thickness of the shell at the same point. The moment of the small magnet whose area is α is $I\alpha t$, hence if θ is the angle which the direction of magnetization makes with PQ, the potential of the small magnet at P is by Art. 123 equal to

$$\frac{I\alpha t \cos \theta}{PQ^2}.$$

If ϕ is the strength of the magnetic shell

$$It = \phi,$$

hence the potential at P is

$$\frac{\phi\alpha \cos \theta}{PQ^2}.$$

This, by Art. 10, is numerically equal to the normal induction over α due to a charge of electricity equal to ϕ

at P.   Hence if ϕ is constant over the shell the potential of the whole shell at P is numerically equal to the total normal electric induction over it due to a charge ϕ at P. This, by Art. 10, is equal to $\phi\omega$, where ω is the area cut off from the surface of a sphere of unit radius with its centre at P by lines drawn from P to the boundary of the shell; ω is called the solid angle subtended by the shell at P; it only depends on the shape of the boundary of the shell.

If the shell is closed, then if P is outside the shell the potential at P is zero, since the total normal electric induction over a closed surface due to a charge at a point outside the surface is zero; if the point P is inside the surface and the negative side of the shell is on the outside, then since the total normal electric induction over the shell due to a charge ϕ at P is $4\pi\phi$, the magnetic potential at P is $4\pi\phi$; as this is constant throughout the shell, the magnetic force vanishes inside the space bounded by the shell.

The signs to be ascribed to the solid angle bounded by the shell at various points are determined in the following way.   Take a fixed point O and with it as centre describe a sphere of unit radius.   Let P be a point at which the magnetic potential of the shell is required.   The contribution to the magnetic potential by any small area round a point Q on the shell, is the area cut off from the surface of the sphere of unit radius by the radii drawn from O parallel to the radii drawn from P to the boundary of the area round Q.   The area enclosed by the lines from O is to be taken as positive or negative according as the lines drawn from P to Q strike first against the positive or negative side of the shell.   By the positive side of the shell we mean the side charged with positive magnetism, by the negative side the side charged with negative magnetism.

With this convention with regard to the signs of the solid angle, let us consider the relation between the potentials due to a shell at two points P and P'; P being close to the shell on the positive side, P' close to P but

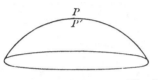

Fig. 66.

on the negative side of the shell. Consider the areas traced out on the unit sphere by radii from O parallel to those drawn from P and P'. The area corresponding to those drawn from P will be the shaded part of the sphere, let this area be ω, the potential at P is $\phi\omega$. The area corresponding to the radii drawn from P' will be the unshaded portion of the sphere whose area is $4\pi - \omega$, but inasmuch as the radii from P' strike first against the negative side of the shell the solid angle subtended at P' will be minus this area, i.e. $\omega - 4\pi$; hence the magnetic potential due to the shell at P' is $\phi(\omega - 4\pi)$. The potential at P thus exceeds that at P' by $4\pi\phi$.

In spite of this finite increment in the potential in passing from P' to the adjacent point P, there will be continuity of potential in passing through the shell if we regard the potential as given in the shell by the same laws as outside.

Consider the potential at a point Q in the shell, and

divide the original shell into two, one on each side of Q. Then as the whole shell is uniformly magnetized the

Fig. 67.

strength of the shells will be proportional to their thicknesses. Thus if ϕ is the strength of the original shell the strength of the shell between P and Q will be $\phi \dfrac{PQ}{PP'}$, and that of the shell between Q and P' will be $\dfrac{QP'}{PP'}$.

The potential at Q due to the shell next to P' is $\omega\phi \dfrac{QP'}{PP'}$, that due to the shell next to P is $(\omega - 4\pi)\,\phi\,\dfrac{QP}{PP'}$, the potential at Q is the sum of these, i.e.

$$\omega\phi - 4\pi\phi\,\frac{QP}{PP'}\,;$$

this changes continuously as we pass through the shell from

$$\phi\,(\omega - 4\pi) \text{ at } P',$$

to $\qquad\qquad\qquad \phi\omega \text{ at } P.$

135. Mutual Potential Energy of the Shell and an external Magnetic System. Let I be the intensity of magnetization at a point Q on the shell; consider a small portion of the shell round Q, α being the area of this portion. Let P, P' be two points on its axis of magnetization, P being on the positive surface of the shell, P' on the negative. Then we have a charge of positive magnetism equal to $I\alpha$ at P, a negative charge $-I\alpha$ at

P'. If V_P, $V_{P'}$ are the potentials at P and P' respectively due to the external magnetic system, then the mutual potential energy of the external system and the small magnet at Q is equal to

$$V_P I \alpha - V_{P'} I \alpha \quad \dots\dots\dots\dots\dots(1).$$

If ϕ is the strength of the shell

$$\phi = I \times PP',$$

hence the expression (1) is equal to

$$\frac{\phi \, (V_P - V_{P'}) \, \alpha}{PP'}.$$

But $(V_P - V_{P'})/PP'$ is the magnetic force due to the external system along PP', the normal to the shell. Let this force be denoted by $-H_n$, the force being taken as positive when it is in the direction of magnetization of the shell, i.e. when the magnetic force passes from the negative to the positive side through the shell, then the mutual potential energy of the external system and the small magnet at Q is equal to

$$- \phi H_n \alpha.$$

Since the strength of the shell is uniform the mutual potential energy of the external system and the whole shell is equal to

$$- \phi \Sigma H_n \alpha,$$

$\Sigma H_n \alpha$ being the sum of the products got by dividing the surface of the shell up into small areas, and multiplying each area by the component along its normal of the magnetic force due to the external system, this component being positive when it is in the direction of magnetization of the shell. This quantity is often called the number of lines of magnetic force due to the external system which pass through the shell.

It is analogous to the total normal electric induction over a surface in Electrostatics, see Art. 9.

136. Force acting on the shell when placed in a magnetic field. If X is the force acting on the shell in the direction x, and if the shell is displaced in this direction through a distance δx, then $X\delta x$ is the work done on the shell by the magnetic forces during the displacement; hence by the principle of the Conservation of Energy, $X\delta x$ must equal the diminution in the energy due to the displacement. Suppose that A, Fig. 68, represents the position of the edge of the shell before,

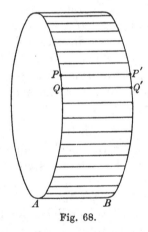

Fig. 68.

B its position after the displacement. The diminution in the energy due to the displacement is, by the last paragraph, equal to

$$\phi (N' - N) \quad \dots\dots\dots\dots\dots\dots\dots(1),$$

where N and N' are the numbers of lines of magnetic force which pass through A and B respectively. Consider the closed surface having as ends the shell in its two positions A and B, the sides of the surface being formed by the lines PP' &c. which join the original position of a point P to its displaced position. We see, as in Art. 10, that unless the closed surface contains an excess of magnetism of one sign $\Sigma H_n \alpha$ taken over its surface must vanish, H_n denoting the magnetic force along the normal to the surface drawn outwards.

But $\Sigma H_n \alpha$ over the whole surface

$$= N' - N + \Sigma H_n \alpha \text{ taken over the sides,}$$

hence $$N' - N = -\Sigma H_n \alpha \dots \dots \dots \dots (2);$$

the summation on the right-hand side of this equation being taken over the sides. Consider a portion of the sides bounded by PQ, $P'Q'$; P', Q' being the displaced positions of P and Q respectively. Since

$$PP' = QQ' = \delta x,$$

the area $PQP'Q'$ is equal to

$$\delta x \times PQ \times \sin \theta,$$

where θ is the angle between PQ and PP'. If H is the magnetic force at P due to the external system, the value of $H_n \alpha$ for the element $PQQ'P'$ is equal to

$$\delta x \times PQ \times \sin \theta \times H \cos \chi,$$

where χ is the angle which the outward-drawn normal to $PQQ'P'$ makes with H. Hence since $X \delta x = \phi (N' - N)$ we have by equation (2)

$$X \delta x = -\phi \Sigma \{\delta x \times PQ \times \sin \theta \times H \cos \chi\},$$

or since δx is the same for all points on the shell

$$X = -\phi \Sigma \{PQ \times \sin \theta \times H \cos \chi\}.$$

Thus the force on the shell parallel to x is the same as it would be if a force parallel to x acted on the boundary of the shell, equal per unit length to

$$- \phi H \sin \theta \cos \chi.$$

Since x is arbitrary this gives the force acting on each element of the boundary in any direction; to find the resultant force on the element, we notice that the component along x vanishes if x is parallel to PQ, for in this case $\theta = 0$, the resultant force is thus at right angles to the element of the boundary. Again, if x is parallel to H, $\chi = \pi/2$, and the force again vanishes, thus the resultant force is at right angles to H. Hence the resultant force on PQ is at right angles both to PQ and H. In order to find the magnitude of this force we have only to suppose that x is parallel to this normal, in this case $\theta = \pi/2$ and $\chi = \dfrac{\pi}{2} - \psi$, where ψ is the angle between PQ and H; the resultant force is therefore

$$- \phi H \sin \psi.$$

Thus the force on the shell may be regarded as equivalent to a system of forces acting over the edge of the shell, the force acting on each element of the edge being at right angles to the element and to the external magnetic force at the element, and equal per unit length to the product of the strength of the shell into the component of the magnetic force at right angles to the element of the edge.

The preceding rule gives the line along which the force acts; the direction of the force is, in any particular case, most easily got from the principle that since the mutual potential energy of the shell and the external magnetic system is equal to $-\phi N$, where N is the number of lines of magnetic force due to the external system which pass through the shell in the direction in which it is magnetized, i.e. which enter the shell on the side with the negative magnetic charge and leave it on the side with the positive charge : the shell will tend to move so as to make N as large as possible, for by so doing it makes the potential energy as small as possible. The force on each element of the boundary will therefore be in such a direction as to tend to move the element of the boundary so as to enclose a greater number of lines of magnetic force passing through the shell in the positive direction.

Thus if the direction of the magnetic force at the element PQ is in the direction PT in Fig. 69, the force on PQ will be outwards along PS as in the figure, for

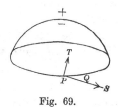

Fig. 69.

if PQ were to move in this direction the shell would catch more lines of force passing through it in the positive direction.

Since $$X\delta x = \phi (N' - N)$$

we get $$X = \phi \frac{dN}{dx}.$$

This expression is often very useful for finding the total force on the shell in any direction.

137. Magnetic force due to the shell. Suppose that the external field is that due to a single unit pole at a point A, the result of the preceding article will give the force on the shell due to the pole, this must however be equal and opposite to the force exerted by the shell on the pole. If however the field is due to a unit pole at A, H the magnetic force due to the external system at an element PQ of the shell is equal to $1/AP^2$ and acts along AP: hence by the last article the magnetic force at A due to the shell is the same as if we supposed each unit of length of the boundary of the shell to exert a force equal to

$$\frac{\phi}{AP^2} \sin \theta,$$

where θ is the angle between AP and the tangent to the boundary at P, ϕ is the strength of the shell. This force acts along the line which is at right angles both to AP and the tangent to the boundary at P. The direction in which the force acts along this line may be found by the rule that it is opposite to the force acting on the element of the boundary at P arising from unit magnetic pole at A; this latter force may be found by the method given at the end of the preceding article.

138. If the external magnetic field in Art. 135 is due to a second magnetic shell, then the mutual potential energy of the two shells is equal to

$$- \phi N,$$

where ϕ is the strength of the first shell, and N the number of lines of force which pass through the first shell, and are produced by the second. It is also equal to

$$- \phi' N',$$

where ϕ' is the strength of the second shell, and N' the number of lines of force which pass through the second shell and are produced by the first. Hence by making $\phi = \phi'$ we see that, if we have two shells α and β of equal strengths, the number of lines of force which pass through α and are due to β is equal to the number of lines of force which pass through β and are due to α.

139. Magnetic Field due to a uniformly magnetized sphere. Let the sphere be magnetized parallel to x, and let I be the intensity of magnetization. We may regard the sphere as made up, as in Fig. 70, of a great number of uniformly magnetized bar magnets of uniform cross section α, the axes of these magnets being parallel to the axis of x. On the ends of each of these magnets we have charges of magnetism equal to $\pm I\alpha$. Now consider a sphere whose radius is equal to that of the magnetized sphere and built up of bars in the same way, each of these bars being however wholly filled with positive magnetism whose volume density is ρ: consider

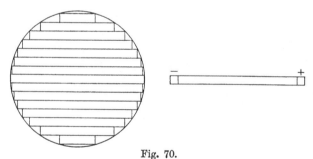

Fig. 70.

also another equal sphere divided up into bars in the same way, each of these bars being however filled with

15—2

negative magnetism whose volume density is $-\rho$; suppose
that these spheres have their centres at O' and O, Fig. 71,
two points very close together, OO' being parallel to the
axis of x. Consider now the result of superposing these
two spheres: take two corresponding bars; the parts of
the bars which coincide will neutralize each other's effects,
but the negative bar will project a distance OO' to the
left, and on this part of the bar there will be a charge of
negative magnetism equal to $OO' \times a \times \rho$: the positive bar
will project a distance OO' to the right, and on this part
of the bar there will be a charge of positive magnetism
equal to $OO' \times a \times \rho$. If OO' is very small we may regard
these charges as concentrated at the ends of the bars, so
that if $OO' \times \rho = I$ the case will coincide with that of the
uniformly magnetized sphere.

We can easily find the effects of the positive and
negative spheres at any point either inside or outside.
Let us first consider the effect at an external point P.

The potential due to the positive sphere is equal to

$$\frac{4}{3} \frac{\pi a^3 \rho}{O'P}$$

if a is the radius of the sphere.

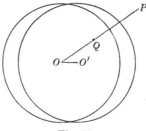

Fig. 71.

The potential due to the negative sphere is equal to

$$-\frac{4}{3}\frac{\pi a^3 \rho}{OP}.$$

Hence the potential due to the combination of the spheres is equal to

$$\frac{4}{3}\pi a^3 \rho \left\{ \frac{1}{O'P} - \frac{1}{OP} \right\}$$

$$= \frac{4}{3}\pi a^3 \rho \cdot \frac{OO' \cos \theta}{OP^2}$$

approximately, if OO' is very small, and θ is the angle which OP makes with OO'.

Now we have seen that this case coincides with that of the uniformly magnetized sphere if $\rho \times OO' = I$, where I is the intensity of magnetization of the sphere; hence the potential due to the uniformly magnetized sphere at an external point P is

$$\frac{4}{3}\pi a^3 I \cdot \frac{\cos \theta}{r^2},$$

where $r = OP$.

Comparing this result with that given in Art. 123 we see that the uniformly magnetized sphere produces the same effect outside the sphere as a very small magnet placed at its centre, the axis of the small magnet being parallel to the direction of magnetization of the sphere, while the moment of the magnet is equal to the intensity of magnetization multiplied by the volume of the sphere.

The magnetic force inside the sphere is indefinite without further definition, since to measure the force on the unit pole, we have to make a hole to receive the pole and the force on the pole depends on the shape of the hole so made: this point is discussed at length in Chapter VIII.

For the sake of completing the solution of this case, we shall anticipate the results of that chapter and assume that the quantity which is defined as 'the magnetic force' inside the sphere is the force which would be exerted on the unit pole if the sphere were regarded as a spherical air cavity over the surface of which there is spread the same distribution of magnetic charge as actually exists over the surface of the magnetized sphere. We may thus in calculating the effect of the charges on the surface suppose that they exert the same magnetic forces as they would in air.

To find the magnetic force at an internal point Q, Fig. 71, we return to the case of the two uniformly charged spheres.

The force due to the uniformly positively charged sphere at Q is equal to

$$\tfrac{4}{3}\pi\rho \,.\, O'Q,$$

and acts along $O'Q$; the force due to the negatively charged sphere is equal to

$$\tfrac{4}{3}\pi\rho \,.\, OQ,$$

and acts along QO.

By the triangle of forces the resultant of the forces exerted by the positive and negative spheres is equal to

$$\tfrac{4}{3}\pi\rho \,.\, OO',$$

and is parallel to OO'. We have seen that the case of the positive and negative spheres coincides with that of the uniformly magnetized sphere if $OO' \times \rho = I$. Hence the force inside the uniformly magnetized sphere is uniform and parallel to the direction of magnetization of the sphere and equal to

$$\tfrac{4}{3}\pi I.$$

The lines of force inside and outside the sphere are given in Fig. 72.

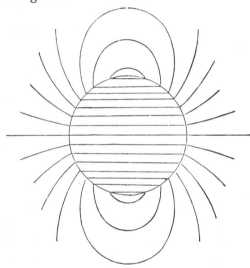

Fig. 72.

CHAPTER VII

TERRESTRIAL MAGNETISM

140. The pointing of the compass in a definite direction was at first ascribed to the special attraction for iron possessed by the pole star. Gilbert, however, in his work *De Magnete,* published in 1600, pointed out that it showed that the earth was itself a magnet. Since Gilbert's time the study of Terrestrial Magnetism, i.e. the state of the earth's magnetic field, has received a great deal of attention and forms one of the most important, and undoubtedly one of the most mysterious departments of Physical Science.

141. To fix the state of the earth's magnetic field we require to know the magnetic force over the whole of the surface of the earth; the observations made at a number of magnetic observatories, scattered unfortunately somewhat irregularly at very wide intervals over the earth, give us an approximation to this.

To determine the magnitude and direction of the earth's magnetic force we require to know three things: the three usually taken are (1) the magnitude of the horizontal component of the earth's magnetic force, usually called the earth's horizontal force; (2) the angle which the direction of the horizontal force makes with the geographical meridian, this angle is called the declination;

the vertical plane through the direction of the earth's horizontal force is called the magnetic meridian; (3) the dip, that is the complement of the angle which the axis of a magnet, suspended so as to be able to turn freely about an axle through its centre of gravity at right angles to the magnetic meridian, makes with the vertical. The fact that a compass needle when free to turn about a horizontal axis would not settle in a horizontal position, but 'dipped,' so that the north end pointed downwards, was discovered by Norman in 1576.

For a full description of the methods and precautions which must be taken to determine accurately the values of the magnetic elements the student is referred to the article on Terrestrial Magnetism in the *Encyclopædia Britannica*: we shall in what follows merely give a general account of these methods without entering into the details which must be attended to if the most accurate results are to be obtained.

The method of determining the horizontal force has been described in Art. 133.

142. Declination. To determine the declination an instrument called a declinometer may be employed; this instrument is represented in Fig. 73. The magnet— which is a hollow tube with a piece of plane glass with a scale engraved on it at the north end and a lens at the south end—is suspended by a single long silk thread from which the torsion has been removed by suspending from it a plummet of the same weight as the magnet: the suspension and the reading telescope can rotate about a vertical axis and the azimuth of the system determined by means of a scale engraved on the fixed horizontal base.

The observer looks through the telescope and observes the
division on the scale at the end of the magnet with which
a cross wire in the telescope coincides; the magnet is
then turned upside down and resuspended and the division
of the scale with which the cross wire coincides again
noted; this is done to correct for the error that would

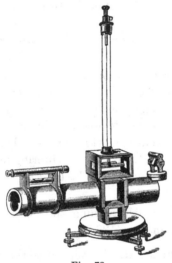

Fig. 73.

otherwise ensue if the magnetic axis of the cylinder did
not coincide with the geometrical axis. The mean of the
readings gives the position of the magnetic axis. If now
we take the reading on the graduated circle and add to
this the known value in terms of the graduations on this
circle of the scale divisions seen through the telescope, we
shall find the circle reading which corresponds to the
magnetic meridian. Now remove the magnet and turn
the telescope round until some distant object, whose

azimuth is known, is in the field of view; take the reading on the graduated circle, the difference between this and the previous reading will give us the angular distance of the magnetic meridian from a plane whose azimuth is known : in other words, it gives us the magnetic declination.

143. Dip. The dip is determined by means of an instrument called the dip-circle, represented in Fig. 74. It

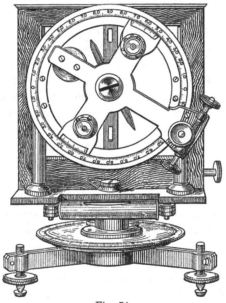

Fig. 74.

consists of a thin magnet with an axle of hard steel whose axis is at right angles to the plane of the magnet, and ought to pass through the centre of gravity of the needle ; this axle rests in a horizontal position on two agate

edges, and the angle the needle makes with the vertical
is read off by means of the vertical circle. The needle
and the vertical circle can turn about a vertical axis.
To set the plane of motion of the needle in the magnetic
meridian, the plane of the needle is turned about the
vertical axis until the magnet stands exactly vertical;
when in this position the plane of the needle must be
at right angles to the magnetic meridian. The instrument
is then twisted through 90° (measured on the horizontal
circle) and the magnet is then in the magnetic meridian;
the angle it makes with the horizontal in this position is
the dip. To avoid the error arising from the axle of the
needle not being coincident with the centre of the vertical
circle, the positions of the two ends of the needle are read;
to avoid the error due to the magnetic axis not being
coincident with the line joining the ends of the needle,
the needle is reversed so that the face which originally
was to the east is now to the west and a fresh set of
readings taken; and to avoid the errors which would arise
if the centre of gravity were not on the axle, the needle
is remagnetized so that the end which was previously
north is now south and a fresh set of readings taken.
The mean of these readings gives the dip.

144. We can embody in the form of charts the deter-
minations of these elements made at the various magnetic
observatories: thus, for example, we can draw a series of
lines over the map of the world such that all points on one
of these lines have the same declination, these are called
isogonic lines: we may also draw another set of lines so
that all the places on a line have the same dip, these are
called isoclinic lines. The lines however which give the

best general idea of the distribution of magnetic force over
the earth's surface are the lines of horizontal magnetic

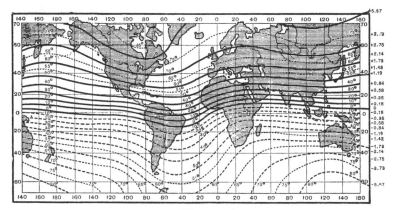

Fig. 75.

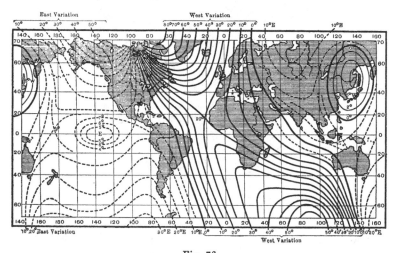

Fig. 76.

force on the earth's surface, i.e. the lines which would be
traced out by a traveller starting from any point and
always travelling in the direction in which the compass
pointed; they were first used by Duperrey in 1836.

The isoclinic lines, the isogonic lines and Duperrey's
lines for the Northern and Southern Hemispheres for 1876
are shown in Figs. 75, 76, 77, and 78 respectively.

145. The points to which Duperrey's lines of force
converge are called 'poles,' they are places where the
horizontal force vanishes, that is where the needle if freely
suspended would place itself in a vertical position.

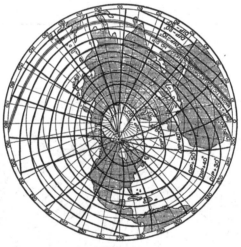

Fig. 77.

Gauss by a very thorough and laborious reduction of
magnetic observations gave as the position in 1836, of
the pole in the Northern Hemisphere, latitude 70° 35',

longitude 262° 1' E., and of the pole in the Southern
Hemisphere, latitude 78° 35', longitude 150° 10' E. The

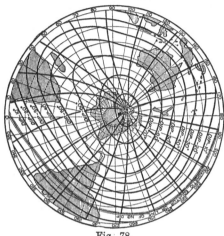

Fig. 78.

poles are thus not nearly at opposite ends of a diameter
of the earth.

146. An approximation, though only a very rough one,
to the state of the earth's magnetic field, may be got by
regarding the earth as a uniformly magnetized sphere.

On this supposition, we have by Art. 139, if θ is the
dip at any place, i.e. the complement of the angle between
the magnetic force and the line joining the place to the
centre of the earth, l the magnetic latitude, i.e. the complement of the angle this line makes with the direction of
magnetization of the sphere,

$$\tan \theta = 2 \tan l,$$

while the resultant magnetic force would vary as

$$[1 + 3 \sin^2 l]^{\frac{1}{2}}.$$

These are only very rough approximations to the truth but are sometimes useful when more accurate knowledge of the magnetic elements is not available.

If M is the moment of the uniformly magnetized sphere which most nearly represents the earth's magnetic field, then in C.G.S. units

$$M = \cdot 323 \text{ (earth's radius)}^3.$$

147. Variations in the Magnetic Elements. During the time within which observations of the magnetic elements have been carried on the declination at London has changed from being 11° 15′ to the East of North as in 1580 to 24° 38′ 25″ to the West of North as in 1818. It is now going back again to the East, but is still pointing between 16° and 17° to the West. The variations in the declination and dip in London are shown in the following table.

Date	Declination	Dip
1576		71° 50′
1580	11° 15′ E.	
1600		72° 0′
1622	6° 0′ E.	
1634	4° 6′ E.	
1657	0° 0′ E.	
1665	1° 22′ W.	
1672	2° 30′ W.	
1676		73° 30′
1692	6° 30′ W.	
1723	14° 17′ W.	74° 42′
1748	17° 40′ W.	
1773	21° 9′ W.	72° 19′
1787	23° 19′ W.	72° 8′

Date	Declination	Dip
1795	23° 57′ W.	
1802	24° 6′ W.	70° 36′
1820	24° 34′ 30″ W.	70° 3′
1830	24°	69° 38′
1838		69° 17′
1860	21° 39′ 51″	68° 19′·29
1870	29° 18′ 52″	67° 57′·98
1880	18° 57′ 59″	
1893	17° 27′	67° 30′
1900	16° 52′·7	

This slow change in the magnetic elements is often called the secular variation. The points of zero declination seem to travel westward.

148. Besides these slow changes in the earth's magnetic force, there are other changes which take place with much greater rapidity.

Diurnal Variation. A freely suspended magnetic needle does not point continually in one direction during the whole of the day. In England in the night from about 7 p.m. to 10 a.m. it points to the East of magnetic North and South (i.e. to the East of the mean position of the needle), and during the day from 10 a.m. to 7 p.m. to the West of magnetic North and South. It reaches the westerly limit about 2 in the afternoon, its easterly one about 8 in the morning, the arc travelled over by the compass being about 10 minutes. This arc varies however with the time of the year, being greatest at midsummer and least at midwinter. There are two maxima in summer, one minimum in winter.

The diurnal variation changes very much from one

T. E.   16

place to another, it is exceedingly small at Trevandrum, a place near the equator.

In the Southern Hemisphere the diurnal variation is of the opposite kind to that in the Northern, i.e. the easterly limit in the Southern Hemisphere is reached in the afternoon, the westerly in the morning.

In the following diagram, due to Prof. Lloyd, the radius vector represents the disturbing force acting on the magnet at different times of the day in Dublin, the

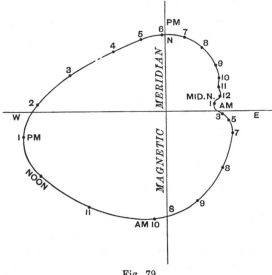

Fig. 79.

forces at any hour are the average of those at that hour for the year. The curve would be different for different seasons of the year.

There is also a diurnal variation in the vertical

component of the earth's magnetic force. In England the vertical force is least between 10 and 11 a.m., greatest at about 6 p.m.

The extent of the diurnal variation depends upon the condition of the sun's surface, being greater when there are many sun spots. As the state of the sun with regard to sun spots is periodic, going through a cycle in about eleven years, there is an eleven-yearly period in the magnitude of the diurnal variation.

149. Effect of the Moon. The magnetic declination shows a variation depending on the position of the moon with respect to the meridian, the nature of this variation varies very much in different localities.

150. Magnetic Disturbances. In addition to the periodic and regular disturbances previously described, rapid and irregular changes in the earth's magnetic field, called magnetic storms, frequently take place ; these often occur simultaneously over a large portion of the earth's surface.

Auroræ are mostly accompanied by magnetic storms, and there is very strong evidence that a magnetic storm accompanies the sudden formation of a sun spot.

151. Very important evidence as to the locality of the origin of the earth's magnetic field, or of its variations, is afforded by a method due to Gauss which enables us to determine whether the earth's magnetic field arises from a magnetic system above or below the surface of the earth. The complete discussion of this method requires the use of Spherical Harmonic Analyses. The principle underlying

16—2

it however can be illustrated by considering a simple case, that of a uniformly magnetized sphere.

Let PQ be two points on a spherical surface concentric with the sphere, then by observation of the horizontal force at a series of stations between P and Q, we can determine the difference between the magnetic potential at P and Q. If Ω_P and Ω_Q are the magnetic potentials at P and Q respectively these observations will give us $\Omega_P - \Omega_Q$. By Art. 139 if θ_1, θ_2 are the angles OP and OQ make with the direction of magnetization of the sphere

$$\Omega_P - \Omega_Q = \frac{M}{r^2} (\cos \theta_1 - \cos \theta_2) \ldots\ldots\ldots\ldots (1),$$

where M is the magnetic moment of the sphere and

$$r = OP = OQ,$$

where O is the centre of the sphere.

If Z_P, Z_Q are the vertical components of the earth's magnetic force, i.e. the forces in the direction OP and OQ respectively, then

$$\left. \begin{aligned} Z_P &= \frac{2M}{r^3} \cos \theta_1 \\ Z_Q &= \frac{2M}{r^3} \cos \theta_2 \end{aligned} \right\} \ldots\ldots\ldots\ldots\ldots\ldots (2),$$

Z_P and Z_Q can of course be determined by observations made at P and Q. By equations (1) and (2), we have

$$\Omega_P - \Omega_Q = \tfrac{1}{2} (Z_P - Z_Q) r \ldots\ldots\ldots\ldots (3),$$

hence if the field over the surface of the sphere through P and Q were due to an internal uniformly magnetized sphere, the relation (3) would exist between the horizontal and vertical components of the earth's magnetic force.

Now suppose that P and Q are points inside a uniformly magnetized sphere, the force inside the sphere is uniform and parallel to the direction of magnetization, let H be the value of this force, then in this case

$$\Omega_P - \Omega_Q = Hr\,(\cos\theta_2 - \cos\theta_1),$$
$$Z_P = H\cos\theta_1,$$
$$Z_Q = H\cos\theta_2,$$

hence in this case

$$\Omega_P - \Omega_Q = -r\,(Z_P - Z_Q)\ldots\ldots\ldots\ldots(4).$$

Thus if the magnetic system were above the places at which the elements of the magnetic field were determined, the relation (4) would exist between the horizontal and vertical components of the earth's magnetic force. Conversely if we found that relation (3) existed between these components we should conclude that the magnets producing the field were below the surface of the earth, while if relation (4) existed we should conclude the magnets were above the surface of the earth; if neither of these relations was true we should conclude that the magnets were partly above and partly below the surface of the earth.

Gauss showed that no appreciable part of the *mean* values of the magnetic elements was due to causes above the surface of the earth. Schuster has however recently shown by the application of the same method that the *diurnal variation* must be largely due to such causes. Balfour Stewart had previously suggested the magnetic action of electric currents flowing through rarefied air in the upper regions of the earth's atmosphere as the probable cause of this variation.

CHAPTER VIII

MAGNETIC INDUCTION

152. When a piece of unmagnetized iron is placed in a magnetic field it becomes a magnet, and is able to attract iron filings; it is then said to be magnetized by *induction*. Thus if a piece of soft iron (a common nail for example) is placed against a magnet, it becomes magnetized by induction, and is able to support another nail, while this nail can support another one, and so on until a long string of nails may be supported by the magnet.

If the positive pole of a bar magnet be brought near to one end of a piece of soft iron, that end will become charged with negative magnetism, while the remote end of the piece of iron will be charged with positive magnetism. Thus the opposite poles of these two magnets are nearest each other, and there will therefore be an attraction between them, so that the piece of iron, if free to move, will move towards the inducing magnet, i.e. it will move from the weak to the strong parts of the magnetic field due to this magnet. If, instead of iron, pieces of nickel or cobalt are used they will tend to move in the same way as the iron, though not to so great an extent. If however we use bismuth instead of iron, we shall find that the bismuth is repelled from the magnet, instead of being attracted towards it; the bismuth tending to move from the strong to the weak parts of the field; the effect is however very small compared with that exhibited by iron; and to make the repulsion evident it is necessary to use a strong electromagnet. When the positive pole of a magnet is brought near a bar of bismuth the end of the bar next the positive pole becomes itself a positive pole, while the further end of the bar becomes a negative pole.

Substances which behave like iron, i.e. which move from the weak to the strong parts of the magnetic field, are called *paramagnetic* substances; while those which behave like bismuth, and tend to move from the strong to the weak parts of the field, are called *diamagnetic* substances.

When tested in very strong fields all substances are found to be para- or dia-magnetic to some degree, though the extent to which iron transcends all other substances is very remarkable.

153. Magnetic Force and Magnetic Induction. The magnetic force at any point in air is defined to be the force on unit pole placed at that point, or—what is equivalent to this—the couple on a magnet of unit moment placed with its axis at right angles to the magnetic force. When however we wish to measure the magnetic force inside a magnetizable substance, we have to make a cavity in the substance in which to place the magnet used in measuring the force. The walls of the cavity will however become magnetized by induction, and this magnetization will affect the force inside the cavity. The magnetic force thus depends upon the shape of the cavity, and this shape must be specified if the expression *magnetic force* is to have a definite meaning.

Let P be a point in a piece of iron or other magnetizable substance, and let us form about P a cylindrical
cavity, the axis of the cylinder being parallel to the
direction of magnetization at P. Let us first take the case
when the cylinder is a very long and narrow one. Then
in consequence of the magnetization at P, there will be
a distribution of positive magnetism over one end of the
cylinder, and a distribution of negative magnetism over
the other. Let I be the intensity of the magnetization
at P, reckoned positive when the axis of the magnet is
drawn from left to right, then when the cylindrical cavity
has been formed round P there will be, if α is the cross
section of the cavity, a charge $I\alpha$ of magnetism on the
end to the left, and a charge $-I\alpha$ on the end to the right.
If $2l$, the length of the cylinder, is very great compared
with the diameter, then the force on unit pole at the
middle of the cylinder due to the magnetism at the ends
of the cylinder will be $2I\alpha/l^2$, and will be indefinitely
small if the breadth of the cylinder is indefinitely small
compared with its length. In this case the force on unit
pole in the cavity is independent of the intensity of
magnetization at P. The force in this cavity is defined
to be '*the magnetic force at P*.' Let us denote it by H.

Let us now take another co-axial cylindrical cavity,
but in this case make the length of the cylinder very
small compared with its diameter, so that the shape of
the cavity is that of a narrow crevasse. On the left end
of this crevasse there is a charge of magnetism of surface
density I, and on the right end of the crevasse a charge
of magnetism of surface density $-I$. If a unit pole be
placed inside the crevasse the force on it due to this
distribution of magnetism will be the same as the force

on unit charge of electricity placed between two infinite plates charged with electricity of surface density $+I$ and $-I$ respectively, i.e. by Art. 14, the force on the unit pole in this case will be $4\pi I$. Thus in a crevasse the total force on the unit pole at P will be the resultant of *the magnetic force* at P and a magnetic force $4\pi I$ in the direction of the magnetization at P. The force on the unit pole in the crevasse is defined to be the '*magnetic induction*' at P, we shall denote it by B. If we had taken a cavity of any other shape the force due to the magnetization at P, would have been intermediate in value between zero for the long cylinder and $4\pi I$ for the crevasse; thus if the cavity had been spherical the force due to the magnetization would (Art. 139) have been $4\pi I/3$.

The magnetic induction is not necessarily in the same direction as the magnetic force, it will only be so when the magnetization at P is parallel to the magnetic force.

154. Tubes of Magnetic Induction. A curve drawn such that its tangent at any point is parallel to the magnetic induction at that point is called a line of magnetic induction: in non-magnetizable substances the lines of magnetic induction coincide with the lines of magnetic force. We can also draw tubes of magnetic induction just as we draw tubes of magnetic force.

We shall choose the unit tube so that the magnetic induction at any place whether in the air or iron is equal to the number of tubes of induction which cross a unit area at right angles to the induction.

Let us consider the case of a small bar magnet, the magnetism being entirely at its ends. Suppose A and B are the ends of the magnet, A being the negative, B the

positive end, then in the air the lines of magnetic induction coincide with those of magnetic force and go
from B to A. To find the lines of magnetic induction
at a point P inside the magnet, imagine the magnet cut
by a plane at right angles to the axis and the two portions
separated by a short distance, the lines of magnetic force
in this short air space will be the lines of magnetic induction in the section through P. If the magnet is cut

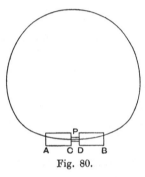

Fig. 80.

as in the figure then the end C will be a positive pole of
the same strength as A, the end D a negative pole of the
same strength as B. Thus through the short air space
between C and D tubes of induction will pass running in
the direction AB. Draw a closed surface passing through
the gap between C and D and enclosing AC or DB. The
magnetic force at any point on this surface is equal to the
magnetic induction at the same point due to the undivided
magnet. Since this surface encloses as much positive as
negative magnetism, we see as in Art. 10 that the total
magnetic force over its surface vanishes. Hence we see
that the tubes of induction inside the magnet are equal
in number at each cross-section and this number is the

same as the number of those which leave the pole B and enter A. In fact the lines of magnetic induction due to the magnet form a series of closed curves all passing through the magnet and then spreading out in the air, the lines running from B to A in the air and from A to B in the magnet.

Thus we may regard any small magnet, whose intensity is I and area of cross-section α, as the origin of a bundle of closed tubes of induction, the number of tubes being $4\pi I\alpha$; every tube in this bundle passes through the magnet, running through the magnet in the direction of the magnetization.

It is instructive to compare the differences between the properties of the tubes of electric polarization in electrostatics and those of magnetic induction in magnetism: the most striking difference is that whereas in electrostatics the tubes are not closed but begin at positive electrification and end on negative, in magnetism the tubes of induction always form closed curves and have neither beginning nor end.

A surface charged with electricity of surface density σ is the origin of σ tubes of electric polarization per unit area. A small magnet whose intensity of magnetization is I is the origin of $4\pi I$ tubes of magnetic induction per unit area of cross-section of the magnet, all these tubes passing through the magnet which acts as a kind of girdle to them.

The properties of these tubes are well summed up by Faraday in the following passage (*Experimental Researches*, § 3117): "there exist lines of force within the magnet, of the same *nature* as those without. What is more, they are exactly equal in *amount* to those without.

They have a relation in *direction* to those without and in fact are continuations of them, absolutely unchanged in their nature so far as the experimental test can be applied to them. Every line of force, therefore, at whatever distance it may be taken from the magnet, must be considered as a closed circuit passing in some part of its course through the magnet, and having an equal amount of force in every part of its course." Faraday's lines of force are what we have called tubes of induction.

155. We shall now proceed to consider the special case, including that of iron and all non-crystalline substances when magnetized entirely by induction, in which the direction of the magnetization and consequently of the magnetic induction is parallel to the magnetic force. Let H be the magnetic force, B the magnetic induction, and I the intensity of magnetization, then we have by Art. 153,

$$B = H + 4\pi I.$$

The ratio of I to H when the magnetization is entirely induced is called the *magnetic susceptibility* and is usually denoted by the letter k. The ratio of B to H under the same circumstances is called the *magnetic permeability* and is denoted by the letter μ.

We thus have

$$I = kH,$$
$$B = \mu H,$$
and since $$B = H + 4\pi I,$$
we have $$\mu = I + 4\pi k.$$

The quantity μ which occurs in magnetism is analogous to the specific inductive capacity in electrostatics; but while as far as our knowledge at present goes, the specific inductive capacity at any time does not depend much, if at all, upon the value of the electric intensity at that time, nor on the electric intensity to which the dielectric has

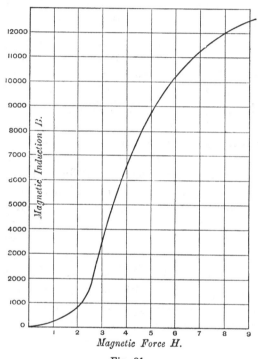

Fig. 81.

previously been exposed; the permeability, on the other hand, if the magnetic force exceeds a certain value (about 1/10 of the earth's horizontal force), depends very greatly upon the magnitude of the magnetic force, and also upon the magnetic forces which have previously been applied to the iron. The variations in the magnetic permeability are most conveniently represented by curves in which the ordinate represents the magnetic induction, the abscissa the corresponding magnetic force. If P be a point on such a curve, PN the ordinate, ON the abscissa, then the magnetic permeability is PN/ON.

Such a curve is shown in Fig. 81, in which the ordinates represent for a particular specimen of iron the values of B, the magnetic induction, the abscissæ the values of H, the magnetic force. For small values of H the curve is straight, indicating that the permeability is independent of the magnetic force. When however the magnetic force increases beyond about $\frac{1}{10}$ of the earth's horizontal force, or about ·018 in C.G.S. units, the curve begins to rise rapidly, and the value of μ is greater than it was for small magnetic forces. The curve rises rapidly for some time, the maximum value of μ occurring when the magnetic force is about 5 C.G.S. units, then it begins to get flatter and there are indications that for very great values of the magnetic force the curve again becomes a straight line making an angle of 45° with the axis along which the magnetic force is measured. The relation between B and H along this part of the curve is

$$B = H + \text{constant}:$$

comparing this with the relation

$$B = H + 4\pi I,$$

we see that it indicates that the intensity of magnetization has become constant. In other words, the intensity of magnetization does not increase as the magnetic force increases. When this is the case the iron or other magnetizable substance is said to be 'saturated.' Thus iron seems not to be able to be magnetized beyond a certain intensity. In a specimen of soft iron examined by Prof. Ewing, saturation was practically reached when the magnetic force was about 2000 in C.G.S. units. For steel the magnetic force required for saturation is very much greater than for soft iron, and in some specimens of steel examined by Prof. Ewing saturation was not attained even when the magnetizing force was as great as 10000.

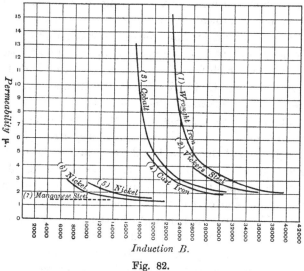

Induction B.

Fig. 82.

For a particular kind of steel called Hadfield's manganese steel the value of μ was practically constant even in the strongest magnetic fields, this steel however is only slightly magnetic, the value of μ being about 1·4. The greatest value of μ which has been observed is 20000 for soft iron, in this case however the iron was tapped when under the influence of the magnetic force. Fig. 82 represents the results of Ewing's experiments on the relation between magnetic permeability and magnetic induction in very intense magnetic fields.

156. Effect of Temperature on the Magnetic Permeability. The permeability of iron depends very much upon the temperature. Dr J. Hopkinson found that as the temperature increases, starting from about 15° C., the magnetic permeability at first slowly increases; this slow rate of increase is however exchanged for an exceedingly rapid one when the temperature approaches a 'critical temperature' which for different samples of iron and steel ranges from 690° C. to 870° C., at this temperature the value of the permeability is many times greater than that at 15° C.: after passing this value the permeability falls even more rapidly than it previously rose. Indeed so fast is the fall that at a few degrees above the critical temperature iron practically ceases to be magnetic. Just below this temperature iron is an intensely magnetic substance, while above that temperature it is not magnetic at all. There are other indications that iron changes its character in passing through this temperature, for here its thermo-electric properties as well as its electrical resistance suffer abrupt changes. This temperature is often called the temperature of recalescence from the fact that a piece of iron wire heated above this temperature to redness and then allowed to cool, will get dull before reaching this temperature and will glow out brightly again when it passes through it.

Though the value of μ at higher temperatures (lower however than that of recalescence) is for small magnetic forces greater than at lower temperatures, still as it is found that at the higher temperatures iron is much more easily saturated than at lower ones, the value of μ for the hot iron might be smaller than for the cold if the magnetic forces were large.

Hopkinson found that some alloys of nickel and iron after being rendered non-magnetic by being raised above the temperature of recalescence remained non-magnetic when cooled below this temperature; it was not until the temperature had fallen far below the temperature of recalescence that they regained their magnetic properties. Thus these alloys can at one and the same temperature exist in both the magnetic and non-magnetic states.

157. Magnet Retentiveness. Hysteresis. When a piece of iron or steel is magnetized in a strong magnetic field it will retain a considerable proportion of its magnetization even after the applied field has been removed and the iron is no longer under the influence of any applied magnetic force. This power of remaining magnetized after the magnetic force has been removed, is called magnetic retentiveness; permanent magnets are a familiar instance of this property. This effect of the previous magnetic history of a substance on its behaviour when exposed to given magnetic conditions has been studied in great detail by Prof. Ewing, who has given to this property the name of hysteresis. To illustrate this properly, let us consider the curve (Fig. 83) which is taken from Prof. Ewing's paper on the magnetic properties of iron (*Phil. Trans.* Part II., 1885), and which represents the relation for a sample of soft iron between the intensity of magnetization (the ordinate) and the magnetizing force (the abscissa), when the magnetic force increases from zero up to ON, then diminishes from ON through zero to $-OM$, and then increases again up to its original value. When the force is first applied we have the state represented by the portion OP of the curve, which begins by being straight, then increases more rapidly, bends round and finally reaches P, the point corresponding to the greatest magnetic force

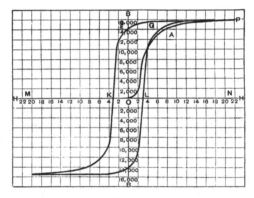

Fig. 83.

applied to the iron. If now the force is diminished it will be found that the magnetization for a given force is greater than it was when the magnet was initially under the action of the same force, i.e. the magnet has retained some of its previous magnetization, thus the curve PE, when the force is diminishing, will not correspond to the curve OP but will be above it. OE is the magnetization retained by the magnet when free from magnetic force;

in some cases it amounts to more than 90 per cent.
of the greatest magnetization attained by the magnet.
When the magnetizing force is reversed the magnet
rapidly loses its magnetization and the negative force
represented by OK is sufficient to deprive it of all
magnetization.   When the negative magnetic force is increased beyond this value, the magnetization is negative.
After the magnetic force is again reversed it requires a
positive force equal to OL to deprive the iron of its
negative magnetization.   When the force is again increased to its original value the relation between the
force and induction is represented by the portion LGP
of the curve.   If after attaining this value the force is
again diminished to $-ON$ and back again, the corresponding curve is the curve PEK.

From the fact that this curve encloses a finite area it
follows that a certain amount of energy must be dissipated
and converted into heat when the magnetic force goes
through a complete cycle.   To show this let us suppose
that we have a small magnet whose intensity is I, cross-
section α, and length l, and that it is moved from a place
where the magnetic force is H to one where it is $H + \delta H$.
We shall show that the work done on the magnet is

$$I\delta H\alpha l.$$

H is considered positive when it acts in the direction of
magnetization of the iron.   For if Ω_1 is the magnetic
potential at A, the negative pole; Ω_2 that at B, the
positive pole, then the potential energy of the magnet
is equal to

$$-I\alpha\Omega_1 + I\alpha\Omega_2$$

$$= I\alpha\,(\Omega_2 - \Omega_1) = -I\alpha l H.$$

17—2

When the magnetic force is $H + \delta H$ the potential energy is equal to

$$- I\alpha l\,(H + \delta H).$$

Thus the diminution in the potential energy when the magnet moves into the stronger field is $I\alpha l\delta H$, this is equal to the work done by the magnet. If the intensity of magnetization changes from I to $I + \delta I$ during the motion of the magnet, the work done is intermediate between $I\alpha l\delta H$ and $(I + \delta I)\alpha l\delta H$; hence neglecting the small terms depending upon $\delta I\delta H$, we may still take $I\alpha l\delta H$ as the expression for the work done. Since $l\alpha$ is the volume of the magnet the work done by the magnet per unit volume is $I\delta H$.

If in Fig. 84 $OS = H$, $OT = H + \delta H$ and $SP = I$, then $I\delta H$ is represented on the diagram by the area $SPQT$.

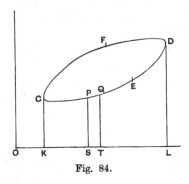

Fig. 84.

Thus the total work done by the magnet when it moves from a place where the force is OK to one where it is OL is represented by the area $CKLDE$. Let the magnet now be pulled back from the place where the force is OL to the place from which it started where the force is OK, work has to be done on the magnet and this work is represented by the area $DFCKL$. Thus the excess of the work done on the magnet over that done by the magnet, when the magnetic force goes through a complete cycle, is represented by the area of the loop $CEDFC$. The area of the loop thus represents the excess of the work spent over that obtained: but since the magnetic force and magnetization at the end of the cycle are the same as at the beginning, this work must have been dissipated and converted into heat. The mechanical equivalent of the amount of heat produced in each unit volume of the iron is represented by the area of the loop.

Another proof of this is given in Chapter XI.

If instead of a curve showing the relation between I and H we use one showing the relation between B and H, there will be similar loops in this second curve and the area of these loops will be 4π times the area of the corresponding loops on the I and H curve.

For the area of a loop on the first curve is

$$- \int I dH,$$

this is equal to

$$- \frac{1}{4\pi} \int (B - H) dH$$

$$= - \frac{1}{4\pi} \int B dH,$$

since $\int H dH = 0$, as the initial and final values of H are equal. The area of a loop on the B and H curve is however equal to

$$- \int B dH.$$

Hence we see that this area is 4π times the area of the corresponding loop on the I and H curve.

158. Conditions which must hold at the boundary of two substances.

At the surface separating two media the magnetic field must satisfy the following conditions.

1.   The magnetic force parallel to the surface must be the same in the two media.

2.   The magnetic induction at right angles to the surface must be the same in the two media.

To prove the first condition, let P and Q be two points close to the surface of separation, Q being in the first, P in the second medium. Now the magnetic force at a point is by definition (see Art. 153) the force on a unit pole placed in a cavity round the point, when the magnetism on the walls of the cavity can be neglected: hence since this magnetism is to be disregarded the difference between the magnetic forces at P and Q must arise from the magnetism on the surface between P and Q: but though the forces at right angles to this portion of the surface due to its magnetism are different at P and Q, the forces parallel to the surface are the same. Hence we see that the tangential magnetic forces will be the same at P as at Q.

We shall now show that the normal magnetic induction is continuous. All the tubes of magnetic induction form closed curves. Hence any tube must cut a closed surface an even number of times; half these times it will be entering the surface, half leaving it. The contributions of each tube to the total normal magnetic induction will be the same in amount but opposite in sign when it enters and when it leaves the surface. Hence the total contribution of each tube is zero, and thus the total normal magnetic induction over any closed surface vanishes. Consider the surface of a very short cylinder whose sides are parallel to the normal at P, one end being in the medium (1), the other in (2). The total normal induction over this surface is zero, but as the area of the sides is negligible compared with that of the ends, this implies that the total normal induction across the end in (1) is equal to that across the end in (2), or that, since the areas of these ends are equal, the induction parallel to the normal in (1) is the same as that in the same direction in (2). This is always true whether the magnet is permanently magnetized or only magnetized by induction.

In Art. 74 we proved that the conditions satisfied at the boundary of two dielectrics are

1.   The tangential electric intensity must be the same in both media.

2.   When there is no free electricity on the surface the normal electric polarization must be the same in both.   That is, if F, F' are the normal electric intensities in the media whose specific inductive capacities are respectively K and K',

$$KF = K'F'.$$

If we compare these conditions with those satisfied at the boundary of two media in the magnetic field and remember that when the magnetization is induced, the magnetic induction is equal to μ times the magnetic force, we see that we have complete analogy between the disturbance of an electric field produced by the presence of uncharged dielectrics and the disturbance in a magnetic field produced by para- or dia-magnetic bodies in which the magnetism is entirely induced.

Hence from the solution of any electrical problem we can deduce that of the corresponding magnetic one by writing magnetic force for electric intensity, and μ for K.

We can prove, as in Art. 75, that if θ_1 is the angle which the direction of the magnetic force in air makes with the normal at a point P on a surface, θ_2 the angle which the magnetic force in the magnetizable substance makes with the normal at the same point, then

$$\mu \tan \theta_1 = \tan \theta_2.$$

Thus when the lines of force go from air to a paramagnetic substance they are bent away from the normal in the substance, since in this case μ is greater than 1; when they go from air to a diamagnetic substance they are bent towards the normal, since in this case μ is less than 1.

The effects produced when paramagnetic and diamagnetic spheres are placed in a uniform field of force are shown in Figs. 39 and 85.

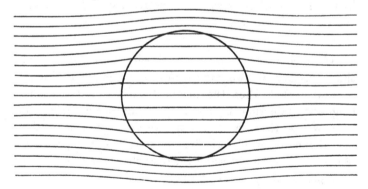

Fig. 85.

159. If μ is infinite $\tan\theta_1$ vanishes, and then the lines of force in air are at right angles to the surface, so that the surface of a substance of infinite permeability is a surface of equi-magnetic potential. The surface of such a substance corresponds to the surface of an insulated conductor without charge in electrostatics, and any problem relating to such conductors can be at once applied to the corresponding case in magnetism. In particular we can apply the principle of images (Chap. v.) to find the effect produced by any distribution of magnetic poles in presence of a sphere of infinite magnetic permeability.

160. Sphere in uniform field. We showed in Art. 104 that if a sphere, whose radius is a, and whose specific inductive capacity is K, is placed in a uniform electric field, and if H is the electric intensity before the introduction of the sphere, then the field when the sphere is present will at a point P outside the sphere, consist of H and an electric intensity whose component along PO is equal to

$$2H\frac{(K-1)}{K+2}\cos\theta\frac{a^3}{r^3},$$

and whose component at right angles to PO in the direction tending to increase θ is

$$H\frac{(K-1)}{K+2}\sin\theta\frac{a^3}{r^3};$$

in these expressions $OP = r$, θ is the angle OP makes with the direction of H, O is the centre of the sphere.

Inside the sphere the electric intensity is constant, parallel to H and equal to

$$\frac{3}{K+2}H.$$

If we write μ for K the preceding expressions will give us the magnetic force when a sphere of magnetic permeability μ is placed in a uniform magnetic field where the magnetic force is H.

A very important special case is when μ is very large compared with unity. In this case the magnetic forces due to the sphere are approximately

$$2H \frac{a^3}{r^3} \cos \theta$$

along PO, and $\qquad H \frac{a^3}{r^3} \sin \theta$

at right angles to it.

Inside the sphere the magnetic force is

$$\frac{3}{\mu} H,$$

and is very small compared with that outside. The magnetic induction inside the sphere is $3H$. Thus through any area in the sphere at right angles to the magnetic force, three times as many tubes of induction pass as through an equal and parallel area at an infinite distance from the sphere.

The resultant magnetic force in air vanishes round the equator of the sphere.

161. Magnetic Shielding. Just as a conductor is able to shield off the electric disturbance which one electrical system would produce on another, so masses of magnetizable material, for which μ has a large value, will shield off from one system magnetic forces due to another.

Inasmuch however as μ has a finite value for all substances the magnetic shielding will not be so complete
as the electrical.

162. Iron Shell. We shall consider the protection
afforded by a spherical iron shell against a uniform magnetic field. We saw in Art. 160 that, when a solid iron
sphere is placed in a uniform magnetic field, the magnetism
induced on the sphere produces outside it a radial magnetic force proportional to $2 \cos \theta / r^3$, and a tangential force
proportional to $\sin \theta / r^3$, and a constant force inside the
sphere. We shall now proceed to show that we can
satisfy the conditions of the problem of the spherical iron
shell by supposing each of the distributions of magnetism
induced on the two surfaces of the shell to give rise to
forces of this character.

Let a be the radius of the inner surface of the shell,
b that of the outer surface. Let H be the force in the
uniform field before the shell was introduced. Let the
magnetic forces due to the magnetism on the outer surface
of the shell consist, at a point P outside the sphere, of a
radial force

$$\frac{2 M_1 \cos \theta}{r^3},$$

a tangential force

$$\frac{M_1 \sin \theta}{r^3},$$

where $r = OP$ and θ is the angle OP makes with the
direction of H. The magnetic force due to this distribution of magnetism will be uniform inside the sphere
whose radius is b, it will act in the direction of H and
be equal to $-M_1/b^3$.

Let the magnetization on the inner surface of the shell give rise to magnetic forces given by similar expressions with M_2 written for M_1 and a for b.

This system of forces, whatever be the values of M_1 and M_2, satisfies the condition that as we cross the surfaces of the shell the tangential components of the magnetic force are continuous. We must now see if we can choose M_1, M_2 so as to make the normal magnetic induction continuous.

The normal magnetic induction (reckoned positive along the outward drawn normal) in the air just outside the outer shell is equal to

$$H \cos \theta + \frac{2M_1}{b^3} \cos \theta + \frac{2M_2}{b^3} \cos \theta,$$

the normal magnetic induction in the iron just inside the outer surface of the shell is equal to

$$\mu \left(H \cos \theta - \frac{M_1}{b^3} \cos \theta + \frac{2M_2}{b^3} \cos \theta \right).$$

These are equal if

$$H + \frac{2M_1}{b^3} + \frac{2M_2}{b^3} = \mu \left(H - \frac{M_1}{b^3} + \frac{2M_2}{b^3} \right),$$

or, if

$$\frac{(\mu + 2) M_1 - 2 (\mu - 1) M_2}{b^3} = (\mu - 1) H \quad \ldots\ldots(1).$$

The normal magnetic induction in the iron just outside the inner surface of the shell is

$$\mu \left(H \cos \theta - \frac{M_1}{b^3} \cos \theta + \frac{2M_2}{a^3} \cos \theta \right),$$

the normal magnetic induction in the air just inside the shell is equal to

$$H \cos \theta - \frac{M_1}{b^3} \cos \theta - \frac{M_2}{a^3} \cos \theta \, ;$$

these are equal if

$$(\mu - 1) \frac{M_1}{b^3} - (2\mu + 1) \frac{M_2}{a^3} = (\mu - 1) H \quad \dots \dots (2).$$

Equations (1) and (2) are satisfied if

$$M_1 = (\mu - 1) H \, \frac{\{b^3 (2\mu + 1) - 2a^3 (\mu - 1)\}}{(2\mu + 1)(\mu + 2) - 2 (\mu - 1)^2 \dfrac{a^3}{b^3}} \, ,$$

$$M_2 = - (\mu - 1) H \, \frac{3a^3}{(2\mu + 1)(\mu + 2) - 2 (\mu - 1)^2 \dfrac{a^3}{b^3}} \, .$$

The magnetic force in the hollow cavity is equal to

$$H - \frac{M_1}{b^3} - \frac{M_2}{a^3} \, .$$

Substituting the values of M_1 and M_2 we see that this is equal to

$$\frac{9\mu H}{9\mu + 2 (\mu - 1)^2 \left(1 - \dfrac{a^3}{b^3}\right)} \, .$$

If μ is very large compared with unity this is approximately equal to

$$\frac{H}{1 + \tfrac{2}{9}\mu \left(1 - \dfrac{a^3}{b^3}\right)} \, .$$

The denominator may be written in the form

$$1 + \tfrac{2}{9}\mu \, . \, \frac{\text{volume of shell}}{\text{volume of outer sphere}} \, .$$

Hence the force inside the shell will not be greatly less than the force outside unless μ is greater than the ratio of the volume of the outer sphere to that of the shell.

In the cases where $\mu = 1000$ and $\mu = 100$, the ratio of H', the force inside the sphere, to H for different values of a/b is given in the following table.

a/b	H'/H $\mu=1000$	H'/H $\mu=100$
·99	3/23	9/15
·9	1/67	1/7
·8	1/109	1/12
·7	1/146	1/15
·6	1/175	1/18
·5	1/195	1/20
·4	1/209	1/22
·3	1/216	1/22
·2	1/221	1/23
·1	1/223	1/23
·0	1/223	1/23

Galvanometers which have to be used in places exposed to the action of extraneous magnets are sometimes protected by surrounding them with a thick-walled tube made of soft iron.

We may regard the shielding effect of the shell as an example of the tendency of the tubes of magnetic induction to run as much as possible through iron; to do this they leave the hollow and crowd into the shell.

163. Expression for the energy in the magnetic field. We shall suppose that the field contains permanent magnets as well as pieces of magnetizable substances magnetized by induction. If the distribution of the permanent magnets is given, the magnetic field will be quite determinate. The forces between magnetic charges follow the same laws as those between electrical ones. Hence the energy due to any system of magnetized bodies will, if the magnetization due to induction is proportional to the magnetic force, i.e. if μ is constant, be equal to the sum of one half the product of the strength of each permanent pole into the magnetic potential at that pole. Thus if Q is the potential energy of the magnetic field,

$$Q = \tfrac{1}{2}\Sigma m\Omega,$$

where m is the strength of the permanent pole and Ω the magnetic potential at that pole. Let us divide each of the permanent magnets up into little magnets and consider the contribution of one of these to the energy. Let I_0 be the intensity of the permanent magnetization, and α the area of the cross section: then the magnet has a pole of permanent magnetism of strength $I_0\alpha$ at A, another pole of strength $-I_0\alpha$ at B. If Ω_A, Ω_B are the values of the magnetic potentials at A and B, the contribution of this magnet to the energy is therefore equal to

$$\tfrac{1}{2}I_0\alpha\,(\Omega_A - \Omega_B).$$

Now the magnet may be regarded as the origin of $4\pi I_0\alpha$ tubes of magnetic induction forming closed curves running through the magnet, leaving it at A and entering it at B; if ds is an element of one of these tubes, and R the resultant magnetic force which acts along this element, then

$$\Omega_A - \Omega_B = \int_A^B R\,ds,$$

the integration being extended over the part of the tube outside the magnet.   Hence the contribution of this magnet to the energy is the same as it would be if each tube of which it is the origin had per unit length at P an amount of energy equal to $1/8\pi$ of the resultant magnetic force at P.   The portion of the tube inside the little magnet in which it has its origin, must not be taken into account.

Now let us consider any small element of volume in the magnetic field, let us take it as cylindrical in shape, the axis of the cylinder being parallel to the resultant magnetic force R at the element.   Let l be the length of this cylinder, ω the area of its cross section.   Now each of the tubes of magnetic induction which pass through the element and have not their origin within it, contributes $R/8\pi$ units of energy for each unit of length of the tube.   Let I_0 be the intensity of the permanent magnetization of the element, I the induced magnetization, then the number of tubes of induction which pass through unit area of the base of the cylinder is equal to the value of the magnetic induction, i.e. it is equal to

$$R + 4\pi (I + I_0);$$

but of these, $4\pi I_0$ have their origin in the element, and hence the number of tubes per unit area which contribute to the energy is equal to

$$R + 4\pi I,$$

and since $I = kR$ and $\mu = 1 + 4\pi k$, this is equal to

$$\mu R,$$

therefore the number passing through the base of the cylinder is equal to

$$\mu R \omega.$$

The energy of the portion of each of the tubes within the element is equal to $Rl/8\pi$, hence the energy contributed by the element is

$$\frac{\mu R^2}{8\pi} l\omega \; ;$$

thus the energy per unit volume is equal to $\mu R^2/8\pi$. We may then regard the energy of the magnetized system as distributed throughout the magnetic field, there being $\mu R^2/8\pi$ units of energy in each unit volume of the field.

164. When a tube of induction enters a paramagnetic substance from air the resultant magnetic force is —when the magnetization is entirely induced—less in the paramagnetic substance than in air, the energy per unit length will be less in the magnetic substance than in the air since the energy per unit length of a tube of induction is proportional to the resultant magnetic force along it. Thus in accordance with the principle that when a system is in equilibrium the potential energy is a minimum, the tubes of induction will tend to leave the air and crowd into the magnet, when this act does not involve so great an increase in their length in the air as to neutralize the diminution of the energy due to the parts passing through the magnet.

Again, when a tube of induction enters a diamagnetic substance the magnetic force inside this substance is greater than it is in the air just outside, the tubes of induction will therefore tend to avoid the diamagnetic substance. Examples of this and the previous effect are seen in Figs. 83 and 39.

A small piece of iron placed in a magnetic field where the force is not uniform will tend to move from the weak to the strong parts of the field, since by doing so it encloses a greater number of tubes of induction and thus produces a greater decrease in the energy. The direction of the force tending to move the iron is in the direction along which the rate of increase of R^2 is greatest. This is not in general the direction of the magnetic force. Thus in the case of a bar magnet AB, the greatest rate of increase in R^2 at C a point equidistant from A and B is along the perpendicular let fall from C on AB, and this is the direction in which a small sphere placed at C will tend to move; it is however at right angles to the direction of the magnetic force at C.

There will be no force tending to move a piece of soft iron placed in a uniform magnetic field.

A diamagnetic substance will tend to move from the strong to the weak parts of the field, since by so doing it will diminish the number of tubes of magnetic induction enclosed by it and hence also the energy, for the tubes of induction have more energy per unit length when they are in the diamagnetic substance than when they are in air.

165. Ellipsoids. We have hitherto only considered the case of spheres placed in a uniform field. Bodies which are much longer in one direction than another have very interesting properties which are conveniently studied by investigating the behaviour of ellipsoids placed in a uniform magnetic field.

We saw in Art. 139 that the magnetic field, due to a sphere uniformly magnetized in the direction of the axis of x, might be regarded as due to two spheres, one of uniform density ρ with its centre at O', the other of uniform density $-\rho$ with its centre at O, the points O and O' being very close together and OO' parallel to the axis of x: the distance OO' is given by the condition that $\rho OO'$ is equal to the intensity of magnetization of the sphere. An exactly similar proof will show that if we have a body of any shape uniformly magnetized, the magnetic potential due to it is the same as that due to two bodies of the shape and size of the magnet, one having the density ρ, the other the density $-\rho$, and so placed that if the negative body is displaced through the distance ξ in the direction of magnetization, it will coincide with the positive body if $\rho \xi = A$, A being the intensity of magnetization of the body.

Let us suppose that the body is uniformly magnetized with intensity A in the direction of the axis of x, and let $\rho \Omega$ be the potential of the positive body at the point P, then the potential of the negative body at P will be equal to $-\rho \Omega'$, where $\rho \Omega'$ is the potential of the positive body at P', if PP' is parallel to the axis of x and equal to ξ.

But since $P'P$ is small,

$$\rho \Omega' = \rho \left(\Omega + \xi \frac{d\Omega}{dx} \right).$$

The potential of the negative body is therefore

$$-\rho \left(\Omega + \xi \frac{d\Omega}{dx} \right).$$

Thus the potential of the positive and negative bodies together, and therefore of the magnetized body, will be

$$-\rho \xi \frac{d\Omega}{dx},$$

$$= -A \frac{d\Omega}{dx},$$

since $\rho \xi = A.$

18—2

If the body instead of being magnetized parallel to x is uniformly magnetized so that the components of the intensity parallel to x, y, z are respectively A, B, C, the magnetic potential is

$$- \left(A \, \frac{d\Omega}{dx} + B \frac{d\Omega}{dy} + C \frac{d\Omega}{dz} \right) \quad \ldots\ldots\ldots(1).$$

We shall now show that if an ellipsoid is placed in a uniform magnetic field it will be uniformly magnetized by induction. To prove this it will be sufficient to show that if we superpose on to the uniform field, the field due to a uniformly magnetized ellipsoid, it is possible to choose the intensity of magnetization so as to satisfy the two conditions, (1) that the tangential magnetic force and (2) that the normal magnetic induction, are continuous at the surface of the ellipsoid. The first of these conditions is evidently satisfied whatever the intensity of magnetization may be: we proceed to discuss the second condition. The forces parallel to the axes of x, y, z (these are taken along the axes of the ellipsoid) due to the attraction of an ellipsoid of uniform unit density, are, see Routh's *Analytical Statics*, vol. II. p. 112, equal to

$$Lx, \quad My, \quad Nz$$

respectively, where L, M, N are constant as long as the point whose coordinates are x, y, z is inside the ellipsoid.

Hence by (1) since

$$\frac{d\Omega}{dx} = - Lx, \text{ &c.,}$$

the magnetic potential inside the ellipsoid due to its magnetization will be

$$(ALx + BMy + CNz),$$

so that the magnetic forces parallel to the axes of x, y, z due to the magnetization of the ellipsoid will be

$$- AL, \quad - BM, \quad - CN$$

respectively.

Hence if N_1 is the component of these forces along the outward drawn normal to the surface of the ellipsoid,

$$N_1 = - (ALl + BMm + CNn),$$

where l, m, n are the direction cosines of the outward drawn normal. If N_2 is the force due to the magnetization on the ellipsoid in the same direction just outside the ellipsoid, then

$$N_2 = N_1 + 4\pi (lA + MB + nC)$$
$$= lA (4\pi - L) + mB (4\pi - M) + nC (4\pi - N).$$

Let X, Y, Z be the components of the force due to the uniform field. Then $N_1{}'$, the total force inside the ellipsoid along the outward drawn normal, will be given by the equation

$$N_1{}' = lX + mY + nZ + N_1,$$

and if $N_2{}'$ is the total force just outside the ellipsoid along the outward drawn normal

$$N_2{}' = lX + mY + nZ + N_2.$$

If μ is the magnetic permeability of the ellipsoid, the normal magnetic induction will be continuous if

$$\mu N_1{}' = N_2{}',$$

that is if

$$l (\mu X - \mu AL) + m (\mu Y - \mu BM) + n (\mu Z - \mu CN)$$
$$= l \{X + A (4\pi - L)\} + m \{Y + B (4\pi - M)\}$$
$$+ n \{Z + C (4\pi - N)\}.$$

But this condition will be satisfied if

$$
\left.
\begin{aligned}
A &= \frac{(\mu - 1)\,X}{4\pi + L\,(\mu - 1)} \\[2mm]
B &= \frac{(\mu - 1)\,Y}{4\pi + M\,(\mu - 1)} \\[2mm]
C &= \frac{(\mu - 1)\,Z}{4\pi + N\,(\mu - 1)}
\end{aligned}
\right\} \quad \ldots\ldots\ldots\ldots(2).
$$

These equations give the intensity of magnetization of an ellipsoid placed in a uniform magnetic field.

The force inside the ellipsoid due to its magnetization has $-AL$, $-BM$, $-CN$ for components parallel to the axes of x, y, z respectively; these components act in the opposite direction to the external field and the force of which these are the components is called the demagnetizing force. We see from equations (2) that the components of the demagnetizing force are

$$
-\frac{(\mu - 1)\,LX}{4\pi + L\,(\mu - 1)},
$$

$$
-\frac{(\mu - 1)\,MY}{4\pi + M\,(\mu - 1)},
$$

$$
-\frac{(\mu - 1)\,NZ}{4\pi + N\,(\mu - 1)}.
$$

We shall now consider some special cases in detail. Let us take the case of an infinitely long elliptic cylinder, let the infinite axis be parallel to z, let $2a$, $2b$ be the axes in the direction of x and y; then (Routh's *Analytical Statics*, vol. II. p. 112)

$$
L = 4\pi\,\frac{b}{a + b}, \quad M = 4\pi\,\frac{a}{a + b}, \quad N = 0.
$$

Thus $A = \dfrac{(\mu-1)X}{4\pi\left\{1+(\mu-1)\dfrac{b}{a+b}\right\}} = \dfrac{kX}{1+(\mu-1)\dfrac{b}{a+b}}$,

$B = \dfrac{(\mu-1)Y}{4\pi\left\{1+(\mu-1)\dfrac{a}{a+b}\right\}} = \dfrac{kY}{1+(\mu-1)\dfrac{a}{a+b}}$,

where k is the magnetic susceptibility.

We see from this equation that A/X is approximately equal to k when $(\mu-1)b/(a+b)$ is very small, but only then. A very common way of measuring k is to measure A/X in the case of an elongated solid, magnetized along the long axis; but we see that in the case of an elongated cylinder this will be equal to k only when $(\mu-1)b/(a+b)$ is very small. Now for some kinds of iron μ is as great as 1000, hence if this method were to give in this case results correct to one per cent., the long axis would have to be 100,000 times as long as the short one. This extreme case will show the importance of using very elongated figures when experimenting with substances of great permeability. Unless this precaution is taken the experiments really determine the value of a/b and not any magnetic property of the body.

When the body is an elongated ellipsoid of revolution the ratio of the long to the short axis need not be so enormous as in the case of the cylinder, but it must still be very considerable. If the axis of x is the axis of revolution, then (Routh's *Analytical Statics*, vol. II. p. 112) we have approximately

$$L = 4\pi\,\frac{b^2}{a^2}\left\{\log\frac{2a}{b} - 1\right\}, \quad M = N = 2\pi.$$

Thus $\quad \dfrac{A}{X} = \dfrac{k}{1 + (\mu - 1)\dfrac{b^2}{a^2}\left\{\log\dfrac{2a}{b} - 1\right\}}$.

Thus if μ were 1000, the ratio of a to b would have to be about 900 to 1 in order that the assumption $A/X = k$ should be correct to one per cent.

166. Couple acting on the Ellipsoid. The moment of the couple tending to twist the ellipsoid round the axis of z, in the direction from x to y, is equal to

$$\text{(volume of ellipsoid)}\,(YA - XB)$$

$$= \tfrac{4}{3}\pi\, \frac{abc\,(\mu - 1)^2\,XY\,(M - L)}{\{4\pi + (\mu - 1)\,L\}\,\{4\pi + (\mu - 1)\,M\}}.$$

If the magnetic force in the external field is parallel to the plane xy and is equal to H and makes an angle θ with the axis of x,

$$X = H\cos\theta, \quad X = H\sin\theta,$$

and the couple is equal to

$$\frac{4\pi H^2}{3}\,\frac{abc\sin\theta\cos\theta\,(\mu - 1)^2\,(M - L)}{\{4\pi + (\mu - 1)\,L\}\,\{4\pi + (\mu - 1)\,M\}}.$$

If $a > b$, M is greater than L. Thus the couple tends to make the long axis coincide in direction with the external force, so that the ellipsoid, if free to turn, will set with its long axis in the direction of the external force. This will be the case whether μ is greater or less than unity, i.e. whether the substance is paramagnetic or diamagnetic, so that in a uniform field both paramagnetic and diamagnetic needles point along the lines of force. It generally happens that a diamagnetic substance places itself athwart the lines of magnetic force, this is due to the want of uniformity in the field, in consequence of

which the diamagnetic substance tries to get as much of
itself as possible in the weakest part of the field. This
tendency varies as $(\mu - 1)$; the couple we are investigating
in this article varies as $(\mu - 1)^2$, and as $(\mu - 1)$ is exceedingly small for bismuth, this couple will be overpowered
unless the field is exceptionally uniform.

167. Ellipsoid in Electric Field. The investigation of Art. 165 enables us to find the distribution of
electrification induced on a conducting ellipsoid when
placed in a uniform electric field. To do this we must
make μ infinite in the expressions of Art. 165. The
quantity $lA + mB + nC$ which occurs in the magnetic
problem corresponds to σ, the surface density of the electrification. Putting $\mu = \infty$ in equations (2) we find

$$\sigma = \left\{ \frac{lX}{L} + \frac{mY}{M} + \frac{nZ}{N} \right\}.$$

If the force in the electric field is parallel to the axis of x

$$\sigma = \frac{lX}{L}.$$

Thus when the electric field is parallel to one of the axes
of the ellipsoid, the density of the electrification is, as in
the case of a sphere, proportional to the cosine of the
angle which the normal to the surface makes with the
direction of the electric intensity in the undisturbed field.

By Coulomb's law the normal electric intensity at the
surface of the ellipsoid is equal to $4\pi\sigma$, i.e. to

$$\frac{4\pi l X}{L}.$$

Thus the electric intensity at the surface of the ellipsoid
is $4\pi/L$ times the electric intensity in the same direction
in the undistributed field.

If the ellipsoid is a very elongated one with its longer axis in the direction of the electric force, then by Art. 165

$$\frac{4\pi}{L} = \frac{a^2}{b^2 \left[\log \dfrac{2a}{b} - 1 \right]}.$$

Thus, when a/b is large, $4\pi/L$ is a large quantity, and the electric intensity at the end of the ellipsoid is very large compared with the intensity in the undisturbed field. Thus if $a/b = 100$, the electric intensity at the end is about 2500 times that in the undisturbed field. This result explains the power of sharply pointed conductors in discharging an electric field, for when these are placed in even a moderate field the electric intensity at the surface of the conductor is great enough to overcome the insulating power of the air, see Art. 37, and the electrification escapes.

If an ellipsoidal conductor is placed in a uniform field of force, at right angles to the axis c and making an angle θ with the axis a, we see from § 166 that the couple round the axis of c tending to make the axis of a move towards the external force is equal to

$$\frac{2\pi}{3} F^2 abc \frac{(M - L)}{LM} \sin 2\theta$$

when F is the external electric force.

When the ellipsoid is one of revolution round the axis of a, and a is large compared with b, the couple is approximately

$$\frac{1}{6} \frac{a^3 F^2 \sin 2\theta}{\left(\log \dfrac{2a}{b} - 1 \right)}.$$

CHAPTER IX.

ELECTRIC CURRENTS.

168. Let two conductors A and B be at different potentials, A being at the higher potential and having a charge of positive electricity, while B is at a lower potential and has a charge of negative electricity; then if A is connected to B by a metallic wire the potential of A will begin to diminish and A will lose some of its positive charge, the potential of B will increase and B will lose some of its negative charge, so that in a short time the potentials of A and B will be equalized. During the time in which the potentials of A and B are changing the following phenomena will occur: the wire connecting A and B will be heated and a magnetic field will be produced which is most intense near the wire. If A and B are merely charged conductors, their potentials are equalized so rapidly, and the thermal and magnetic effects are in consequence so transient, that it is somewhat difficult to observe them. If, however, we maintain A and B at constant potentials by connecting them with the terminals of a voltaic battery the thermal and magnetic effects will persist as long as the connection with the battery is maintained, and are then easily observed.

The wire connecting the two bodies A and B at different potentials is said to be conveying a current of electricity, and when A is losing its positive charge and B its negative charge the current is said to flow from A to B along the wire.

Let us consider the behaviour of the Faraday tubes during the discharge of the conductors A, B. Before the conductors were connected by the wire these tubes may be supposed to be distributed somewhat as in the figure.

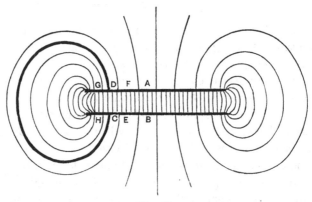

Fig. 86.

When the conducting wire CD is inserted, the tubes which were previously in the region occupied by the wire cannot subsist in the conductor, they therefore shrink, their ends travelling along the wire until the ends which were previously on A and B come close together and the effect of these tubes is annulled. The distribution of the tubes in the field before the wire was inserted was one in which there was equilibrium between the tensions along the tubes and the lateral repulsion they exert on each other:

now after the tubes in the wire have shrunk the lateral repulsion they exerted is annulled and there will therefore be an unbalanced pressure tending to push the surrounding tubes such as *EF*, *GH* into the wire, where they will shrink like those previously in the wire. This process will go on until all the tubes which originally stretched from *A* to *B* have been forced into the wire and their effects annulled.

The discharge of the conductors is thus accompanied by the movement of the tubes in towards the wire and the sliding of the ends of these tubes along the wire. The positive ends of the tubes move on the whole from *A* towards *B* along the wire, the negative ends from *B* towards *A*.

169. Strength of the current. If we consider any cross-section of the wire at *P*, and if in the time δt *N* units of positive electricity cross it from *A* towards *B* and *N'* units of negative electricity from *B* towards *A*, $(N + N')/\delta t$ is called the strength of the current at *P*. When the wire is in a steady state the strength of the current must be the same at all points along the wire, for if it were not the same at *P* as at *Q* a positive or negative charge would accumulate between *P* and *Q* and the state of the wire would not be steady.

170. Electrodes. Anode. Cathode. If the ends *R*, *S* of a body through which a current is flowing are portions of equipotential surfaces, then *R* and *S* are called the electrodes, and if the current is in the direction *RS*, *R* is called the anode and *S* the cathode.

171. Electrolysis. In addition to the thermal and magnetic effects mentioned in Art. 168, there is another effect characteristic of the passages of the current through a large class of substances called electrolytes. Suppose for example that a current passes between platinum plates immersed in a dilute solution of sulphuric acid, then the solution suffers chemical decomposition to some extent and oxygen is liberated at the platinum anode, hydrogen at the platinum cathode. There is no liberation of hydrogen or oxygen in the portions of the liquid not in contact with the platinum plates however far apart these plates may be. Substances whose constituents are separated in this way by the current are called *electrolytes,* and the act of separation is called *electrolysis.* Electrolytes may be solids, liquids, or, as recent experiments have shown, gases. Iodide of silver is an example of a solid electrolyte, while as examples of liquid electrolytes we have solutions of a great number of mineral salts or acids as well as many fused salts.

The constituents into which the electrolyte is separated by the current are called the *ions :* the constituent which is deposited at the anode is called the *anion,* that which is deposited at the cathode the *cation.* With very few exceptions, an element, or such a group of elements as is called by chemists a 'radical,' is deposited at the same electrode from whatever compound it is liberated; thus for example hydrogen and the metals are cations from whatever compounds they are liberated, while chlorine is always an anion.

The amount of the ions deposited by the passage of a current through an electrolyte was shown by Faraday to be connected by a very simple relation with the quantity of electricity which passes through the electrolyte.

172. Faraday's First Law of Electrolysis. *The quantity of an electrolyte decomposed by the passage of a current of electricity is directly proportional to the quantity of electricity which passes through it.*

Thus as long as the quantity of electricity passing through an electrolyte remains the same, it is immaterial whether the electricity passes as a very intense current for a short time or as a very weak current for a long time.

173. Faraday's Second Law of Electrolysis. *If the same quantity of electricity passes through different electrolytes the weights of the different ions deposited will be proportional to the chemical equivalents of the ions.*

Thus, if the same current passes through a series of electrolytes from which it deposits as ions, hydrogen, oxygen, silver, and chlorine, then for every gramme of hydrogen deposited, 8 grammes of oxygen, 108 grammes of silver and 35·5 grammes of chlorine will be deposited.

If we define the *electro-chemical equivalent* of a substance as the number of grammes of that substance deposited during the passage of the unit charge of electricity, we see that Faraday's Laws may be comprised in the statement that the number of grammes of an ion deposited during the passage of a current through an electrolyte is equal to the number of units of electricity which have passed through the electrolyte multiplied by the electro-chemical equivalent of the ion.

Elements which form two series of salts, such as copper, which forms cuprous and cupric salts, or iron, which forms ferrous and ferric salts, have different electro-chemical equivalents according as they are deposited from solutions of the cuprous or cupric, ferrous or ferric salts. The

electro-chemical equivalents of a few substances are given
in the following table; the numbers represent the weight
in grammes of the substance deposited by the passage of
one electro-magnetic unit of electricity (see Chap. XII.).

Hydrogen	·00010352.
Oxygen	·000828.
Chlorine	·003675.
Iron (from ferrous salts)	·002898.
„ (from ferric salts)	·001932.
Copper (from cuprous salts)	·006522.
„ (from cupric salts)	·003261.
Silver	·01118.

The chemical composition of the portions of the electrolyte situated between the electrodes is unchanged by
the passage of the current. Imagine a plane drawn across
the electrolyte, there must pass in any time towards the
cathode across the plane an amount of the cation chemi-
cally equivalent to that of the anion deposited in the
same time at the anode; while a corresponding amount
of the anion must cross the plane towards the anode.
Thus in every part of the electrolyte the cation is moving
in the direction of the current, the anion in the opposite
direction.

Faraday's laws of electrolysis give a method of
measuring the quantity of electricity which has passed
through a conductor in any time and hence of measuring
the average current. For if we place an electrolyte in
circuit with the conductor in such a way that the current
through the electrolyte is always equal to that through the
conductor, then the amount of the electrolyte decomposed
will be proportional to the quantity of electricity which

has passed through the conductor; if we divide the weight in grammes of the deposit of one of the ions by the electro-chemical equivalent of that ion we get the number of electro-magnetic units of electricity which has passed through the conductor, dividing this by the time we get the average current in electro-magnetic units.

An electrolytic cell used in this way is called a volta-meter; the forms most frequently used are those in which we weigh the amount of copper deposited from a solution of copper sulphate, or of silver from a solution of silver nitrate, or measure the amount of hydrogen liberated by the passage of the current through acidulated water.

174. Relation between Electromotive Force and Current. Ohm's Law. The work done by the electric forces on unit charge of electricity in going from a point A to another point B is called the *electromotive force* from A to B. It is frequently written as the E.M.F. from A to B.

Ohm's Law. The relation between the electromotive force and the current was enunciated by Ohm in 1827, and goes by the name of Ohm's Law.

This law states that if E is the electromotive force between two points A and B of a wire, I the current passing along the wire between these points, then

$$E = RI,$$

where R is a quantity called the resistance of the wire. The point of Ohm's Law is that the quantity R defined by this equation is independent of the strength of the current flowing through the wire, and depends only upon the shape and size of the wire, the material of which it is made, and upon its temperature and state of strain.

The most searching investigations have been made as to the truth of this law when currents pass through metals or electrolytes; these have all failed to discover any exceptions to it, though from the accuracy with which resistances can be measured (in several investigations an accuracy of one part in 100,000 has been attained) the tests to which it has been subjected are exceptionally severe.

Ohm's Law does not however hold when the currents pass through rarefied gases.

175. Resistance of a number of Conductors in Series. Suppose we have a number of wires AB, CD,

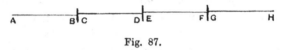

Fig. 87.

EF... (Fig. 87) connected together so that B is in contact with C, D with E, F with G and so on. This method of connection is called putting the wires in series.

Let r_1, r_2, r_3 ... be the resistances of the wires AB, CD, EF... and let i be the current entering the circuit AB, CD... at A, then the current i will flow through each of the conductors. Let us consider the case when the field is steady, then if v_A, v_B, v_C, &c. denote the potentials at A, B, C, &c. respectively, the E.M.F. from A to B is $v_A - v_B$; thus we have by Ohm's Law,

$$v_A - v_B = r_1 i,$$
$$v_C - v_D = r_2 i,$$
$$v_E - v_F = r_3 i,$$
$$\dots\dots\dots\dots$$

But since B and C are in contact they will, if the wires are made of the same substance, be at the same potential; hence $v_B = v_C$, $v_D = v_E$, and so on; hence adding the preceding equations we get

$$v_A - v_F = (r_1 + r_2 + r_3 + \ldots) \, i.$$

But if R is the resistance between A and F, then by Ohm's Law we have

$$v_A - v_F = Ri.$$

Comparing this expression with the preceding, we see that

$$R = r_1 + r_2 + r_3 + \ldots.$$

Hence when a system of conductors are put in series, the resistance of the series is equal to the sum of the resistances of the individual conductors.

176. Resistance of a number of Conductors arranged in Parallel. If the wires instead of being arranged so that the end of one coincides with the beginning of the next, as in the last example, are arranged as in Fig. 88, the beginnings of all the wires being in

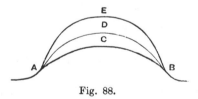

Fig. 88.

contact, as are also their ends, the resistances are said to be arranged in parallel, or in multiple arc.

We proceed now to find the resistance of a system of wires so arranged. Let i be the current flowing up to A, let this divide itself into currents $i_1, i_2, i_3 \ldots$ flowing through

19—2

the circuits ACB, ADB, AEB... whose resistances are
r_1, r_2, r_3... respectively. Then if v_A, v_B are the potentials
of A and B respectively, we have by Ohm's Law

$$v_A - v_B = r_1 i_1,$$
$$v_A - v_B = r_2 i_2,$$
$$v_A - v_B = r_3 i_3,$$
$$\dotsb$$

Now $\quad i = i_1 + i_2 + i_3 + \dots$

$$= (v_A - v_B) \left(\frac{1}{r_1} + \frac{1}{r_2} + \frac{1}{r_3} + \dots \right).$$

But if R is the resistance of the system of conductors,
then by Ohm's Law,

$$Ri = (v_A - v_B);$$

hence comparing this expression with the preceding one
we see that

$$\frac{1}{R} = \frac{1}{r_1} + \frac{1}{r_2} + \frac{1}{r_3} + \dots,$$

or the reciprocal of the resistance of a number of conductors in parallel is equal to the sum of the reciprocals
of the individual resistances. The reciprocal of the resistance of a conductor is called its *conductivity*, hence we
see that we may express the result of this investigation
by saying that the conductivity of a number of conductors
in parallel is equal to the sum of the conductivities of
the individual conductors.

In the special case when all the wires connected up in
multiple arc have the same resistance, and if there are n
wires, their resistance when in multiple arc is $1/n$ of the
resistance of one of the individual wires.

177. Specific resistance of a substance. If we have a wire whose length is l and whose cross section is uniform and of area α, we may regard it as built up of cubes whose edges are of unit length, in the following way; take a wire formed by placing l of these cubes in series and then place α of these filaments in parallel; the resistance of this system is evidently the same as that of the wire under consideration. If σ is the resistance of one of the cubes the resistance of the filament formed by placing l such cubes in series is $l\sigma$, and when α of these filaments are placed in parallel the resistance of the system is $l\sigma/\alpha$; hence the resistance of the wire is

$$\frac{l\sigma}{\alpha}.$$

Since σ only depends on the material of which the wire is made we see that the resistance of a wire of uniform cross section is proportional to the length and inversely proportional to the area of the cross section.

The quantity denoted by σ in the preceding expression is called the specific resistance of the substance of which the wire is made; it is the resistance of a cube of the substance of which the edge is equal to the unit of length, the current passing through the cube parallel to one of its edges.

178. Heat generated by the passage of a current through a conductor. Let A and B be two points connected by a conductor, let E be the electromotive force from A to B. By the definition of electromotive force, work equal to E is done on unit positive charge when it goes from A to B, and on unit negative charge when it goes from B to A ; hence if in unit time N units of positive charge go from A to B and N' units of negative charge from B to A, the work done is E $(N+N')$. But $N+N'$ is equal to C, the strength of the current flowing from A to B, thus the work done is equal to EC. If R is the resistance of the conductor between A and B, $E = RC$; thus the work done in unit time is equal to RC^2. We see that the same amount of work would be spent in driving a current of the same intensity in the reverse direction, viz. from B to A.

By the principle of the Conservation of Energy the work spent by the electric forces in driving the current cannot be lost, it must give rise to an equivalent amount of energy of some kind or other. The passage of the current heats the conductor, but if the heat is caused to leave the conductor as soon as produced the state of the conductor is not altered by the passage of the current. The mechanical equivalent of the heat produced in the conductor was shown by Joule to be equal to the work spent in driving the current through the conductor, so that the work done in driving the current is in this case entirely converted into heat. Thus if H is the mechanical equivalent of the heat produced in time t,

$$H = RC^2t.$$

The law expressed by this equation is called Joule's Law. It states that the heat produced in a given time is proportional to the square of the strength of the current.

Since by Ohm's Law $E = RC$, the heat produced in the time t is also equal to

$$\frac{E^2}{R}t = ECt.$$

179. Voltaic Cell. We have seen that in an electric field due to any distribution of positive and negative electricity, the work done when unit charge is taken round a closed circuit vanishes; the electric intensity due to such a field tending to stop the unit charge in some parts of its course and to help it on in others. Hence such a field cannot produce a steady current round a closed circuit. To maintain such a current other forces must come into play by which work can be done; this work may be supplied from chemical sources, as in the voltaic battery; from thermal sources, as in the thermoelectric circuit; or by mechanical means, as when currents are produced by dynamos. We shall consider here the case of the voltaic circuit. Let us consider the simple form of battery consisting of two plates, one of zinc, the other of copper, dipping into a vessel containing dilute sulphuric acid. If the zinc and copper plates are connected by a wire, a current will flow round the circuit, flowing from the zinc to the copper through the acid, and from the copper to the zinc through the wire. When the current flows round the circuit the zinc is attacked by the acid and zinc sulphate is formed. For each unit of electricity that flows round the circuit one electro-chemical equivalent of zinc and sulphuric acid disappears and equivalent amounts of zinc sulphate and hydrogen are formed. Now if a piece of pure zinc is placed in dilute acid very little chemical action goes on, but if a piece of copper is attached to the zinc the latter is immediately attacked by the acid and zinc sulphate and hydrogen are produced; this action is accompanied by a considerable heating effect, and we find that for each gramme of zinc consumed a definite

amount of heat is produced. Now let us consider two
vessels (α) and (β), such that in (α) the zinc and copper
form the plates of a battery, while in (β) the zinc has
merely got a bit of copper fastened to it : let a definite
amount of zinc be consumed in the latter and then let the
current run through the battery until the same amount
of zinc has been consumed in (α) as in (β). The same
amount of chemical combination has gone on in the two
cells, hence the loss of chemical energy is the same in
(α) as in (β). This energy has been converted into heat
in both cases, the difference being that in the cell (β)
the heat is produced close to the zinc plate, while in (α)
the places where heat is produced are distributed through
the whole of the circuit, and if the wire connecting the
plates has a much greater resistance than the liquid
between them, by far the greater portion of the heat
is produced in the wire, and not in the liquid in the
neighbourhood of the zinc. Though the distribution of
the places in which the heat is produced is different in the
two cases, yet, since the same changes have gone on in the
two cells, it follows from the principle of the Conservation
of Energy, that the total amount of heat produced in the
two cases must be the same. Thus the total amount of
heat produced by the battery cell (α) must be equivalent
to that developed by the combination of the amount of
zinc consumed in the cell while the current is passing
with the equivalent amount of sulphuric acid.

180. Electromotive Force of a Cell. If C is the
current, R the resistance of the wire between the plates,
r that of the liquid between the plates, t the time the
current has been flowing, then by Joule's law the mechanical

equivalent of the heat generated in the wire is RC^2t, that of the heat generated in the liquid is rC^2t. We shall see in Chapter XIII, that when a current flows across the junction of two different metals, heat is produced or absorbed at the junction; this effect is called the Peltier effect. The laws governing the thermal effects at the junction of two metals differ very materially from Joule's law. The heat developed in accordance with Joule's law in a conductor AB is, as long as the strength of the current remains unaltered, the same whether the current flows from A to B or from B to A. The thermal effects at the junction of two metals C and D depend upon the direction of the current; thus if there is a development of heat when the current flows across the junction from C to D, there will be an absorption of heat at the junction, if the current flows from D to C. These heat effects which change sign with the current are called reversible heat effects. The heat developed at the junction of two substances in unit time is directly proportional to the strength of the current and not to the square of the strength.

In the case of the voltaic cell formed of dilute acid and zinc and copper plates, the current passes across the junctions of the zinc and acid and of the acid and copper as well as across the metallic junctions which occur in the wire used to connect up the two plates. Let P be the total heat developed at these junctions when traversed by unit current for unit time. Then the total amount of heat developed in the voltaic cell is

$$RC^2t + rC^2t + PCt.$$

Since a current C has passed through the cell for a time t, the number of units of electricity which have passed through the cell is Ct, hence, if e is the electro-chemical equivalent of zinc, eCt grammes of zinc have been converted into zinc sulphate. Let w be the mechanical equivalent of the heat produced when one gramme of zinc is turned into zinc sulphate, then the mechanical equivalent of the heat which would be developed by the chemical action which has taken place in the cell is $eCtw$; but this must be equal to the mechanical equivalent of the heat developed in the cell, and hence we have

$$RC^2t + rC^2t + PCt = eCtw,$$

or $\qquad (R + r)\, C = ew - P.$

The quantity on the right-hand side is called the electro-motive force of the cell.

We see that it is equal to the sum of the products of the current through the external circuit and the external resistance and the current through the battery and the battery resistance.

We shall now prove that if the zinc and copper plates instead of being joined by a wire are connected to the plates of a condenser, then if these plates are made of the same material, they will be at different potentials, and the difference between their potentials will equal the electromotive force of the battery. For when the system has got into a state of equilibrium, and any change is made in the electrical conditions, the increase in the electrical energy must equal the energy lost in making the change. Suppose that the potential of the plate of the condenser in connection with the copper plate in the battery exceeds by E the potential of the other plate of the condenser in connection with the zinc plate of the battery; and suppose now that the electrical state is altered by a quantity of electricity equal to δQ passing from the plate of the condenser at low potential to the plate at high potential through the battery from the zinc to the copper. The electrical energy of the condenser is increased by $E\delta Q$, while the passage of this quantity of electricity will develop at the junctions of the different substances in the cell a quantity of heat whose mechanical equivalent is equal to $P\delta Q$. If t is the time this charge takes to pass from the one plate to the other, the average current will be equal to $\delta Q/t$, hence the heat developed in accordance with Joule's law will be proportional to $(\delta Q/t)^2 \times t$ or to $(\delta Q)^2/t$; by making δQ small enough, we can make this exceedingly small compared with either $E\delta Q$ or $P\delta Q$ which depend on the first powers of δQ. The loss of chemical energy is $e\delta Q \times w$, and this must be equal to the heat produced plus the increase in the electrical energy, hence we have

$$E\delta Q + P\delta Q = e\delta Q \times w,$$

or
$$E = ew - P,$$

that is, the difference of potential between the plates of the condenser is equal to the electromotive force of the battery. Hence we can determine this electromotive force by measuring the difference of potential.

The simple form of voltaic cell just described does not give a constant electromotive force, as the hydrogen produced by the chemical action does not all escape from the cell; some of it adheres to the copper plate, forming a gaseous film which increases the resistance and diminishes the electromotive force of the cell.

The copper plate with the hydrogen adhering to it is said to be polarized and to be the seat of a back electromotive force which makes the electromotive force of the battery less than its maximum theoretical value. We shall perhaps get a clearer view of the condition of the copper plate with its film of hydrogen from the following considerations. The hydrogen in an electrolyte follows the current and thus behaves as if it had a positive charge of electricity; if now the hydrogen ions when they come up to the copper plate, do not at once give up their charges to the plate, but remain charged at a small distance from it; we shall have what is equivalent to a charged parallel plate condenser at the copper plate, the positively charged hydrogen atoms corresponding to the positive plate of the condenser, and the copper to the negative plate. If the positively charged hydrogen ions charge up the positive plate of this condenser, driving off by induction an equal positive charge from the copper plate, instead of giving up their charge directly to this plate; the flow of electricity through the battery increases the charge in the condenser. To increase this charge work has to be done. If V is the potential difference between the plates of the condenser, the rest of the notation being the same as on p. 298, we have

$$(R + r)\, C^2 t + VCt + PCt = ewCt$$

or $$(R + r)\, C = ew - P - V;$$

thus the electromotive force of the cell is diminished by the potential difference between the plates of the condenser.

Another cause of inconstancy is that the zinc sulphate formed acts as an electrolyte and carries some of the current; the zinc, travelling with the current, is deposited against the copper plate and alters the electromotive force of the cell.

The deposition of hydrogen against the positive plate of the battery, and its liberation as free hydrogen, can be avoided in several ways; in the Bichromate Battery the copper plate is replaced by carbon, and potassium bichromate is added to the sulphuric acid; as the bichromate is an active oxidising agent it oxidises the hydrogen as soon as it is formed, and thus prevents its accumulation on the positive plate.

181. Daniell's Cell. In Daniell's cell, the zinc and sulphuric acid are enclosed in a porous pot (Fig. 89) made

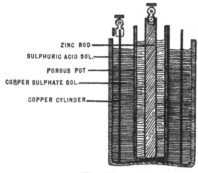

ZINC ROD
SULPHURIC ACID SOL.
POROUS POT
COPPER SULPHATE SOL.
COPPER CYLINDER

Fig. 89.

of unglazed earthenware; the copper electrode usually takes the shape of a cylindrical copper vessel, in which the porous pot is placed. The space between the porous pot and the copper is filled with a saturated solution of copper sulphate, in which crystals of copper sulphate are placed to replace the copper sulphate used up during the working of the cell. When the sulphuric acid acts upon the zinc, zinc sulphate is formed and hydrogen gas liberated; the hydrogen following the current, travels through the porous pot, where it meets with the copper sulphate, chemical action takes place and sulphuric acid is formed and copper set free. This copper travels to the copper cylinder and is there deposited. Thus in this cell instead of hydrogen being deposited on the copper, we have copper deposited, so that no change takes place in the condition of the positive pole and there is no polarization.

182. Calculation of E.M.F. of Daniell's Cell. The chemical energy lost in the cell during the passage of one unit of electricity may be calculated as follows: in the porous pot we have one electro-chemical equivalent of zinc sulphate formed while one equivalent of sulphuric acid disappears; in the fluid outside this pot one equivalent of sulphuric acid is formed and one equivalent of copper sulphate disappears, thus the chemical energy lost is that which is lost when the copper in one electro-chemical equivalent of copper sulphate is replaced by the equivalent quantity of zinc.

Now the electro-chemical equivalent of copper is ·003261 grammes, and when 1 gramme of copper is dissolved in sulphuric acid the heat given out is 909·5 thermal units, or $909·5 \times 4·2 \times 10^7$ mechanical units, since the mechanical equivalent of heat on the C.G.S. system is $4·2 \times 10^7$. Thus the heat given out when one electro-chemical equivalent of copper is dissolved in sulphuric acid is $·003261 \times 909·5 \times 4·2 \times 10^7 = 1·245 \times 10^8$ mechanical units.

The electro-chemical equivalent of zinc is ·003364 grammes, and the heat developed when 1 gramme of zinc is dissolved in sulphuric acid is $1670 \times 4·2 \times 10^7$ mechanical units. Hence the heat developed when one electro-chemical equivalent of zinc is dissolved in sulphuric acid is $\cdot003364 \times 1670 \times 4\cdot2 \times 10^7 = 2\cdot359 \times 10^8$ mechanical units.

Thus the loss of chemical energy in the porous pot is $2\cdot359 \times 10^8$ while the gain in the copper sulphate is $1\cdot245 \times 10^8$, thus the total loss is $1\cdot114 \times 10^8$. Thus ew in Art. $180 = 1\cdot114 \times 10^8$. The electromotive force of a Daniell's cell is about $1\cdot028 \times 10^8$. We see from the near agreement of these values that the reversible thermal effects (see Art. 180) are of relatively small importance, though if we ascribe the difference between the two numbers to this cause these effects would be much greater than those observed when a current flows across the junction of two metals.

183. In **Grove's cell** the hydrogen at the positive pole is got rid of by oxidising it by strong nitric acid. The zinc and sulphuric acid are placed in a porous pot, and this is placed in a larger cell of glazed earthenware containing nitric acid ; the positive pole is a strip of platinum foil dipping into the nitric acid. This cell has a large electromotive force, viz. $1\cdot97 \times 10^8$.

Bunsen's cell is a modification of Grove's, in which the platinum is replaced by hard gas carbon.

184. **Clark's cell**, which on account of its constancy is very useful as a standard of electromotive force, is made as follows. The outer vessel (Fig. 90) is a small test-tube containing a glass tube down which a platinum wire passes ; a quantity of pure redistilled mercury sufficient to cover the end of this wire is then poured into the tube ; on the mercury rests a paste made by mixing mercurous sulphate, saturated zinc sulphate and a little zinc oxide to neutralize it; a rod of pure zinc dips into the paste and is held in position by passing through a

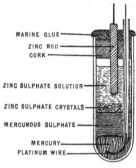

Fig. 90.

cork in the mouth of the test-tube. The electromotive force of this cell is $1\cdot434 \times 10^8$ at $15°$ Centigrade.

Cadmium cell. In this cell the zinc of the Clark cell is replaced by Cadmium, the negative electrode instead of being zinc is an amalgam containing twelve parts by weight of Cadmium in 100 of the amalgam; the zinc sulphate solution is replaced by a saturated solution of Cadmium sulphate; the rest of the cell is the same as in the Clark cell. This all has a smaller temperature coefficient than the Clark cell and is the one now most frequently used as a standard; its electromotive force at $t°$ C. is

$$1\cdot0185 - 0\cdot000038\,(t - 20) - 0\cdot00000065\,(t - 20)^2 \text{ volts.}$$

185. Polarization. When two platinum plates are immersed in a cell containing acidulated water, and a current from a battery is sent from one plate to the other through the water, we find that the current for some time after it begins to flow is not steady but keeps diminishing. If we observe the condition of the plates, we shall find that oxygen adheres to the plate A, at which the current enters the cell, while hydrogen adheres to the other plate B, by which the current leaves the cell. If these plates are now disconnected from the battery and connected by a wire, a current will flow round the circuit so formed, the current going from the plate B to the plate A through the electrolyte and from A to B through the wire. This current is thus in the opposite direction to that which originally passed through the cell. The plates are said to be polarized, and the E.M.F. round the circuit, when they are first connected by the wire, is called the electromotive force of polarization. When the plates are disconnected from the battery and connected by the wire the hydrogen and oxygen gradually disappear from the plates as the current passes. In fact we may regard the polarized plates as forming a voltaic battery, in which the chemical action maintaining the current is the combination of hydrogen and oxygen to form water. Though hydrogen and oxygen do not combine at ordinary temperatures if merely mixed together, yet the oxygen and hydrogen condensed on the platinum plates combine readily as soon as these plates are connected by a wire so as to make the oxygen and hydrogen parts of a closed electrical circuit. There are numerous other examples of the way in which the formation of such a circuit facilitates chemical combination.

186. A Finite Electromotive Force is required to liberate the Ions from an Electrolyte. This follows

at once by the principle of the Conservation of Energy
if we assume the truth of Faraday's Law of Electrolysis.
Thus suppose for example that we have a single Daniell's
cell placed in series with an electrolytic cell containing
acidulated water; then if this arrangement could produce
a current which would liberate hydrogen and oxygen from
the electrolytic cell, for each electro-chemical equivalent of
zinc consumed in the battery an electro-chemical equivalent of water would be decomposed in the electrolytic cell.
Now when one electro-chemical equivalent of hydrogen
combines with oxygen to form water, $1\cdot47 \times 10^8$ mechanical
units of heat are produced, and the decomposition of one
electro-chemical equivalent of water into free hydrogen
and oxygen would therefore correspond to the gain of this
amount of energy.   But for each electro-chemical equivalent of zinc consumed in the battery the chemical energy
lost is (Art. 182) equal to $1\cdot114 \times 10^8$ mechanical units.
Hence we see that if the water in the electrolytic cell
were decomposed, $3\cdot56 \times 10^7$ units of energy would be
gained for each unit of electricity that passed through
the cell: as this is not in accordance with the principle
of the Conservation of Energy the decomposition of the
water cannot go on.   We see that electrolytic decomposition can only go on when the loss of energy in the
battery is greater than the gain of energy in the electrolytic cell.

If we attempt to decompose an electrolyte, acidulated
water for example, by an insufficient electromotive force
the following phenomena occur.   When the battery is
first connected to the cell a current of electricity runs
through the cell, hydrogen travelling with the current
to the plate where the current leaves the cell, oxygen

travelling up against the current to the other plate. Neither the hydrogen nor the oxygen, however, is liberated at the plates, but adheres to the plates, polarizing them and producing a back E.M.F. which tends to stop the current; as the current continues to flow the amount of gas against the plates increases, and with it the polarization, until the E.M.F. of the polarization equals that of the battery, when the current sinks to an excessively small fraction of its original value. The current does not stop entirely, a very small current continues to flow through the cell. This current has however been shown by v. Helmholtz to be due to hydrogen and oxygen dissolved in the electrolytic cell and does not involve any separation of water into free hydrogen and oxygen. The way in which the residual current is carried is somewhat as follows. Suppose that the battery with its small E.M.F. has caused the current to flow through the cell until the polarization of the plates is just sufficient to balance the E.M.F. of the battery; the oxygen dissolved in the water near the hydrogen coated plate will attack the hydrogen on this plate, combining with it to form water, and will, by removing some of the hydrogen, reduce the polarization of the plate; similarly the hydrogen dissolved in the water or it may be absorbed in the plate, will attack the oxygen on the oxygen coated plate and reduce its polarization. The E.M.F. of the polarization being reduced in this way no longer balances the E.M.F. of the battery; a current therefore flows through the cell until the polarization is again restored to its original value, to be again reduced by the action of the dissolved gases. Thus in consequence of the depolarizing action of the dissolved gases there will be a continual current tending to keep the E.M.F.

of the polarization equal to that of the battery; the current however is not accompanied by the liberation of free hydrogen and oxygen and its production does not violate the principle of the Conservation of Energy.

187. Cells in Series. When a series of voltaic cells, Daniell's cells for example, are connected so that the zinc pole of the first is joined up to the copper pole of the second, the zinc pole of the second to the copper pole of the third, and so on, the cells are said to be connected up in series. In this case the total electromotive force of the cells so connected up is equal to the sum of the electromotive forces of the individual cells. We can see this at once if we remember (see Art. 180) that the electomotive force of any system is equal to the difference between the chemical energy lost, when unit of electricity passes through the system, and the mechanical equivalent of the reversible heat generated at junctions of different substances: when the cells are connected in series the same chemical changes and reversible heat effects go on in each cell when unit of electricity passes through as when the same quantity of electricity passes through the cell by itself, hence the E.M.F. of the cells in series is the sum of the E.M.F.'s of the individual cells.

The resistance of the cells when in series is the sum of their resistances when separate. Thus if E is the E.M.F. and r the resistance of a cell, the E.M.F. and resistance of n such cells arranged in series are respectively nE and nr.

188. Cells in parallel. If we have n similar cells and connect all the copper terminals together for a new terminal and all the zincs together for the other terminal the cells are said to be arranged in parallel. In this case we form what is equivalent to a large cell whose E.M.F. is equal to E, that of any one of the cells, but whose resistance is only r/n.

189. Suppose that we have N equal cells and wish to arrange them so as to get the greatest current through a given external resistance R. Let the cells be divided into m sets, each of these sets consisting of n cells in series, and let these m sets be connected up in parallel. The E.M.F. of the battery thus formed will be nE, its resistance nr/m, where E and r are respectively the E.M.F. and resistance of one of the cells. The current through the external resistance R will be equal to

$$\frac{nE}{R+\dfrac{nr}{m}} = \frac{E}{\dfrac{R}{n}+\dfrac{r}{m}}.$$

Now $nm = N$, hence the denominator of this expression is the sum of two terms whose product is given, it will therefore be least when the terms are equal, i.e. when

$$\frac{R}{n} = \frac{r}{m},$$

or

$$R = \frac{n}{m}\, r.$$

Since the denominator in this case is as small as possible the current will have its maximum value. Since nr/m is the resistance of the battery we see that we must arrange the battery so as to make, if possible, the resistance of the battery equal to the given external resistance. This arrangement, though it gives the largest current, is not economical, for as much heat is wasted in the battery as is produced in the external circuit.

190. Distribution of a steady current in a System of Conductors.

Kirchhoff's Laws. The distribution of a steady current in a network of linear conductors can be readily determined by means of the following laws, which were formulated by Kirchhoff.

1. The algebraical sum of the currents which meet at any point is zero.

2. If we take any closed circuit the algebraical sum of the products of the current and resistance in each of the conductors in the circuit is equal to the electromotive force in the circuit.

The first of these laws expresses that electricity is not accumulating at any point in the system of conductors; this must be true if the system is in a steady state.

The second follows at once from the relation (see Art. 180)

$$RI + rI = E,$$

where R is the external resistance, r the resistance of the battery whose E.M.F. is E, and I the current through the battery. For RI is the difference of potential between the terminals of the battery, and by Ohm's law this is equal to the sum of the products of the strength of the current and the resistance for a series of conductors forming a continuous link between the terminals of the battery.

191. Wheatstone's Bridge. We shall illustrate these laws by applying them to the system known as the

Wheatstone's Bridge. In this system a battery is placed in a conductor AB, and five other conductors AC, BC, AD, BD, CD are connected up in the way shown in Fig. 91.

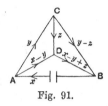

Fig. 91.

Let E be the electromotive force of the battery, B the resistance of the battery circuit AB, i.e. the resistance of the battery itself plus the resistance of the wires connecting its plates to A and B. Let G be the resistance of CD, and b, a, α, β the resistances of AC, BC, AD, BD respectively. Let x be the current through the battery, y the current through AC, z that through CD. By Kirchhoff's first law the current through AD will be $x - y$, that through CB $y - z$, and that through DB $x - y + z$.

Since there is no electromotive force in the circuit ACD we have by Kirchhoff's second law,

$$by + Gz - \alpha (x - y) = 0 ;$$

the negative sign is given to the last term because travelling round the circuit in the direction ACD the current $x - y$ flows in the direction opposite to that in which we are moving; rearranging the terms we get

$$(b + \alpha) y + Gz - \alpha x = 0 \quad \dots\dots\dots\dots\dots(1).$$

Since there is no electromotive force in the circuit CDB, we have

$$Gz + \beta (x - y + z) - a (y - z) = 0,$$
or $\qquad - (a + \beta) y + (G + a + \beta) z + \beta x = 0 \quad \dots\dots(2).$

From (1) and (2) we get

$$\frac{x}{G\,(a+b+\alpha+\beta)+(b+\alpha)\,(a+\beta)}=\frac{y}{G\,(\alpha+\beta)+\alpha\,(a+\beta)}$$

$$=\frac{z}{a\alpha-b\beta}\ \ldots\ldots\ldots\ldots(3).$$

Since the electromotive force round the circuit ACB is E, we have

$$Bx+by+a\,(y-z)=E\,;$$

hence by (3), we have

$$\left.\begin{aligned}
x &= \{G\,(a+b+\alpha+\beta)+(b+\alpha)\,(a+\beta)\}\frac{E}{\Delta}\\[4pt]
y &= \{G\,(\alpha+\beta)+\alpha\,(a+\beta)\}\,\frac{E}{\Delta}\\[4pt]
z &= (a\alpha-b\beta)\,\frac{E}{\Delta}\\[4pt]
x-y &= \{G\,(a+b)+b\,(a+\beta)\}\,\frac{E}{\Delta}\\[4pt]
y-z &= \{G\,(\alpha+\beta)+\beta\,(\alpha+b)\}\,\frac{E}{\Delta}\\[4pt]
x-y+z &= \{G\,(a+b)+a\,(b+\alpha)\}\,\frac{E}{\Delta}
\end{aligned}\right\}\ \ldots(4),$$

where

$$\begin{aligned}
\Delta &= BG\,(a+b+\alpha+\beta)+B\,(b+\alpha)\,(a+\beta)\\
&\quad+G\,(a+b)\,(\alpha+\beta)+\alpha\,(a+\beta)\,(a+b)-a\,(a\alpha-b\beta)\\
&= BG\,(a+b+\alpha+\beta)+B\,(b+\alpha)\,(a+\beta)\\
&\quad+G\,(a+b)\,(\alpha+\beta)+ab\alpha+ab\beta+a\alpha\beta+b\alpha\beta,
\end{aligned}$$

Δ is the sum of the products of the six resistances B, G, a, b, α, β, taken three at a time, omitting the product of any three which meet in a point.

In the expressions given in equations (4) for the currents through the various branches of the network of resistances, we see that the multiplier of E/Δ in the expression for the current through an arm (P) (other than CD) is the sum of the products of the resistances other than the battery resistance and the resistance of P taken two and two, omitting the product of any two which meet at either of the extremities of the battery arm or at either of the extremities of the arm P.

From these expressions we see at once that if we keep *all* the resistances the same, then the current in one arm (A) due to an electromotive force E in another arm (B), is equal to the current in (B) when the electromotive E is placed in the arm A. This reciprocal relation is not confined to the case of six conductors, but is true whatever the number of conductors may be.

We may write the expression for x given by equation (4) in the form

$$x = \frac{E}{B + R},$$

where

$$R = \frac{G\,(a+b)\,(\alpha+\beta) + a\alpha\beta + a\alpha b + a\beta b + \alpha\beta b}{G\,(a+b+\alpha+\beta) + (b+\alpha)\,(a+\beta)}.$$

R is the resistance, between A and B, of the crossed quadrilateral $ACBD$.

We see that $R =$ (sum of products of the 5 resistances of this quadrilateral taken 3 at a time, leaving out the product of any three that meet in a point): divided by the sum of the products of the same resistances taken two at a time, leaving out the product of any pair that meet in A or B.

192. Conjugate Conductors. The current through CD will vanish if

$$a\alpha = b\beta \; ;$$

in this case AB and CD are said to be conjugate to each other, they are so related that an electromotive force in AB does not produce any current in CD: it follows from the reciprocal relation that when this is the case an electromotive force in CD will not produce any current in AB.

The condition that CD should be conjugate to AB may be got very simply in the following way. If no current flows down CD, C and D must be at the same potential; hence since $z = 0$, we have by Ohm's law

$$by = \alpha \, (x - y),$$

since the difference of potential between A and C is equal to that between A and D.

Since the difference of potential between C and B is equal to that between D and B, we have

$$ay = \beta \, (x - y);$$

hence eliminating y and $x - y$, we get

$$\frac{b}{a} = \frac{\alpha}{\beta},$$

or $\qquad\qquad b\beta = a\alpha.$

When this relation holds we may easily prove that

$$\Delta = (a + b + \alpha + \beta) \left\{ G + \frac{(b + \alpha)\,(a + \beta)}{a + b + \alpha + \beta} \right\}$$

$$\left\{ B + \frac{(\alpha + \beta)\,(a + b)}{a + b + \alpha + \beta} \right\},$$

which we may write as

$$\Delta = S\,(G + P')\,(B + P);$$

where S is the resistance of ADB, ACB placed in series, P the resistance of the same conductors when in parallel, and P' the resistance of CAD, CBD in parallel.

When AB is conjugate to CD, then in whatever part of the network an electromotive force is placed, the current through one of these arms is independent of the resistance in the other. We may deduce this from the preceding expressions for the currents in various arms of the circuit; it can also be proved in the following way, which is applicable to any number of conductors. Suppose that an electromotive force in some branch of the system produces a current through AB, then we may introduce any E.M.F. we please into AB without altering the current through its conjugate CD. We may in particular introduce such an electromotive force as would make the current through AB vanish, without altering the current in CD, but the effect of making the current in AB vanish would be the same as supposing AB to have an infinite resistance; hence we may make the resistance of AB infinite without altering the current through CD.

193. We may use Wheatstone's Bridge to get a difference of potential which is a very small fraction of that of the battery in the Bridge.   The difference of potential between C and D is equal to Gz, i.e. to

$$\frac{G\left(a\alpha - b\beta\right)E}{\Delta};$$

it thus bears to E the ratio of $G\left(a\alpha - b\beta\right)$ to Δ.   By making $a\alpha - b\beta$ small we can without using either very small or very large resistances make the ratio of the potential difference between C and D to E exceedingly small;

for example, let $a = 101$, $\alpha = 99$, $b = \beta = 100$, $B = G = 1$. Thus we find that this ratio is nearly equal to $1/4 \times 10^6$, or the potential difference between C and D is only about one four-millionth part of the E.M.F. of the battery.

194. Heat produced in the System of Conductors. Assuming Joule's law (see Art. 178) we shall show that for all possible distributions consistent with Kirchhoff's first law, the one that gives the minimum rate of heat production is that given by the second law.

For, consider any closed circuit in a network of conductors, let u, v, w... be the currents through the arms of this circuit as determined by Kirchhoff's laws, and r_1, r_2, ... the corresponding resistances. The rate of heat production in this closed circuit is by Joule's law equal to

$$r_1 u^2 + r_2 v^2 + \ldots \quad \ldots\ldots\ldots\ldots\ldots(1).$$

Now suppose that the currents in this circuit are altered in the most general way possible consistent with leaving the currents in the conductors not in the closed circuit unaltered, and consistent also with the condition that the algebraical sum of the currents flowing into any point should vanish: we see that these conditions require that all the currents in the closed circuit should be increased or diminished by the same amount. Let them all be increased by ξ; the rate of heat production in the circuit is now by Joule's law

$$r_1 (u + \xi)^2 + r_2 (v + \xi)^2 + \ldots$$
$$= r_1 u^2 + r_2 v^2 + \ldots + 2\xi (r_1 u + r_2 v + \ldots) + (r_1 + r_2 + r_3 + \ldots)\, \xi^2.$$

Now since the currents u, v, w are supposed to be determined by Kirchhoff's laws

$$r_1 u + r_2 v + \ldots = 0,$$

if there is no electromotive force in the closed circuit
Hence the rate of heat production is equal to

$$r_1 u^2 + r_2 v^2 + \dots + (r_1 + r_2 + r_3 + \dots)\, \xi^2 \ \dots\dots(2).$$

Of the two expressions (1) and (2) for the rate of heat
production (2) is always the greater; hence we see that
any deviation of the currents from the values determined
by Kirchhoff's law would involve an increase in the rate
of heat production.

195. Use of the Dissipation Function. We may
often conveniently deduce the actual distribution of the
currents by writing down F the expression for the rate of
heat production and making it a minimum, subject to the
condition that the algebraical sum of the currents which
meet in a point is zero. Or we may by the aid of this
condition express, as in the example of the Wheatstone's
Bridge, the current through the various arms in terms of
a small number of currents x, y, z, then express the rate
of heat production in terms of x, y, z.

F is often called the *Dissipation Function.*

When there are electromotive forces E_p, E_q in the
arms through which currents u_p, u_q are flowing respectively, then the actual distribution of current is that
which makes

$$F - 2\,(E_p u_p + E_q u_q + \dots)$$

a minimum. Thus in the case of the Wheatstone's
Bridge (Art. 191)

$$F = Bx^2 + by^2 + a\,(y - z)^2 + Gz^2 + \alpha\,(x - y)^2 + \beta\,(x - y + z)^2,$$

and equations (4) of Art. 191 are equivalent to

$$\frac{d}{dy}\left(F - 2Ex\right) = 0,$$

$$\frac{d}{dz}\left(F - 2Ex\right) = 0,$$

$$\frac{d}{dx}\left(F - 2Ex\right) = 0,$$

which are the conditions that $F - 2Ex$ should be a minimum.

A very important example of the principle that steady currents distribute themselves so as to make the rate of heat production as small as possible, is that of the flow of a steady current through a uniform wire; in this case the rate of heat production is a minimum when the current is uniformly distributed over the cross section of the wire.

196. It follows from Art. 194 that if two electrodes are connected by any network of conductors, the equivalent resistance is in general increased, and is never diminished, by an increase in the resistance of any arm of this network.

If R is the resistance between the electrodes, i the current flowing in at one electrode and out at the other, then Ri^2 is the rate of heat production. Let A and B respectively denote the network before and after the increase in resistance in one or more of its arms. Without altering the resistance alter the currents until the distribution of currents through A is the same as that actually existing in B. The rate of heat production for the new distribution is by Art. 194 greater than that in A. Now take this constrained system and without altering the currents suppose that the resistances are increased until they are the same as in B. Since the resistances are increased without altering the currents the rate of heat production is increased, so that as this rate was greater than in A before the resistances were increased it will *à fortiori* be greater afterwards. But after the resistances were increased the currents and resistances are the same as B, hence the rate of heat production in B and therefore its resistance is greater than that of A.

197. The following proof of the reciprocal relations between the currents and the electromotive forces in a network of conductors is due to Professor Wilberforce. Let A, B be two of the points in a network of conductors, let R_{AB} denote the resistance of the wire joining AB; V_A the potential of A, V_B that of B, C_{AB} the current flowing along the wire from A to B, E_{AB} the electromotive force of a battery in AB, tending to make the current flow through the battery in the direction AB; let currents from an external source be led into the network, the current entering at a point A being denoted by I_A. Then if ΣI_A denotes the sum of all these currents $\Sigma I_A = 0$.

We have by Ohm's law,
$$R_{AB} C_{AB} = V_A - V_B + E_{AB} \quad \ldots\ldots\ldots\ldots(1).$$

Let us suppose that another distribution of currents, potentials, and electromotive forces is denoted by dashed letters. We have by (1),
$$R_{AB} C_{AB} C'_{AB} = (V_A - V_B) C'_{AB} + E_{AB} C'_{AB}.$$

Taking the whole network of conductors we have

$$\Sigma R_{AB} C_{AB} C'_{AB} = \Sigma (V_A - V_B) C'_{AB} + \Sigma E_{AB} C'_{AB}.$$

The coefficient of V_A is the sum of all the currents that flow outwards from A, this must be equal to I'_A, hence

$$\Sigma R_{AB} C_{AB} C'_{AB} = \Sigma V_A I'_A + \Sigma E_{AB} C'_{AB}.$$

Since the left-hand side is symmetrical with respect to the accented and unaccented letters we have

$$\Sigma V_A I'_A + \Sigma E_{AB} C'_{AB} = \Sigma V'_A I_A + \Sigma E'_{AB} C_{AB} \dots (2).$$

Now suppose that all the I's and I''s are zero, and that all the E's are zero except E_{AB}, all the E''s except E'_{CD}; (2) becomes

$$E_{AB} C'_{AB} = E'_{CD} C_{CD},$$

i.e. that when unit electromotive force acts in AB, the current sent through another branch CD of the network is the same as the current through AB when unit electromotive force acts in CD. Again in equation (2) suppose that all the E's and E''s are zero, that a current I_A is led in at A and out at B, all the other I's vanishing, and that in the distribution represented by the dashed letters a current I'_C is led in at C and out at D, all the other I''s vanishing, then by (2)

$$I'_C (V_C - V_D) = I_A (V'_A - V'_B).$$

Thus if unit current be led in at A and out at B the potential difference between C and D is the same as the potential difference between A and B when unit current is led in at C and out at D.

198. Distribution of Current through an infinite Conductor. We shall now consider the case when the currents instead of being constrained to flow along wires are free to distribute themselves through an unlimited conductor whose conductivity is constant throughout its volume.   We shall suppose that the current is introduced into this conductor by means of perfectly conducting electrodes, i.e. electrodes made of a material whose specific resistance vanishes. The currents will enter and leave the conductor at right angles to the electrodes, for a tangential current in the conductor would correspond to a finite tangential electric intensity in the conductor and therefore in the electrode, but in the perfectly conducting electrode a finite electric intensity would correspond to an infinite current. Let A and B be the electrodes, i the current which enters at A and leaves at B; then we shall prove that the current at any point P in the conductor is in the same direction as, and numerically equal to, the *electric intensity* at the same point, if we suppose the conducting material between the electrodes to be replaced by air, and the electrodes A and B to have charges of electricity equal to $i/4\pi$ and $-i/4\pi$ respectively. For the current is determined by the conditions (1) that it is at right angles to the surfaces A and B, and (2) that since the current is steady, and there is no accumulation of electricity at any part of the conductor, the quantity of electricity which flows into any region equals the quantity which flows out.   Hence we see that the outward flow over any closed surface enclosing A and not B is equal to i, over any closed surface enclosing B and not A is equal to $-i$, and over any closed surface enclosing neither or both of these surfaces is zero. But the electric intensity, when the conductor is replaced by air and A has a charge $i/4\pi$ of positive electricity, while B has an equal charge of negative electricity, satisfies exactly the same conditions, which are sufficient to determine it without ambiguity; hence the current in the conductor is equal to the electric intensity in the air and is in the same direction. A line such that the tangent to it at any point is in the direction of the current at that point is called a stream-line. The stream-lines coincide with the lines of force in the electrostatic problem.

199. If q is the intensity of the current at any point P (i.e. the current flowing through unit area at right angles to the stream-line at P), σ the specific resistance of the conductor, ds an element of the stream-line, then by Ohm's law the E.M.F. between the electrodes A and B is equal to

$$\int \sigma q ds,$$

the integral being extended from the surface of A to that of B. As σ is constant, this is equal to

$$\sigma \int q ds.$$

If F is the electric intensity at P in the electrostatic problem, since $F = q$, the E.M.F. between A and B is equal to

$$\sigma \int F ds;$$

but if V is the difference of potential between A and B in the electrostatic problem,

$$V = \int F ds.$$

Hence the E.M.F. between A and B is equal to σV. But if C is the electrostatic capacity of the two conductors, since these have the charges $i/4\pi$ and $-i/4\pi$ respectively,

$$\frac{i}{4\pi} = CV.$$

Hence the E.M.F. between A and $B = \dfrac{\sigma i}{4\pi C},$

or the resistance between A and B is equal to

$$\frac{\sigma}{4\pi C}.$$

We see from this that the resistance of a shell bounded by concentric spherical surfaces, whose radii are a and b, is equal to

$$\frac{\sigma}{4\pi ab}(b-a).$$

The resistance per unit length of a shell of conducting material bounded by two coaxial cylindrical surfaces whose radii are a and b is equal to

$$\frac{\sigma}{2\pi}\log\frac{b}{a}.$$

The resistance between two spherical electrodes whose radii are a and b and whose centres are separated by a distance R, where R is very large compared with either a or b, is equal to

$$\frac{\sigma}{4\pi}\left\{\frac{1}{a}+\frac{1}{b}-\frac{2}{R}\right\},$$

approximately.

The resistance per unit length between two straight parallel cylindrical wires whose radii are a and b, and whose axes are at a distance R apart, where R is very large compared with a or b, is approximately

$$\frac{\sigma}{2\pi}\log\frac{R^2}{ab}.$$

If we have two infinite cylinders, one with a charge of electricity E per unit length, the other with the charge $-E$; then if A and B are the centres of the sections of these cylinders by a plane perpendicular to the axis and

21—2

P a point in this plane, the electrostatic potential at P will be equal to

$$2E \log \frac{BP}{AP},$$

if the cylinders are so far apart that the electricity may be regarded as uniformly distributed over them. Thus the lines along which the electrostatic potential is constant are those for which

$$\frac{BP}{AP} = \text{a constant quantity.}$$

That is, they are the series of circles for which A and B are inverse points. The lines of force are the lines which cut these circles at right angles, i.e. they are the series of circles passing through A and B. But the lines of force in the electrostatic problem coincide with the lines along which the currents flow between two parallel cylinders as electrodes; hence these currents flow in planes at right angles to the axes of the cylinders, along the circles passing through the two points in which these planes intersect the axes of the cylinders.

Since the resistance of unit length of the cylinders is

$$\frac{\sigma}{2\pi} \log \frac{R^2}{ab},$$

the resistance of a length t is

$$\frac{\sigma}{2\pi t} \log \frac{R^2}{ab}.$$

This will be the resistance of a thin lamina whose thickness is t when the current is led in by circular electrodes radii a and b, if the thickness of the lamina is so small that the currents are compelled to flow parallel to the lamina. The lines of flow in this case are circles passing through A and B; they are represented in Fig. 92.

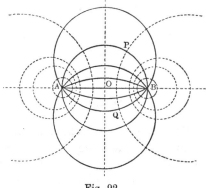

Fig. 92.

Since the currents flow along these circles we shall not alter the distribution of current if we imagine the lamina cut along one or other of these circles; hence if the lamina is bounded by two circular areas such as APB, BQA the lines of flow will be circular arcs passing through A and B.

To find the resistance of a lamina so bounded, consider for a moment the flow through the unlimited lamina. The current will flow from out of each electrode approximately uniformly in all directions; hence if we draw a series of circles intersecting at the constant angle a at A and B, we may regard the whole lamina as made up of the conductors between the stream-lines placed in multiple arc; the number of these conductors is $\dfrac{2\pi}{a}$, and since the same current flows through each, the resistance of any one of them is $2\pi/\alpha$ of the whole resistance; thus the resistance of one of these conductors is

$$\frac{\sigma}{\alpha t}\log\frac{R^2}{ab}.$$

Thus, for example, if the electrodes are placed on the circumference of a complete circle, $\alpha = \pi$ and the resistance of the lamina is

$$\frac{\sigma}{\pi t}\log\frac{R^2}{ab}.$$

200. Conditions satisfied when a current flows from one medium to another. Let AB be a portion of the surface of separation of two media, σ_1 the specific resistance of the upper medium, σ_2 that of the lower, let θ and ϕ be the angles which the directions of the current in the upper and lower media respectively make with the normal to the surface. Let q_1, q_2 be the intensities of the currents in the two media, i.e. the amount of current flowing across unit areas drawn at right angles to the direction of flow. Then since, when things are in a steady state, there is no increase or decrease in the electricity at the junction of the two media, the currents along the normal must be equal in the two media.

Thus $$q_1 \cos\theta = q_2 \cos\phi \quad\dotfill(1).$$

Again, the electric intensity parallel to the surface must be equal in the two media, and since the electric intensity in any direction is equal to the specific resistance of the medium multiplied by the intensity of the current in that direction, we have

$$\sigma_1 q_1 \sin\theta = \sigma_2 q_2 \sin\phi \quad\dotfill(2),$$

hence from (1) and (2) we have

$$\sigma_1 \tan \theta = \sigma_2 \tan \phi.$$

This relation between the directions of the currents in the two media is identical in form with that given in Arts. 74 and 157, for the relation between the directions of the lines of electric intensity and of magnetic force when these lines pass from one medium to another.

We see that if σ_1 is greater than σ_2, then ϕ is greater than θ; hence when the current flows from a poor conductor into a better one the current is bent away from the normal.

The bending of the current as it flows from one medium into another is illustrated in Fig. 93, which is taken from a paper by Quincke. The figure represents

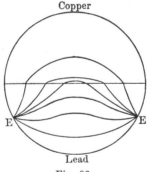

Fig. 93.

the current lines in a circular lamina, one half of which is lead, the other half copper, the electrodes E, E being placed on the circumference. It shows how the currents in going from the worse conductor (the lead) to the better one (the copper) get bent away from the normal to the surface of separation.

The electric intensity parallel to the normal in the medium whose specific resistance is σ_1 is

$$\sigma_1 q_1 \cos \theta,$$

that in the medium whose specific resistance is σ_2 is $\sigma_2 q_2 \cos \phi$. Since $q_1 \cos \theta$ is by equation (1), equal to $q_2 \cos \phi$, we see that if σ_1 differs from σ_2 the normal electric intensity will be discontinuous at the surface of separation.

If the normal electric intensity is discontinuous there must be a distribution of electricity over the surface such that 4π times the surface density of this distribution is equal to the discontinuity in the normal electric intensity; hence if s is the surface intensity of the electricity on the surface, and if the current is flowing from the first medium to the second

$$4\pi s = \sigma_2 q_2 \cos \phi - \sigma_1 q_1 \cos \theta$$

$$= (\sigma_2 - \sigma_1)\, q_1 \cos \theta.$$

CHAPTER X

MAGNETIC FORCE DUE TO CURRENTS

201. It was not known until 1820 that an electric current exerted any mechanical effect on a magnet in its vicinity. In that year however Oersted, a Professor at Copenhagen, showed that a magnet was deflected when placed near a wire conveying an electric current.

When a long straight wire with a current flowing through it was held near the magnet, the magnet tended to place itself at right angles both to the wire and the perpendicular let fall from the centre of the magnet on the wire.

The lines of magnetic force due to a long straight wire may be readily shown by making the wire pass through a hole in a card-board disc over which iron filings are sprinkled. When the disc is at right angles to the wire, the iron filings will arrange themselves in circles when the current is flowing; these circles are concentric, having as their centre the point where the wire crosses the plane of the disc.

The connection between the direction of the current and that of the magnetic force is such that if the axis

of a right-handed screw (i.e. an ordinary corkscrew) coincides with the direction of the current, then if the screw
is screwed forward into a fixed nut in the direction of the
current the magnetic force at a point P due to the current
is in the direction in which P would move if it were rigidly
attached to the screw.

Many students will find that they can remember the
connection between the direction of the current and
the magnetic force more easily by means of a figure
than by a verbal rule. The following figure exhibits
this relation.

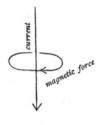

Fig. 94.

**202. Ampère's law for the magnetic field due
to any closed linear circuit.** This may be stated as
follows: At any point P, not in the wire conveying the
current, the magnetic forces due to the current can be
derived from a potential Ω where $\Omega = Ci\omega$, i being the
current flowing round the circuit, ω the solid angle subtended by the circuit at P, and C a constant which
depends on the unit in which the current is expressed.

When the unit of current is what is known as the
'electromagnetic unit,' see Chap. XII., C is unity. We

shall in the following investigations suppose that the current is measured in terms of this unit.

We see from Art. 134 that this is equivalent to saying that the magnetic field due to a current is the same as that due to a magnetic shell whose strength is i, the boundary of the shell coinciding with the circuit conveying the current. The direction of magnetization of the shell is related to the direction of the current in such a way that if the observer stands on the side of the shell which is charged with positive magnetism and looks at the current, the current in front of him flows from right to left.

The best proof of the truth of Ampère's law is that though its consequences are being daily compared with the results of experiments, no discrepancy has ever been detected.

The potential due to the magnetic shell at a point in the substance of the shell is not the same as that due to the electric circuit, nor is the magnetic force at such a point the same in the two cases. This however does not cause any difficulty in determining the magnetic force due to a circuit at any point P, for, since only the boundary of the equivalent magnetic shell is fixed, we can always arrange the shell in such a way that it does not pass through P.

We can easily prove, however, that at any point, whether in the substance of the shell or not, the magnetic *force* due to the circuit is equal to the magnetic *induction* due to the shell. For let P be a point in the substance of the shell, then though the magnetic force due to the shell will not be the same as at P', a point just outside the shell, yet the force due to the current at P' will differ from that at P by an amount which vanishes when the distance PP' is indefinitely diminished. The magnetic force at P' due to the current is the same as the magnetic force at P' due to the shell. Since the shell is magnetized along the normal, the tangential magnetic force in the shell is equal to the tangential magnetic induction. Now, by Art. 158, the tangential magnetic force at P', a point just outside the shell, is equal to the tangential magnetic force at P, a point just inside the shell, and this, as we have just seen, is equal to the tangential magnetic induction at P. Again, by Art. 158, the normal magnetic force at P' is equal to the normal magnetic induction at P. Thus since the normal force at P' is equal to the normal induction at P, and the tangential force at P' is equal to the tangential induction at P, the magnetic force at P' is equal in magnitude and direction to the magnetic induction at P. Since the magnetic force at P due to the current is equal to the magnetic force at P' due to the shell, we see that the magnetic force due to the current at P is equal to the magnetic induction due to the shell at P.

Thus since the lines of magnetic induction due to the shell form a series of closed curves passing through the shell, the lines of magnetic force due to the current flowing round a closed linear circuit will be a series of closed curves threading the circuit.

203. Work done in taking a magnetic pole round a closed curve in a magnetic field due to electric currents. Let $EFGH$ be the closed curve traversed by the magnetic pole; if this curve threads the circuit traversed by the current, then the magnetic shell whose magnetic effect is equivalent to that of the current must cut the curve, let it do so in PQ. Let a, b, c be the components of magnetic induction due to the shell at any point, α, β, γ the components of the magnetic force at the same point, and A, B, C the components of the intensity of magnetization. Since the magnetic force due to the

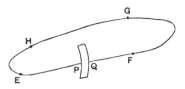

Fig. 95.

circuit is the same as the magnetic induction due to the shell, W, the work done on the unit pole when it traverses the closed curve $EFGH$ under the influence of the electrical currents, is given by the equation

$$W = \int (a\,dx + b\,dy + c\,dx),$$

the integral being taken round the closed curve.

Hence we have by Art. 153

$$W = \int \{(\alpha + 4\pi A)\,dx + (\beta + 4\pi B)\,dy + (\gamma + 4\pi C)\,dz\},$$

or since by Art. 134 the line integral of the magnetic force due to the shell vanishes when taken round a closed circuit, we have

$$\int (\alpha\,dx + \beta\,dy + \gamma\,dz) = 0;$$

hence $$W = 4\pi \int (A\,dx + B\,dy + C\,dz),$$

where the integral is now taken from P to Q, the points where the shell cuts the curve $EFGH$, since it is only between P and Q that A, B, C do not vanish.

If ϕ is the strength of the magnetic shell, and the direction of integration is from the negative to the positive side of the shell

$$\int (A\,dx + B\,dy + C\,dz) = \phi\,;$$

hence $W = 4\pi\phi.$

If i is the strength of the current which the shell replaces

$$\phi = i,$$

see Art. 202; hence

$$W = 4\pi i.$$

Thus the work done on unit pole when it travels round a closed curve which threads the circuit once in the positive direction, i.e. when the pole enters at the negative side of the equivalent shell and leaves at the positive, is constant whatever be the path, and is equal to

$$4\pi i.$$

If the closed curve along which the unit pole travels does not thread the circuit of the current, the work done on the unit pole vanishes, for we can draw the equivalent shell so as to be wholly outside the path of the pole, and in this case A, B, C vanish at all points of the path.

If the path along which the unit pole is taken threads the circuit n times in the positive direction (the positive direction being when the pole in its path enters the equivalent magnetic shell at the negative side and leaves it at the positive), and m times in the negative direction, the work done on the pole on its path is equal to

$$4\pi i\,(n - m).$$

The value of $\int (\alpha\,dx + \beta\,dy + \gamma\,dz)$ taken round a closed circuit is independent of the nature of the material which is traversed by the circuit; it is the same, if the currents are unaltered, whether the circuit lies entirely in air, entirely in iron or any other magnetizable medium, or partly in air and partly in iron. For the field may be regarded as made up of two parts, one, in which the components of the magnetic force are α_1, β_1, γ_1 due to the magnetic action of the currents when there is nothing but air in the neighbourhood; the other, a field whose components are α_0, β_0, γ_0 due to the magnetization induced or permanent of the iron.

Hence

$$\int(\alpha dx + \beta dy + \gamma dz)$$
$$= \int\{(\alpha_1 + \alpha_0)\,dx + (\beta_1 + \beta_0)\,dy + (\gamma_1 + \gamma_0)\,dz\}.$$

Since α_0, β_0, γ_0 are the forces due to a distribution of *magnets* the work done by these forces on a unit pole taken round a closed circuit must vanish, hence

$$\int(\alpha_0 dx + \beta_0 dy + \gamma_0 dz) = 0,$$

when the integral is taken round any closed circuit. Thus

$$\int(\alpha dx + \beta dy + \gamma dz) = \int(\alpha_1 dx + \beta_1 dy + \gamma_1 dz),$$

and

$$\int(\alpha_1 dx + \beta_1 dy + \gamma_1 dz)$$
$$= 4\pi \text{ (sum of currents embraced by the circuits)}.$$

Thus $\int(\alpha dx + \beta dy + \gamma dz)$ depends merely upon the currents in the field and not upon the nature of the material intersected by the circuit.

204. Magnetic force due to an infinitely long straight current, in a field in which there are no magnetizable substances. In this case the magnetic force is numerically equal to the magnetic induction, and hence the total normal magnetic force taken over any closed surface vanishes. Take as the closed surface a

right circular cylinder with the current for axis, and let
R be the radial magnetic force at any point of the curved
surface of this cylinder; by symmetry R is constant over
the curved surface. Since the current is infinitely long
the magnetic force will not vary as we move parallel to
the wire conveying the current; hence the normal magnetic force taken over one of the plane ends will cancel
that taken over the other. Thus, if S is the curved
surface of the cylinder, the total magnetic force taken
over the cylinder is RS, and since this vanishes, R must
vanish; hence there is no radial magnetic force due to
the current.

To find T the tangential magnetic force, let P be
any point, and OP the perpendicular let fall from P
on the current; T is the magnetic force at right angles
to OP and to the direction of the current. With O as
centre and radius OP describe in a plane at right angles
to the current a circle; at each point on the circumference of this circle the tangential magnetic force will
by symmetry be constant, and equal to T. The work
done when unit pole is taken round this circle is $2\pi r T$,
and since the path encircles the current once this must
by Art. 203 be equal to $4\pi i$, if i is the strength of the
current; hence we have

$$T = \frac{2i}{r},$$

or the tangential magnetic force varies inversely as the
distance from the current.

We shall now show that the magnetic force parallel to
the current vanishes.

We can do this by regarding the straight circuit as
the limit of a circular one with a very large radius.

Consider the magnetic force at a point P due to the circular current. Through P draw a circle in a plane parallel to that of the current, so that the line joining O the centre of this circle, to the centre of the circle in which the current is flowing, is perpendicular to the planes of these circles. Then if T is the magnetic force along the tangent to this circle at P, T will be, by symmetry, the tangential force at each point of this circle. Hence the work done in taking unit pole round the circumference of this circle is $2\pi OP \cdot T$, this must however vanish as the circle does not enclose any current, thus T must be zero. Proceeding to the limit when the radius of the circle is indefinitely increased we see that the magnetic force due to a straight current has no component parallel' to the current.

Thus the lines of magnetic force due to the long straight current are a series of circles whose centres are on the axis of the current and their planes at right angles to the current. The direction of the magnetic force is related to that of the current in the way shown in the diagram, Fig. 92; i.e. the directions of current and magnetic force are related in the same way as the directions of translation and rotation in a right-handed screw.

The magnetic force at a point P not in the current itself is thus derivable from a potential Ω, where

$$\Omega = 2i\theta \pm 4\pi ni,$$

where θ is the angle PO, the perpendicular let fall from P on the axis of the current, makes with a fixed line in the plane through O at right angles to the current: n is an integer. The potential is a multiple-valued function having at each point an infinite series of values differing from each other by multiples of $4\pi i$, which is the work done in taking unit magnetic pole round a closed circuit embracing the current. This indeterminateness in the potential arises from the fact that the work done on unit pole as it goes from one point P to another point Q, depends not merely on the relative positions of P and Q but also on the number of times the pole in its path from P to Q encircles the current.

205. Magnetic force inside the conductor conveying the current. When the current is flowing symmetrically through a circular cylinder, we can easily find the magnetic force at a point inside the cylinder. Let O be the centre of a cross section of the conductor, and P a point at which the tangential force T is required; in the plane of the section draw a circle whose centre is O and radius OP. The work done in taking unit pole round this circle is $2\pi OP . T$, this by Art. 203 is equal to 4π times the current enclosed by the circle. Hence we have

$2\pi OP . T = 4\pi$ (current enclosed by the circle with

centre O and radius OP).

If the current is all outside this circle, the right-hand side of this equation vanishes: hence T vanishes and there is no magnetic force. Thus there is no magnetic force in the interior of a cylindrical tube conveying a current.

If the current is uniformly distributed over the cross section, and i is the total current flowing through the cylinder whose radius we shall denote by a, the current through the circle whose radius is OP is equal to

$$i . \frac{OP^2}{a^2} .$$

Hence $2\pi OP \cdot T = 4\pi i \cdot \dfrac{OP^2}{a^2},$

$$T = 2i \cdot \dfrac{OP}{a^2}.$$

Thus when the current is uniformly distributed, the magnetic force inside the cylinder varies directly as the distance from the axis; outside the cylinder it varies inversely as this distance.

206. The total normal magnetic induction through any cylindric surface passing through two lines parallel to the current is the same whatever be the shape of the

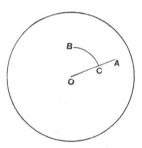

Fig. 96.

surface connecting these lines. This follows at once from the principle that the total magnetic induction over any closed surface is zero. To find an expression for the induction through the cylindric surface, let A and B be the points where the two lines intersect a plane at right angles to the current, O the point where the axis of the current intersects this plane. Take the cylindrical surface such that if B is the point nearest to O, the normal section of the surface is the circular arc BC and the radial portion CA. Since the magnetic force is everywhere tangential to

BC no tube of force passes through the portion corresponding to BC; if r is the distance of any point P on CA from O, the magnetic force at P is

$$\frac{2i}{r};$$

hence the number of tubes of magnetic force passing through the portion corresponding to AC is

$$\int_{oc}^{OA} \frac{2i}{r}\, dr = 2i \log \frac{OA}{OC}$$

$$= 2i \log \frac{OA}{OB},$$

and this represents the number passing through each unit of length of any cylindric surface passing through A and B.

207. Two infinitely long straight parallel currents flowing in opposite directions. Let A and B, Fig. 97, be the points where the axes of the currents intersect a plane drawn at right angles to the direction of the currents. Let the direction of the current at A

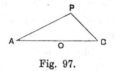

Fig. 97.

be downwards through the paper, that at B upwards; if i is the strength of either current, the magnetic potential at a point P is, Art. 204, equal to

$$2i \left\{ \angle PAB \pm 2\pi n \right\} - 2i \left\{ \pi - \angle PBA \pm 2\pi m \right\}.$$

This may be written

$$4\pi i \, (n + m) - \angle\, APB \times 2i\,;$$

thus along an equipotential line the angle APB is constant, hence the equipotential lines are the series of circles passing through AB.

The lines of magnetic force are at right angles to the equipotential lines, they are therefore the series of circles having their centres along AB such that the tangents to them from O, the middle point of AB, are of the constant length OA.

The lines of magnetic force and the equipotential lines are represented in Fig. 98.

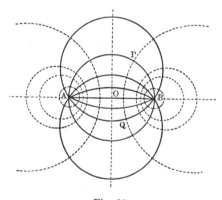

Fig. 98.

The direction of the magnetic force is easily found as follows. If PT is the direction of the magnetic force at P, then since PT is the normal to the circle round

APB, the angle BPT is equal to the complement of the angle PAB.

The magnetic force R at P is the resultant of the forces $2i/AP$ at right angles to AP and $2i/BP$ at right angles to BP. Resolving these along PT, we have

$$R = \frac{2i}{AP} \cos ABP + \frac{2i}{BP} \cos BAP$$

$$= \frac{2iAB}{AP.BP}.$$

Thus the intensity of the magnetic force at P varies inversely as the product of the distances of P from A and B.

At a point on the line bisecting AB at right angles $AP = BP$, and along this line, which may be called the axis of the current, the magnetic force is inversely proportional to the square of the distance from A or B; the direction of the force is parallel to the axis.

At a point whose distances from A and B are large compared with AB we may put $AP = BP = OP$, in this case the magnetic force varies inversely as OP^2, and the direction of the force makes with OP the same angle as OP makes with the line at right angles to AB.

208. Number of tubes of magnetic force due to the two currents which pass through a circuit consisting of two parallel wires. Let A, B be the points where the two currents intersect a plane drawn at right angles to them, C, D the points where the wires of the circuit cut the same plane. Then, Art. 206, the number of tubes of magnetic force due to A which pass through CD per unit length $= 2i \log \dfrac{AC}{AD}$. Similarly the number which pass through CD and are due to the current B is

$$- 2i \log \frac{BC}{BD};$$

hence the number through CD per unit length due to the current i at A and $-i$ at B, is

$$2i \left\{ \log \frac{AC}{AD} - \log \frac{BC}{BD} \right\}$$

$$= 2i \log \frac{AC \cdot BD}{AD \cdot BC}.$$

We see from the symmetry of the expression that this is the number which would pass through the circuit AB due to currents $+i$ and $-i$ at C and D respectively.

When the circuits AB, CD are so situated that the total number of tubes passing through CD due to the current in A, B is zero, the circuits AB, CD are said to be *conjugate* to each other. The condition for this is that $\log \dfrac{AC \cdot BD}{AD \cdot BC}$ should vanish, or that

$$\frac{AC}{BC} = \frac{AD}{BD};$$

another way of stating this result is that C and D must be two points on the same line of magnetic force due to the currents at A and B; this is equivalent to the condition that A and B should be points on a line of magnetic force due to equal and opposite currents at C and D. Since the lines of magnetic force due to the currents A and B are a series of circles with their centres on AB it follows that if CD is conjugate to AB it will remain conjugate however CD is rotated round the point O', O' being the point where the line bisecting CD at right angles intersects AB.

A case of considerable practical importance is when we have two equal circuits AB and CD, the current through A being in the same direction as that through C and that through B in the same direction as that through D.

Let us consider the case when AB and CD are equal and parallel and so placed that the points A, B, D, C are at the corners of a rectangle. Then if i is the current flowing round each of the circuits, Ω the magnetic potential at a point P will, by Art. 204, be given by the equation

$$\Omega = -2i\theta - 2i\phi + \text{constant},$$

where θ and ϕ are the angles subtended respectively by AB and CD at P.

The lines of magnetic force are the curves which cut these at right angles; along such a line

$$\frac{r_1}{r_2}\frac{r_3}{r_4}$$

is constant, where r_1, r_2, r_3, r_4 are the distances of a point on the line from A, B, C, D respectively.

The lines of magnetic force are represented in Fig. 99. There are two points E, F where the magnetic force vanishes; these points are on the line drawn through O, the centre of the rectangle, parallel to the sides AB and CD; we can easily prove that OE is equal to OA.

At a point P on the axis of the current, i.e. on the line through O at right angles to AB, the magnetic force is parallel to the axis and is by Art. 207 equal to

$$\frac{2i \cdot AB}{AP^2} + \frac{2i \cdot CD}{CP^2};$$

if $OP = x$, $AB = 2a$, $AC = 2d$, the magnetic force at P is equal to

$$\frac{4ia}{a^2 + (x+d)^2} + \frac{4ia}{a^2 + (d-x)^2}.$$

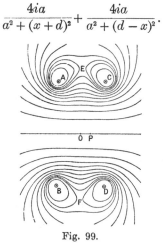

Fig. 99.

This is, neglecting the fourth and higher powers of x, equal to

$$\frac{8ia}{a^2 + d^2}\left\{1 + \frac{3d^2 - a^2}{(a^2 + d^2)^2}\, x^2\right\},$$

thus, if $\sqrt{3}d = a$, the term in x^2 disappears and the lowest power of x which appears in the expression for the magnetic force is the fourth. Thus with this relation between the size of the coils and the distance between them the force near O varies very slowly as we move along the axis.

The number of tubes of magnetic force which pass through one circuit when a current i flows round the other may, by using the result given on page 343, easily be proved to be equal to

$$4i \log \frac{BC}{AC}.$$

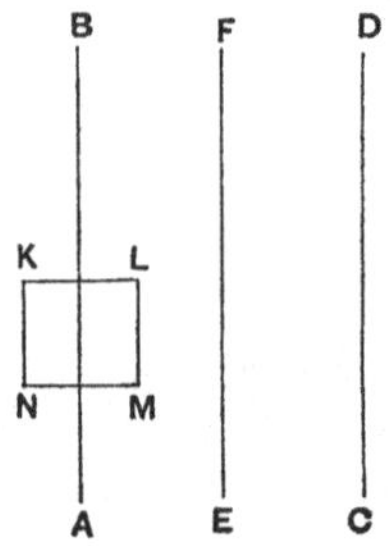

Fig. 100.

209. Direct and return currents flowing uniformly through two parallel and infinite planes.

Let the two parallel planes be at right angles to the plane of the paper and let this plane intersect them in the lines AB, CD, Fig. 100. Let a current i flow upwards at right angles to the plane of the paper through each unit length of AB and downwards through each unit length of CD. Let EF be the section of the plane parallel to AB and CD and midway between them. We shall prove that the magnetic force between the planes is uniform and parallel to EF, being thus parallel to the planes in which the currents are flowing and at right angles to the currents.

We shall begin by proving that the magnetic force has no component at right angles to the planes in which the currents are flowing. This is evidently true by

symmetry at all points in the plane midway between
AB and CD; we can prove it is true at all points in
the following way. Take a rectangular parallelepiped one
of whose faces is in the plane whose section is EF, let
another pair of faces be parallel to the plane of the paper
and the third pair perpendicular to the line EF. The
total normal magnetic induction over this closed surface
vanishes. Since the currents are uniformly distributed
in the infinite planes, the magnetic induction will be the
same at all points in a plane parallel to those in which the
currents are flowing. Hence the total magnetic induction
over the pairs of faces of the parallelepiped which are at
right angles to the parallel planes will vanish : for the
induction at a point on one face will be equal to that at
a corresponding point on the opposite face, and in the
one case it will be along the inward normal, in the other
along the outward. Hence since the total induction over
the parallelepiped is zero the induction over one of the
faces parallel to the planes must be equal and opposite to
that over the opposite face. But one of these faces is
in the plane EF where the magnetic induction normal
to the face vanishes; hence the total normal induction
over the other face must vanish, and since the induction
is the same at each point on the face the induction can
have no component at right angles to this face, i.e. at
right angles to the planes in which the currents are
flowing. This proof applies to all parts of the field,
whether between the planes or outside them.

To prove that the force parallel to the currents
vanishes, we take a rectangle $PQRS$ with two sides PQ,
RS parallel to the currents, the other sides PS, QR being
at right angles to the planes of the currents. No current

flows perpendicularly through this rectangle, hence (Art. 203) the work done when unit magnetic pole is taken round its circumference is zero. But since the magnetic force parallel to PS, RQ vanishes, the work done on unit pole, if F is the force along PQ, F' that along RS, is equal to

$$(F - F')\,PQ.$$

Since this vanishes $F = F'$, i.e. F is constant throughout the field, and since by symmetry it vanishes along EF it must vanish throughout the field.

We have now proved that throughout the field the components of the magnetic force in two directions at right angles to each other vanish, hence the magnetic force, where it exists, must be parallel to EF, Fig. 100.

By drawing a rectangle in the space outside the planes with one pair of its sides parallel to EF we can prove that the force parallel to EF also vanishes outside the planes, so that in this region there is no magnetic force. To find the magnitude of the magnetic force H between the planes, take a rectangle such as $LMNK$, Fig. 100, cutting one of the planes, the sides of the rectangle being respectively parallel and perpendicular to EF. The quantity of current flowing through this rectangle is $i \times LM$, since i flows through each unit of length of the plane; hence $4\pi i \times LM$ is equal to the work done in taking unit magnetic pole round the rectangle. But this work is $H \times LM$, since no work is done when the pole is moving along MN, NK and KL, hence we have

$$H \times LM = 4\pi i \times LM,$$

or $\qquad\qquad\qquad H = 4\pi i.$

Thus the magnetic force is independent of the distance between the planes.

210. Solenoid. We can apply exactly the same method to the very important case of an infinitely long right circular solenoid, i.e. an infinitely long right circular cylinder round which currents are flowing in planes perpendicular to the axis. Such a solenoid may be constructed by winding a right circular cylinder uniformly with wire, the planes of the winding being at right angles to the axis of the cylinder, so that between any two planes at right angles to the axis and at unit distance apart there are the same number of turns of wire. We can show by the same method as in Art. 209, that inside the cylinder the radial magnetic force vanishes, and that the force parallel to the axis of the cylinder is uniform, that outside the cylinder the magnetic force vanishes: and that if H is the magnetic force inside the cylinder parallel to the axis

$H = 4\pi$ (current flowing between two planes separated by unit distance).

If there are n turns of wire wound round each unit length of the cylinder and i is the current flowing through the wire, this equation is equivalent to

$$H = 4\pi ni.$$

The preceding result is true whatever be the shape of the cross section of the cylinder on which the wire is wound, provided the number of turns of wire between two parallel planes at unit distance apart perpendicular to the axis of the cylinder is uniform.

Endless Solenoids. Near the ends of a straight solenoid the magnetic field is not uniform and ceases to be parallel to the axis of the cylinder and equal to $4\pi ni$. We can, however, avoid this irregularity if we wind the wire on a ring instead of on a straight cylinder. Suppose the ring is generated by the revolution of a plane area about an axis in its own plane which does not cut it, and let the ring be wound with wire so that the windings are in planes through the axis of the ring and so that the number of windings between two planes which make an angle θ with each other is equal to $n\theta/2\pi$; n is thus the whole number of windings on the ring. Then we can prove as in Art. 209 that the magnetic force vanishes outside the solenoid, and that inside the solenoid the lines of magnetic force are circles having their centres on the axis of the solenoid and their planes at right angles to the axis. Let H be the magnetic force at a distance r from this axis; the work done on unit pole when taken round a circle whose radius is r and whose centre is on the axis and plane perpendicular to it is $2\pi rH$; this by Art. 203 is equal to 4π times the current flowing through this circle, and is thus equal to $4\pi ni$, if i is the current flowing through one of the turns of wire. Hence

$$2\pi rH = 4\pi ni$$

or
$$H = \frac{2ni}{r}.$$

Thus the force is inversely proportional to the distance from the axis.

The preceding proof will apply if the solenoid is wound round a closed iron ring; if however there is a gap in the iron it requires modification.

Let Fig. 101 represent a section of the solenoid and suppose that $ABDC$ is a gap in the iron, the faces of the iron being planes passing through the axis of the solenoid. Let this axis cut the plane of the paper in O.

Let P be a point on the face of one of the gaps, B the magnetic induction in the iron at right angles to OP, then since the normal magnetic induction is continuous B will also be the magnetic induction in the air. Hence if μ is the magnetic permeability of the iron, the magnetic force in the iron is B/μ while that in the air is B. If

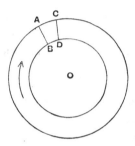

Fig. 101.

$OP = r$, the work done in taking unit pole round a circle whose radius is r is

$$\frac{Br}{\mu}(2\pi - \theta) + Br\theta,$$

where θ is the angle subtended by the air gap at the axis of the solenoid.   Hence by Art. 203 we have

$$rB\left\{\frac{2\pi - \theta}{\mu} + \theta\right\} = 4\pi ni$$

or

$$B = \frac{2n\mu i}{r\left\{1 + \dfrac{\theta}{2\pi}(\mu - 1)\right\}}.$$

This formula shows the great effect produced by even a very small air gap in diminishing the magnetic induction.

Let us take the case of a sample of iron for which $\mu - 1 = 1000$, then if $\theta/2\pi = 1/100$, i.e. if the air is only one per cent. of the whole circuit, the value of B is only one-eleventh of what it would be if the iron circuit were complete, while even though $\theta/2\pi$ were only equal to $1/1000$ the magnetic induction would be reduced one-half by the presence of the gap.

We can explain this by the tendency which the tubes of magnetic induction have to leave air and run through iron. If the magnetic force in the solenoid due to the current circulating round it is in the direction of the arrow, the face AB of the gap will be charged with positive magnetism, the face CD with negative. If this distribution of magnetism existed in air, tubes of magnetic induction starting from AB and running through the air to CD would be pretty uniformly distributed in the field; in this case they would only be in the solenoid for a short part of their course. But as soon as the solenoid is filled with soft iron these tubes forsake the air and run through the iron, and as they are in the opposite direction to the tubes due to the current they diminish the magnetic induction in the iron.

Problems like the one just discussed can be easily solved by making use of the analogy between the distribution of magnetic induction in a field containing magnetic and non-magnetic substances, and the distribution of electric current in a field containing substances of different electrical conductivity. This analogy is shown by the following table, the properties stated on the left-hand side relating to the magnetic field due to a magnetizing circuit traversed by a current i, those on the right relating to the

distribution of current produced by a battery of electromotive force E.

<table>
<tr><td align="center">MAGNETIC SYSTEM.</td><td align="center">CURRENT SYSTEM.</td></tr>
<tr><td>1. The line integral of the magnetic force round any closed curve threading the magnetizing circuit is $4\pi i$, while round any other closed curve it vanishes.</td><td>1. The line integral of the electric force round any closed curve passing through the battery is E, while round any other closed curve it vanishes.</td></tr>
<tr><td>2. The lines of magnetic induction are closed curves threading the magnetizing circuit.</td><td>2. The lines of flow of the current are closed curves passing through the battery.</td></tr>
<tr><td>3. The magnetic induction is μ times the magnetic force, where μ is the magnetic permeability.</td><td>3. The intensity of the current is by Ohm's Law c times the electric force, where c is the specific conductivity of the substance, i.e. the reciprocal of the specific resistance.</td></tr>
<tr><td>4. At the junction of two different media the normal magnetic induction and the tangential magnetic force are continuous.</td><td>4. At the junction of two different media the normal electric current and the tangential electric force are continuous.</td></tr>
</table>

From these results we see that the magnetic induction
due to a magnetizing circuit carrying a current i will be
numerically equal to the current produced by a battery
coinciding with the circuit, if the electromotive force of
the battery is $4\pi i$, and if the specific conductivity of the
medium at any point in the surrounding field is numerically equal to the magnetic permeability at that point.

Since the magnetic permeability of iron is so much
greater than that of air or other non-magnetic substances,
we may, when we use the analogy of the current, regard
the magnetic substances as good conductors, the non-
magnetic substances as very bad ones.

Thus in the case of a magnetizing coil round an iron ring, the current analogue is a battery inserted in a ring of high conductivity, the ring being surrounded by a very bad conductor; in this case practically all the current will go round the ring, very little escaping into the surrounding medium. If E is the E.M.F. of a battery, c the specific conductivity of the ring, l its length, a the radius of its cross section, the resistance of the ring is $l/c\pi a^2$, the current through the ring is $Ec\pi a^2/l$, the average intensity of the current is Ec/l: hence the magnetic induction in an iron ring of length l due to a magnetizing circuit traversed by a current is $4\pi i \mu/l$. Suppose now that there is a gap in this circuit, in the electric analogue this would correspond to cutting the ring, inserting a disc of a bad conductor in the opening, this would evidently greatly reduce the current; if d is the width of the slit, c_1 the specific conductivity of the material with which it is filled, then the resistance of the ring is $\dfrac{l-d}{c\pi a^2} + \dfrac{d}{c_1\pi a^2}$, the current through the ring is equal to

$$\frac{E\pi a^2 c}{l + d\left(\dfrac{c}{c_1} - 1\right)},$$

the average intensity of current is equal to

$$\frac{Ec}{l + d\left(\dfrac{c}{c_1} - 1\right)}.$$

The magnetic induction in the slit iron ring will therefore since the magnetic permeability of air is unity, be

$$\frac{4\pi i\mu}{l + d\left(\mu - 1\right)}.$$

Any problem in the distribution of currents has a magnetic analogue. Thus take the problem of the Wheatstone Bridge (Art. 191), in the magnetic analogue we have six iron bars AB, BC, CA, AD, BD, CD (Fig. 89) with a magnetizing circuit round AB; if l_1, l_2, l_3, l_4 are the lengths, a_1, a_2, a_3, a_4 the areas of the cross sections, and μ_1, μ_2, μ_3, μ_4 the magnetic permeability of AC, CB, AD, BD respectively, we see from the theory of the Wheatstone Bridge that there will be no lines of magnetic induction down CD if

$$\frac{l_1}{\mu_1 a_1}\frac{l_4}{\mu_4 a_4}=\frac{l_2}{\mu_2 a_2}\frac{l_3}{\mu_3 a_3},$$

a result which may be applied to the comparison of the magnetic permeabilities of various samples of iron.

The student will find the use of this analogy between magnetic and current problems of great assistance in dealing with the former, and he will find it profitable to take a number of simple cases of distribution of current and find their magnetic analogues.

211. Ampère's Formula. We saw, Art. 137, that the magnetic force exerted by a magnetic shell of uniform strength ϕ, is that which would be produced if each unit of length at a point P on the boundary of the shell exerted a magnetic force at Q equal to $\phi \sin \theta / PQ^2$, where θ is the angle between PQ and the tangent at P to the boundary of the shell: the direction of the magnetic force at Q is at right angles to both PQ and the tangent to the boundary at P. Since the magnetic force due to the shell is by Ampère's rule the same as that due to a current flowing round the boundary of the shell, the intensity of the current being equal to the strength of the shell, it follows that the magnetic force due to a linear current may be calculated by supposing an element of current of length ds at P to exert at Q a magnetic force equal to $i\,ds \sin \theta / PQ^2$, where i is the strength of the current, and θ the angle between PQ and the direction of the current at P: the direction of the magnetic force being at right angles both to PQ and to the direction of the current at P.

The direction of the magnetic force is related to the direction of the current, like rotation to translation in a right-handed screw working in a fixed nut.

212. Magnetic force due to a circular current. The preceding rule will enable us to find the magnetic force along the axis of a circular current.

Let the plane of the current be at right angles to the plane of the paper. Let the current intersect this plane

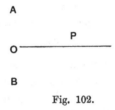

Fig. 102.

in the points A, B, Fig. 102, flowing upwards at A and downwards at B. Let O be the centre of the circle round which the current is flowing, P a point on the axis of the circle. The force at P will by symmetry be along OP. If i is the intensity of the current, then the force at P due to an element ds of the current at A will be at right angles to the current at A, i.e. it will be in the plane of the paper, it will also be at right angles to AP: the magnitude of this force is ids/AP^2, hence the component along OP is equal to

$$ids\,\frac{OA}{AP^3}.$$

By symmetry each unit length of the current will furnish the same contribution to the magnetic force along the axis at P: hence the magnetic force due to the circuit is equal to

$$2\pi i\,\frac{OA^2}{AP^3}.$$

Thus the force varies inversely as the cube of the distance from the circumference of the circle. At the

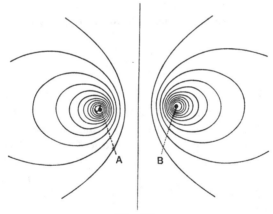

Fig. 103.

centre of the circle $AP = OA$, hence the magnetic force at the centre is equal to

$$\frac{2\pi i}{OA},$$

and thus, if the current remains of the same intensity, varies inversely as the radius of the circle.

The lines of magnetic force round a circular current are shown in Fig. 103. The plane of the current is at right angles to the plane of the paper and the current passes through the points A and B.

213. A case of some practical importance is that of two equal circular circuits conveying equal currents and placed with their axes coincident. Let A, B; C, D be the points in which the currents, which are supposed to flow in planes at right angles to the plane of the paper, cut this plane, the currents flowing upwards at A and C, downwards at B and D: let P be a point on the common axis of the two circuits. The magnetic force at P is, if i is the intensity of the current through either circuit, equal to

$$\frac{2\pi i a^2}{AP^3} + \frac{2\pi i a^2}{CP^3},$$

where a is the radius of the circuits. If $2d$ is the distance between the planes of the circuits, and $x = OP$, where O is the point on the axis midway between the planes of the currents, the magnetic force at P is

$$2\pi i a^2 \left\{ \frac{1}{(a^2 + (d+x)^2)^{\frac{3}{2}}} + \frac{1}{(a^2 + (d-x)^2)^{\frac{3}{2}}} \right\}$$

$$= \frac{4\pi i a^2}{(a^2 + d^2)^{\frac{3}{2}}} \left\{ 1 - \tfrac{3}{2}(a^2 - 4d^2)\frac{x^2}{(a^2 + d^2)^2} \right.$$

$$\left. + \text{ terms in } x^4 \text{ and higher powers of } x \right\}.$$

Thus if $a = 2d$, that is if the distance between the currents is equal to the radius of either circuit, the lowest power of x in the expression for the magnetic force will be the fourth. Thus near O where x is small the magnetic force will be exceedingly uniform.

This disposition of the coils is adopted in Helmholtz's Galvanometer.

214. Mechanical Force acting on an electric current placed in a magnetic field.

The mechanical forces exerted by currents on a magnetic system are equal and opposite to the forces exerted by the magnetic system on the currents. Since the forces exerted by the currents on the magnets are the same as those exerted by Ampère's system of magnetic shells, it follows that the mechanical forces on the currents must be the same as those on the magnetic shells; hence the determination of the mechanical forces on a system of currents can be effected by the principles investigated in Art. 136. Introducing the intensity of the current instead of strength of the magnetic shell we see from that article that the force in any direction acting on a circuit conveying a current i is equal to i times the rate of increase of the number of unit tubes of magnetic induction passing through the circuit, when the circuit is displaced in the direction of the force. In many cases the deduction from this principle given on page 219 is useful, as it shows that the forces on the current are equivalent to a system of forces acting on each element of the circuit. If i is the strength of the current, ds the length of an element at P, B the magnetic induction at P, θ the angle between ds and B, then the force on the element is equal in magnitude to $idsB \sin \theta$, and its direction is at right angles both to ds and B. The relation between the direction of the mechanical force and the directions of the current and the magnetic induction is shown in the accompanying figure, where the magnetic induction is supposed drawn upwards from the plane of the paper.

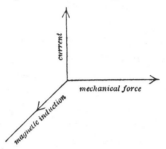

Fig. 104.

215. Couple acting on a plane circuit placed in a uniform magnetic field. Let A be the area of the circuit, i the intensity of the current, ϕ the angle between the normal to the plane of the circuit and the direction of the magnetic induction. The number of unit tubes of magnetic induction due to the uniform field passing through the circuit is $iAB \cos \phi$, where B is the strength of the magnetic induction in the uniform field, and this does not change as the circuit is moved parallel to itself; there are therefore no translatory forces acting on the system. The number of tubes passing through the circuit changes however as the circuit is rotated, and there will therefore be a couple acting on the circuit; the moment of the couple tending to increase ϕ is by the last article equal to the rate of increase with ϕ of the number of unit tubes passing through the circuit, that is to

$$\frac{d}{d\phi} (iAB \cos \phi)$$
$$= - iAB \sin \phi.$$

The couple vanishes with ϕ, and hence the circuit tends to place itself with its normal along the direction of the magnetic induction, and in such a way that the direction of the magnetic induction through the circuit and the direction in which the current flows round it are related like translation and rotation in a right-handed screw working in a fixed nut.

216. Force between two infinitely long straight parallel currents. Let the currents be at right angles to the plane of the paper, intersecting this plane in A and B, let the intensity of the currents be i, i' respectively, and let the currents come from below upwards through the paper. Then, by Art. 204, the magnetic force at B due to the current through A is equal to

$$\frac{2i}{AB},$$

and is at right angles to AB; hence, by Art. 214, the mechanical force per unit length on the current at B is equal to

$$\frac{2ii'}{AB},$$

and since it acts at right angles both to the current and to the magnetic force, it acts along AB. By the rule given in Art. 214, we see that if the currents are in the same direction the force between them is an attraction, if the currents are in opposite directions the force between them is a repulsion. Hence, we see that straight parallel currents attract or repel each other, according as they are flowing in the same or opposite directions, with a force which varies inversely as the distance between them.

217. Mechanical force between two circuits, each circuit consisting of a pair of infinitely long parallel straight conductors. Let the currents be all perpendicular to the plane of the paper and let the currents of the first and second pairs intersect the plane of the paper in A, B and C, D respectively: we shall consider the case when the circuits are placed symmetrically and so that the line EF bisects both AB and CD at right angles. Let the current i flow upwards through

Fig. 105.

the paper at A, downwards at B, the current i' upwards through the paper at C, downwards at D. The force between the circuits will by symmetry be parallel to EF. Between the currents at A and C there is an attraction along CA equal per unit length to

$$\frac{2ii'}{AC};$$

the component of this parallel to EF is

$$\frac{2ii'}{AC^2}EF.$$

Between the currents B and C there is a repulsion along BC equal per unit length to

$$\frac{2ii'}{BC};$$

the component of this parallel to EF is

$$-\frac{2ii'}{BC^2}EF.$$

Hence on each unit length of C there is a force parallel to FE, and equal to

$$2ii'EF\left\{\frac{1}{AC^2}-\frac{1}{BC^2}\right\};$$

there is an equal force acting in this direction on each unit length of D; hence the total force per unit length on the circuit CD is an attraction parallel to EF equal to

$$4ii'EF\left\{\frac{1}{AC^2}-\frac{1}{BC^2}\right\}.$$

If $EF=x$, $AE=a$, $CF=b$, this is equal to

$$4ii'x\left\{\frac{1}{(a-b)^2+x^2}-\frac{1}{(a+b)^2+x^2}\right\},$$

this vanishes when $x=0$ and when x is infinite. Hence there must be some intermediate value of x when the attraction is a maximum. This value of x is easily found to be given by the equation

$$x^2=\tfrac{1}{3}\left\{2\sqrt{a^4+b^4-a^2b^2}-(a^2+b^2)\right\}:$$

when $a-b$ is very small this gives

$$x=a-b,$$

when b/a is very small

$$x=\frac{a}{\sqrt{3}}.$$

218. Force between two coaxial circular circuits.

The solution of the general case requires the use of more analysis than is permissible in this work: there are however two important cases which can be solved by elementary considerations. The first of these is when the radii of the circuits are nearly equal, and the circuits are so close together that the distance between their planes is a very small fraction of the radius of either circuit. In this case the force per unit length of each circuit is approximately the same as that between two infinitely long straight parallel circuits, the distance between the straight circuits being equal to the shortest distance between the circular ones. Thus if i, i' are the currents through the circular circuits, whose radii are respectively a and b, and x is the distance between the planes of the circuits, the attraction between the parallel circuits is at right angles to the planes of the circuits and is approximately equal to

$$\frac{4\pi a i i' x}{(a-b)^2 + x^2}.$$

This is a maximum when $x = a - b$; that is, when the distance between the planes of the circuits is equal to the difference of their radii.

Another case which is easily solved is that of two coaxial circular circuits, the radius of one being small compared with that of the other. Let i be the intensity of the current flowing round the large circuit whose radius is a, i' the current round the small circuit whose radius is b; let x be the distance between the planes of the circuits. Then since b is very small compared with a, the magnetic force due to the large circuit will be approximately uniform over the second circuit and equal to $2\pi i a^2/(a^2 + x^2)^{\frac{3}{2}}$, its value at the centre of that circuit. Thus the number of

unit tubes of magnetic induction due to the first circuit
which pass through the second circuit is equal to

$$\frac{2\pi^2 i a^2 b^2}{(a^2 + x^2)^{\frac{3}{2}}}.$$

Hence by Art. 214 the force on the second circuit
in the direction in which x increases, i.e. the repulsion
between the circuits, is equal to

$$2\pi^2 i i' a^2 b^2 \frac{d}{dx} \frac{1}{(a^2 + x^2)^{\frac{3}{2}}}.$$

Thus the attraction between the circuits is equal to

$$6\pi^2 i i' a^2 b^2 \frac{x}{(a^2 + x^2)^{\frac{5}{2}}}.$$

This is a maximum when $x = a/2$, so that the attraction
between the circuits is greatest when the distance between
their planes is half the radius of the larger circuit.

In the more general case when the radii have any
values, there is, unless the radii are equal, a position in
which the attraction is a maximum. When we use the
attraction between currents as a means of measuring
their intensities, the currents ought to be placed in this
position, for not only is the force to be measured greatest
in this case, but it is also practically independent of any
slight error in the proper adjustment of the distance
between the coils.

219. Coefficient of Self and Mutual Induction.
The coefficient of self-induction of a circuit is defined
to be the number of unit tubes of magnetic induction
which pass through the circuit when it is traversed by
unit current, there being no other current or permanent
magnet in its neighbourhood.

The coefficient of mutual induction of two circuits A and B is defined to be the number of unit tubes of magnetic induction which pass through B when unit current flows round A, there being no current except that through A, or permanent magnet in the neighbourhood of the circuits.

We see from Art. 138 that the coefficient of mutual induction is also equal to the number of unit tubes of induction which pass through A when unit current flows round B.

If the circuit consist of several turns of wire, then in the preceding definitions we must take as the number of tubes of magnetic induction which pass through the circuit, the sum of the number of tubes of magnetic induction which pass through the different turns of the circuit.

We see from the preceding definitions that if we have two circuits A and B, and if the currents i, j flow respectively through these circuits, then the numbers of tubes of magnetic induction which pass through the circuits A and B are respectively,

$$Li + Mj, \text{ and } Mi + Nj,$$

where L and N are the coefficients of self-induction of the circuits A and B respectively, and M is the coefficient of mutual induction between the circuits. The results given in the preceding articles enable us to calculate the coefficient of self-induction in some simple cases.

In the case of the long straight solenoid discussed in Art. 210, when unit current flows through the wire the magnetic force in the solenoid is $4\pi n$, where n is the number of turns per unit length; hence if A is the area of the core of the solenoid, and if the core is filled with

air, the number of unit tubes of magnetic induction
passing through each turn of wire is equal to $4\pi nA$, and
since there are n turns per unit length, the coefficient of
self-induction of a length l of the solenoid is equal to

$$4\pi n^2 l A.$$

If the core were filled with soft iron of permeability μ,
then the number of unit tubes of magnetic induction
which pass through each turn of wire is $4\pi n\mu A$, and the
coefficient of self-induction of a length l is $4\pi n^2 l\mu A$.

If the iron instead of completely filling the core only
partially fills it, then if B is the area of the core occupied
by the iron, the coefficient of self-induction of a length
l is $4\pi n^2 l \{\mu B + A - B\}$.

Consider now the coefficient of mutual induction of
two solenoids α and β with parallel axes. The coefficient
of mutual induction will vanish unless one of the solenoids
is inside the other, for the magnetic force due to a current
through a solenoid vanishes outside the solenoid. Hence
when a current flows through α no lines of induction will
pass through β unless β is either inside α or completely
surrounds it.

Let β be inside α. Let B be the area of the solenoid β,
and let m be the number of turns of wire per unit length.
Then if unit current flows through α, the magnetic force
inside is $4\pi n$, where n is the number of turns per unit
length. Hence if there is no iron inside the solenoids, the
number of tubes of magnetic induction passing through
each turn of β is $4\pi nB$, and since there are m turns
per unit length, the coefficient of mutual induction of
a length l of the two solenoids is $4\pi nmlB$.

We see, by Art. 218, that the coefficient of mutual

induction between a large circle of radius a and a small one of radius b, with their planes parallel and the line joining their centres at right angles to their planes, is equal to

$$\frac{2\pi^2 a^2 b^2}{(a^2 + x^2)^{\frac{3}{2}}},$$

where x is the distance between the planes.

If we have two circuits α, β, each consisting of two infinitely long, parallel straight conductors, the current flowing up one of these and down the other, then by Art. 208, the coefficient of mutual induction between α and β is, per unit length, equal to

$$2 \log \frac{AC \cdot BD}{AD \cdot BC},$$

where A, B, C, D are respectively the points where the wires of the circuits α and β intersect a plane at right angles to their common direction. The current through the conductor intersecting this plane in A is in the same direction as that through the conductor passing through C.

220. We can express the energy in the magnetic field due to a system of currents very easily in terms of the currents and the coefficients of self and mutual induction of the circuits. We proved, Art. 163, that the energy per unit length in a unit tube of *induction* at P is equal to $R/8\pi$, where R is the magnetic *force* at P. The tube of induction is a closed curve, and the total amount of energy in this tube is equal to

$$\frac{1}{8\pi} \Sigma Rds,$$

where ds is an element of length of the tube and ΣRds denotes the sum of all the products Rds for the tube.

But ΣRds is the work done on unit pole when it is taken round the closed curve formed by the tube of induction, and this by Art. 203 is equal to 4π times the sum of the currents encircled by the curve. Hence the energy in a tube of induction is equal to

$\frac{1}{2}$ (the sum of the currents encircled by the tube).

Hence the whole energy in the magnetic field is equal to half the sum of the products obtained by multiplying the current in each circuit by the number of tubes of magnetic induction passing through that circuit.

Thus if we have two circuits A and B, and if i, j are the currents through A and B respectively, L, N the coefficients of self-induction of A and B, M the coefficient of mutual induction between these circuits, then the numbers of tubes of magnetic induction passing through A and B respectively are

$$Li + Mj,$$

and $$Mi + Nj.$$

Hence the energy in the magnetic field around this circuit is

$$\tfrac{1}{2}i(Li + Mj) + \tfrac{1}{2}j(Mi + Nj)$$
$$= \tfrac{1}{2}Li^2 + Mij + \tfrac{1}{2}Nj^2.$$

If we have only one circuit carrying a current i, then if L is its coefficient of self-induction, the energy in the magnetic field is

$$\tfrac{1}{2}Li^2.$$

Thus the coefficient of self-induction is equal to twice the energy in the magnetic field due to unit current.

We may use this as the definition of coefficient of self-induction, and this definition has a wider application than the previous one. The definition in Art. 219 is only applicable when the currents flow through very fine wires, the present one however is applicable when the current is distributed over a conductor with a finite cross section. Thus let us consider the case where we have a current flowing through an infinitely long cylinder whose radius is OA, the direction of flow being parallel to the axis of the cylinder, and where the return current flows down a thin tube, whose radius is OB, coaxial with this cylinder.

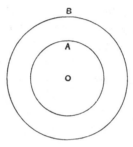

Fig. 106.

Let i be the current which flows up through the cylinder and down through the tube, let us suppose that the current through the cylinder is uniformly distributed over its cross section. The magnetic force will vanish outside the tube, for since as much current flows up through the cylinder as down through the tube, the total current flowing through any curve enclosing them both vanishes, and therefore the work done in taking unit pole round a circle with centre O and radius greater than that of the tube will vanish. Since the magnetic force due to the currents must by symmetry be tangential to this circle and have the same value at each point on its circumference, it follows that the magnetic force vanishes outside the tube.   We can prove as in Art. 204 that at a point P between the cylinder and the tube the magnetic force is equal to

$$\frac{2i}{r},$$

where $r = OP$.

At a point P inside the cylinder the magnetic force is

$$\frac{2ir}{a^2},$$

where $a = OA$, the radius of the cylinder.

By Art. 163 the energy per unit volume is equal to $\mu H^2/8\pi$, where H is the magnetic force; hence if μ is the magnetic permeability of the cylinder, the magnetic energy between two planes at right angles to the axis of the cylinder and at unit distance apart is equal to

$$\frac{4i^2}{8\pi} \int_{OA}^{OB} \frac{2\pi r dr}{r^2} + \frac{4i^2\mu}{8\pi} \int_{0}^{OA} \frac{r^2}{a^4} 2\pi r dr$$

$$= i^2 \log \frac{OB}{OA} + \frac{i^2}{4} \mu.$$

Hence, since the coefficient of self-induction per unit length is twice the energy when the current is unity, it is equal to

$$2 \log \frac{OB}{OA} + \tfrac{1}{2}\mu.$$

In this case the coefficient of self-induction will be very much greater when the cylinder is made of iron than when it is made of a non-magnetic metal like copper.   For take the case when $OB = e.OA$, where $e = 2\cdot718$, the base of the Napierian logarithms; then the self-induction for copper, for which μ is equal to unity, is equal to $2\cdot5$ per unit length, but if the cylinder is made of a sample of iron whose magnetic permeability is 1000, the coefficient of self-induction per unit length is 502. Thus in this case the material of the conductor through which the current flows produces an enormous effect, much greater than it does in the case of the solenoids.

The self-induction depends upon the way in which the current is distributed in the cylinder; thus if the current instead of spreading uniformly across the section of the cylinder were concentrated on the surface, the magnetic force inside the cylinder would vanish, while that in the space between the tube and the cylinder would be the same as before, hence the energy would now be

$$\frac{4i^2}{8\pi} \int_{OA}^{OB} \frac{2\pi r dr}{r^2} = i^2 \log \frac{OB}{OA},$$

so that the coefficient of self-induction would now be $2 \log (OB/OA)$, thus it would be less than before and independent of the material of which the cylinder is made.

221. Rational Current Element. In Ampère's expression for the magnetic force due to a current, the current is supposed to be divided up into elements, an element ds giving rise to a magnetic force equal to $ids \sin \theta/r^2$. Each of those elements when regarded as a separate unit corresponds to an unclosed electric current, whereas on the Modern Theory of Electricity such currents do not exist. Thus the mathematical unit does not correspond to a physical reality. To obviate this inconvenience Mr Heaviside has proposed another interpretation of the element of current; he points out that the magnetic force $ids \sin \theta/r^2$ is that due to a system of *closed* currents distributed through space like the lines of magnetic *induction* due to a small magnet, PQ, PQ being the element of current ds, and i representing the number of lines of magnetic induction running through PQ, i.e. passing through each cross section of the magnet ; the current at any point in the field round the element of current is represented in magnitude and direction by the *magnetic induction* at that point due to the little magnet. The reader will have no difficulty in proving this result, if he applies the principle that the work done in taking the unit magnetic pole round any closed circuit is equal to 4π times the current passing through the circuit. The element, PQ, *with its associated system of currents*, Mr Heaviside calls the *rational current element*, it has the advantage of corresponding to a possible physical system. It is important to notice that this view of the element of current gives us for closed circuits the same result as the old one, i.e. the closed current is entirely confined to the closed circuit and does not spread out at all into the surrounding space ; for let PQ, RS be two elements, then if we place these together so that the end Q of one coincides with the beginning, R, of the other, then the analogy with the lines of magnetic induction shows that the currents which when PQ was alone in the field diverged from Q now run through QS and diverge from S, hence if we put a number of such elements together so as to form a closed circuit the current will never leave the circuit.

We shall see that the magnetic force produced at P when a charged particle O moves with a velocity v, is $ev \sin \theta / OP^2$, where e is the charge on the particle and θ the angle between v and OP ; the direction of the force is at right angles to the plane containing v and OP. Thus another interpretation of the element ds of a circuit is, that it is a place where n charged particles are moving in the direction of the element with the velocity v; n, e, v and i being connected by the relation $nev = ids$.

MEASUREMENT OF CURRENT AND RESISTANCE.

Galvanometers.

222. The magnetic force produced by a current may be used to measure the intensity of the current. This is most frequently done by means of the tangent galvanometer, which consists of a circular coil of wire placed with its plane in the magnetic meridian. If the magnetic field is not wholly due to the earth, the plane of the coil must contain the resultant magnetic force. At the centre of the coil there is a magnet which can turn freely about a vertical axis. When the magnet is in equilibrium its axis will lie along the horizontal component of the magnetic force at the centre of the coil, thus when no current is flowing through the coil the axis of the magnet will be in the plane of the coil. A current flowing through the coil will produce a magnetic force at right angles to the plane of the coil, proportional to the intensity of the current. Let this magnetic force be equal to Gi where i is the intensity of the current flowing through the coil and G a quantity depending upon the dimensions of the coil. G is called the 'Galvanometer constant.' Let H be the horizontal component of the magnetic force at the centre of the coil. Then the resultant magnetic force at the centre of the coil has a component H in the plane of the coil and a component Gi at right angles to it, hence if θ is the angle which the resultant magnetic force makes with the plane of the coil,

$$\tan \theta = \frac{Gi}{H} \quad \dots\dots\dots\dots\dots\dots(1).$$

When the magnet is in equilibrium its axis will lie along the direction of the resultant magnetic force, hence the passage of the current will deflect the magnet through an angle θ given by equation (1). As the current is proportional to the tangent of the angle of deflection, this instrument is called the *tangent* Galvanometer.

The smaller we can make H, the external magnetic force at the centre of the coil, the larger will be the angle through which a given current will deflect the magnet. By placing permanent magnets in suitable positions in the neighbourhood of the coil we can partly neutralize the earth's magnetic field at the centre of the coil: in this way we can reduce H and increase the sensitiveness of the galvanometer. A magnet for this purpose is shown in Fig. 107, which represents an ordinary type of galvanometer.

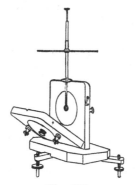

Fig. 107.

Another method of increasing the sensitiveness of the instrument is employed in the 'astatic galvanometer.' In this galvanometer (Fig. 108) we have two coils A and B in series, so arranged that the current circulates round

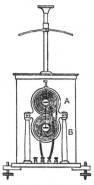

Fig. 108.

them in opposite directions. Thus, if the magnetic force at the centre of the upper coil is upwards from the plane of the paper, that at the centre of the lower coil will be downwards. Two magnets α, β, mounted on a common axis, are placed at the centres of the coils A and B respectively, the axes of magnetization of these magnets point in opposite directions; thus as the magnetic forces at the centres of the two coils due to the currents are also in opposite directions, the couples due to the currents acting on the two magnets will be in the same direction. The couples arising from the external magnetic field will however be in opposite directions: if the external magnetic field is uniform and the moments of the two magnets very nearly equal, the couple tending to restore the magnet to its position of equilibrium will be very small, and the galvanometer will be very sensitive.

The larger we make G the greater will be the sensitiveness of the galvanometer. If the galvanometer consists of a single circle of radius a, then (see Art. 212) $G = 2\pi/a$. If there are n turns close together and arranged so that the distance between any two turns is a very small fraction of the radius of the turns, then G is approximately $2\pi n/a$. If the galvanometer consists of a circular coil of rectangular cross section, the sides of the rectangle being in and at right angles to the plane of the coil, and if $2b$ is the breadth of this rectangle (measured at right angles to the plane of the coil), $2a$ the depth in the plane of the coil, n the number of turns of wire passing through unit area, then taking as axis of x the line through the centre of the coil at right angles to its plane, and as axis of y a line through the centre at right angles to this, we have

$$G = 2\pi n \int_{-b}^{b} \int_{c-a}^{c+a} \frac{y^2 dx dy}{(x^2 + y^2)^{\frac{3}{2}}},$$

where c is the mean radius of the coil.

If 2θ, 2ϕ are the angles subtended at the centre by AB, CD, Fig. 109, this reduces to

$$G = 4\pi n b \log \frac{\cot \dfrac{\theta}{2}}{\cot \dfrac{\phi}{2}}.$$

In sensitive galvanometers the hole in the centre for the magnet is made as small as possible, so that the inner windings have very small radii; when this is the case, we may put $\phi = \dfrac{\pi}{2}$, and then

$$G = 4\pi n b \log \cot \frac{\theta}{2}.$$

In this case when the area of the cross section of the coil is given, i.e. when $2b^2 \cot \theta$ is given, we can prove that G is a maximum when

$$\log \cot \frac{\theta}{2} = 2 \cos \theta,$$

the solution of this equation is $\theta = 16° 46'$: this makes the breadth bear to the depth the ratio of 1 to 1·61.

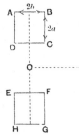

Fig. 109.

The sensitiveness of modern galvanometers is very great, some of them will detect a current of 10^{-13} ampères. It would take a current of this magnitude centuries to liberate 1 c.c. of hydrogen by electrolysis.

Since

$$i = \frac{H}{G} \tan \theta,$$

$$\frac{\delta \theta}{\delta i} = \frac{G}{H} \cos^2 \theta,$$

while

$$\frac{\delta \theta}{(\delta i / i)} = \sin \theta \cos \theta.$$

Thus for a given absolute increment of i, $\delta \theta$ will be greatest when θ is zero, and for a given relative increment,

$\delta\theta$, or the change in deflection, will be greatest when $\theta = 45°$.

In some cases it is important to have the magnetic field near the magnet as uniform as possible. This can be attained (see Art. 213) by using two equal coils placed parallel to one another and at right angles to the line joining their centres, the distance between the coils being equal to the radius of either. The magnet is then placed on the common axis of the two coils and midway between them.

223. Sine Galvanometer. In this galvanometer, Fig. 110, the coil itself can move about a vertical axis, its

Fig. 110.

position being determined by means of a graduated horizontal circle. In using the instrument the coil is placed so that when no current goes through it the magnetic axis of the magnet at its centre is in the plain of the coil. When a current passes through the coil, the magnet is deflected out of this plane, and the coil is now moved

round until the axis of the magnet is again in the plane
of the coil. When this is the case the components of
the magnetic force at right angles to the plane of the
coil due respectively to the current and to the external
magnetic field must be equal and opposite. If H is
the external magnetic force, ϕ the angle through which
the coil has been twisted when the axis of the magnet
is again in the plane of the coil, the external force at right
angles to the plane of the coil is $H \sin \phi$. If i is the
current through the coil, G the magnetic force at its
centre when the wires of the coil are traversed by unit
current, then the magnetic force at right angles to the
coil due to the current is Gi; hence when this is in equilibrium with the component due to the external field,

$$H \sin \phi = Gi,$$

or
$$i = \frac{H}{G} \sin \phi.$$

The advantage of this form of galvanometer is that the
magnet is always in the same position with respect to the
coil. For the same coils and magnetic field the deflection
is greater for the sine than for the tangent galvanometer.

224. Desprez-d'Arsonval Galvanometer. In this
galvanometer the coil carrying the current moves while
the magnets are fixed. The galvanometer is represented
in Fig. 111. A rectangular coil is suspended by very fine
metal wires which also serve to convey the current to the
coil. The coil moves between the poles of a horse-shoe
magnet, and the magnetic field is concentrated on the coil
by a fixed soft iron cylinder placed inside the coil. When
a current flows round the coil, the coil tends to place itself

so as to include as many tubes of magnetic induction
as possible (Art. 215).   It therefore tends to place itself so
that its plane is at right angles to the lines of magnetic
induction.   The motion of the coil is resisted by the
torsion of the wire which suspends it, and the coil takes a
position in which the couple due to the torsion of the wire
just balances that due to the magnetic field.   When the
magnetic field is uniform the relation between the de-

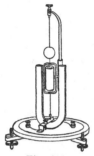

Fig. 111.

flection and the current is as follows.   Let A be the area
of the coil, n the number of turns of wire, i the current
through the wire, B the magnetic induction at the coil.
When the plane of the coil makes an angle ϕ with the
direction of magnetic induction the number of tubes of
magnetic induction passing through it is

$$BAn \sin \phi,$$

hence, by Art. 215, the couple tending to twist the coil is

$$iBAn \cos \phi.$$

If the torsional couple vanishes when ϕ is zero, the
couple when the coil is twisted through an angle ϕ

will be proportional to ϕ; let it equal $\tau\phi$, then when there is equilibrium, we have

$$iBAn \cos\phi = \tau\phi,$$

or

$$i = \frac{\tau\phi}{BAn \cos\phi};$$

if ϕ is small this equation becomes approximately

$$i = \frac{\tau\phi}{BAn}.$$

225. Ballistic Galvanometer. A galvanometer may be used to measure the total quantity of electricity passing through its coil, provided the electricity passes so quickly that the magnet of the galvanometer has not time to appreciably change its position while the electricity is passing. Let us suppose that when no current is passing the axis of the magnet is in the plane of the coil, then if i is the current passing through the plane of the coil, G the galvanometer constant, i.e. the magnetic force at the centre of the coil when unit current passes through it, m the moment of the magnet, the couple on the magnet while the current is passing is

$$Gim.$$

If the current passes so quickly that the magnet has not time sensibly to depart from the magnetic meridian while the current is flowing, the earth's magnetic force will exert no couple on the magnet. Thus if K is the moment of inertia of the magnet, θ the angle the axis of the magnet makes with the magnetic meridian, the equation of motion of the magnet during the flow of the current is

$$K\frac{d^2\theta}{dt^2} = Gim,$$

thus if the magnet starts from rest the angular velocity after a time t is given by the equation

$$K \frac{d\theta}{dt} = Gm \int_0 i\,dt.$$

If the total quantity of electricity which passes through the galvanometer is Q and the angular velocity communicated to the magnet ω, we have therefore

$$K\omega = GmQ.$$

This angular velocity makes the magnet swing out of the plane of the coil : if H is the external magnetic force at the centre of the coil, the equation of motion of the magnet is, if there is no retarding force,

$$K \frac{d^2\theta}{dt^2} + mH \sin\theta = 0.$$

Integrating this equation we get

$$K \left\{ \left(\frac{d\theta}{dt}\right)^2 - \omega^2 \right\} + 2mH (1 - \cos\theta) = 0.$$

If ϑ is the angular swing of the magnet, the angular velocity vanishes when $\theta = \vartheta$, hence

$$K\omega^2 = 2mH (1 - \cos\vartheta) = 4mH \sin^2 \frac{\vartheta}{2}.$$

On substituting for ω the value previously found we get

$$Q = 2 \sin \tfrac{1}{2}\vartheta \, \frac{1}{Gm} \sqrt{mH \cdot K}.$$

If T is the time of a small oscillation of the magnet,

$$T = 2\pi \sqrt{\frac{K}{mH}},$$

hence

$$Q = \sin \tfrac{1}{2}\vartheta \cdot \frac{TH}{\pi G}.$$

We have neglected any retarding force such as would arise from the resistance of the air. Galvanometers which are used for the purpose of measuring quantities of electricity are called 'ballistic galvanometers,' and are constructed so as to make the effects of the frictional forces as small as possible. This is done either by making the moment of inertia of the magnet very large, or by making the magnet so symmetrical about its axis of rotation that the frictional forces are but small. The correction to be applied when the frictional forces are not negligible is investigated in Maxwell's *Electricity and Magnetism,* Vol. II. p. 386.

226. Measurement of Resistance. The arrangement of conductors in the Wheatstone's Bridge (Art. 191)

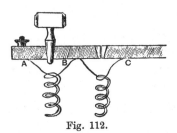

Fig. 112.

enables us to determine the resistance of one arm of the bridge, say *BD*, Fig. 89, in terms of the resistances of the arms *AC, CB* and *AD*. For the measurement of resistances by this method wires having a known resistance are used. These are called resistance coils, and are made in the following way. A piece of silk-covered German-silver wire is taken and doubled back on itself (to avoid effects due to electromagnetic induction, see Chap. XI.) and then wound in a coil. Its length is then carefully adjusted until its resistance is some multiple of the standard resistance, the ohm. Each end of this coil is soldered to a stout piece of brass such as A, B, or C, Fig. 112; these pieces are attached to an ebonite board to insulate them from each other. Two adjacent pieces of brass can be put in electrical connection by inserting stout well-fitting brass plugs between them. When the plug is out the resistance between B and C is that of the wire, while when the plug is in there is practically no resistance between these places.

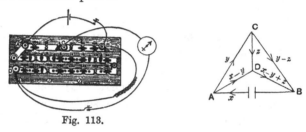

Fig. 113.

When there is no current through the arm CD of the Wheatstone's Bridge there is, by Art. 191, a certain relation between four resistances: hence to measure a resistance by this method we require three known resistances. These resistances are conveniently arranged in what is known as the Post-Office Resistance Box. This is a box of coils arranged as in Fig. 113, and provided with screws at A, B, C, D, to which wires can be attached. To determine the resistance of a conductor such as R connect one end to B and the other end to D; connect one terminal of a galvanometer to C and the other to D, and one electrode of a battery to A, the other to B. The arrangement of the conductors is the same as that in the diagram in Art. 191, which is reproduced here by the

side for convenience. To measure the resistance of R:
take one or more plugs out of CA and CB and then proceed to take plugs out of AD until there is no deflection
of the galvanometer, when the battery circuit is completed.
As the current through CD vanishes, we must have by
Art. 191

resistance of BD × resistance of AC

$$= \text{resistance of } BC \times \text{resistance of } AD.$$

As the resistances of AC, BC, AD are known, that of BD
is determined by this equation.

227. Resistance of a Galvanometer. A method
due to Lord Kelvin for measuring the resistance of a
galvanometer is an interesting example of the property
of conjugate conductors. We saw (Art. 192) that if CD
is conjugate to AB, then the current sent through any
arm of the bridge by a battery in AB is independent of
the resistance in CD, and the converse is also true. To
apply this to measure the resistance of a galvanometer,
place the galvanometer in the arm BD of the bridge and
replace the galvanometer in CD by a key by means of
which the circuit CD can be completed or broken at
pleasure. Then adjust the resistance of AD until the
deflection of the galvanometer is the same when the
circuit CD is completed as when it is broken. As in
this case the current through BD is independent of the
resistance of CD, CD must be conjugate to AB, and we
have therefore (Art. 191),

resistance of galvanometer × resistance of AC

$$= \text{resistance of } BC \times \text{resistance of } AD.$$

CHAPTER XI

ELECTROMAGNETIC INDUCTION

228. Electromagnetic Induction, of which the laws were unravelled by Faraday, may be illustrated by the following experiment. Two circuits A and B, Fig. 114, are placed near together, but completely insulated from each

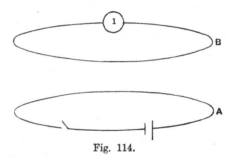

Fig. 114.

other; a galvanometer is in the circuit B, and a battery and key in A. Suppose the circuit A at the beginning of the experiment to be interrupted, press down the key and close the circuit, the galvanometer in B will be deflected, indicating the passage of a current through B, although B is completely insulated from the battery. The deflection of the galvanometer is not a permanent one, but is of the same kind as that of a ballistic galvanometer when a finite

25—2

quantity of electricity is quickly discharged through it, that is, the magnet of the galvanometer is set swinging, but is not permanently deflected, as it oscillates symmetrically about its old position of equilibrium. This indicates that an electromotive force, acting for a very short time, has acted round B. The direction of the deflection of the magnet of the galvanometer in B indicates that the direction of the momentary current induced in B was opposite to that started in A. After a time the motion of the magnet subsides and the magnet remains at rest, although the current continues to flow through A. If, after the magnet has come to rest, we raise the key in A, so as to stop the current flowing through the circuit, the galvanometer in B is again affected, the direction of the first swing in this case being opposite to that which occurred when the current in A was started, indicating that when the current in A is stopped, an electromotive force is produced round B tending to start a current through B in the same direction as that which previously existed in A. This electromotive force, like the one produced when the circuit A was completed, is but momentary.

These experiments show that the starting or the stopping of a current in a circuit A is accompanied by the production of another current in a neighbouring circuit B, the current in B being in the opposite direction to that in A when the current is started and in the same direction when the current is stopped.

If instead of making or breaking the current in A, this current is kept steadily flowing in the circuit, while the circuit itself is moved about, then when A is moving away from B an electromotive force is produced tending to send round B a current in the same direction as that round A, while if A is moved towards B an electromotive force acts round B tending to produce a current in the opposite direction to that round A. These electromotive forces in B only occur when A is moving, they stop as soon as it is brought to rest. If we replace the circuit A, with the current flowing through it, by its equivalent magnet, then we shall find that the motion of the magnet will induce the same currents in B as the motion of the circuit A. If we keep the circuit A, or the magnet, fixed and move B, we also get currents produced in B.

The currents started in B by the alteration in intensity or position of the current in A, or by the alteration of the position of B with respect to magnets in its neighbourhood, are called induced currents ; and the phenomenon is called electromagnetic induction.

A good deal of light is thrown on these phenomena if we interpret them in terms of the tubes of magnetic induction. Let us first take the case when the induction is produced by starting a current in A. Then before the current circulates through A no tubes of magnetic induction pass through B; when the current is started through A this circuit is at once threaded by a number of tubes of magnetic induction, some of which pass through B. The induced current through B also causes B to be threaded by tubes of magnetic induction, which since the induced current is in the opposite direction to the primary one in A, pass through the circuit in the opposite direction to those sent through it by the current in A ; thus the effect of the induced current in B is to tend to make the total number of tubes of magnetic induction passing through B zero; that is, to keep the total number of tubes of magnetic induction through B the same as it was before the current was started in A. We shall find, when we investigate the laws of induction more closely, that the tubes of magnetic induction passing through B, due to the induced current, are at the moment of making the primary circuit equal in number and opposite in direction to those sent through B by the current in A. The laws of the induction of currents may thus be expressed by saying that the number of tubes of magnetic induction passing through B does not change abruptly.

Again, take the case when currents are induced in B by stopping the current in A. Initially the current flowing through A sends a number of tubes of magnetic induction through B: when the current in A is stopped these tubes cease, but the current induced in B in the same direction as that in A causes a number of tubes of magnetic induction to pass through B in the same direction as those due to the original current in A. Thus the action of the induced current is again to tend to keep the number of tubes of magnetic induction passing through B constant.

The same tendency to keep the number of tubes of magnetic induction through B constant is shown by the induction of a current in B when A is moved away from or towards B. When A is moved away from B, the number of tubes of magnetic induction due to A which pass through B is diminished, but there is a current induced in B in the same direction as that through A, which causes additional tubes of magnetic induction to pass through B in the same direction as those due to A: the production of these tubes counterbalances the diminution due to the recession of A, and thus the induced current again tends to keep the number of tubes of magnetic induction passing through B constant. The same thing occurs when A is moved towards B, or when currents are induced in B by the motion of permanent magnets in its neighbourhood.

Not only is there a tendency to keep the number of tubes of magnetic induction passing through any circuit in the neighbourhood of A constant, there is also the same tendency with respect to the circuit A itself. Let us suppose that A is alone in the field, then, when a current is flowing round A, tubes of magnetic induction pass through it. If the circuit is broken, and the current stopped, the number of tubes would fall to zero; the tendency, however, to preserve unaltered the number of tubes passing through the circuit, will under suitable circumstances, cause the current, in its effort to continue flowing in the same direction, to spark across an air-gap when the circuit is broken, even though the original E.M.F., applied to send the current through A, was totally

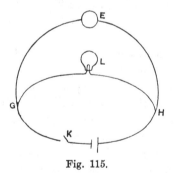

Fig. 115.

inadequate to produce a spark. To show this effect experimentally it is desirable to wind the coil A round a core of soft iron, so as, with a given current, to increase the number of tubes of magnetic induction passing through the circuit; the coil of an electro-magnet shows this effect very well. The effect of this tendency is shown very clearly in the following experiment. The coil of an electro-magnet E, Fig. 115, is placed in parallel with an electric lamp L, the resistance of the lamp being very large compared with that of an electro-magnet; in consequence of this, when the two are connected up to a battery, by far the greater part of the current will flow through the coil, comparatively little through the lamp, too little indeed to raise the lamp to incandescence. If however the circuit is broken at K, the tendency to keep the number of tubes of magnetic induction passing through the circuit constant, will send a current momentarily round the circuit $HLGE$, which will be larger than that flowing through the lamp when the battery is kept continuously connected up to the circuit; and thus though the lamp remains quite dark when the current is steady, it can be raised to bright incandescence by repeatedly making and breaking the circuit.

229. The electromotive force round a circuit due to induction does not depend upon the metal of which the circuit is made. This may be proved by taking two equal circuits of different metals, iron and copper, say, placed close together and arranged so that the electro-motive forces due to induction in the two circuits tend to oppose each other. When this circuit, connected up to a galvanometer, is placed in a varying magnetic field, no current passes, showing that the electromotive forces in the two circuits are equal and opposite.

Faraday proved that in a magnetic field varying at

an assigned rate, the electromotive force round a circuit
due to induction is proportional to the number of tubes of
magnetic induction passing through the circuit, by taking
a coil made of several turns of very fine wire, and inserting in it a galvanometer whose resistance was small
compared with that of the coil: when this coil was placed
in a varying field the deflection of the galvanometer was
found to be independent of the number of turns in the
coil.   As all the resistance in the circuit is practically in
the coil, the resistance of the circuit will be proportional to
the number of turns in the coil.   Since the quantity of
electricity passing through the circuit is independent of
the number of turns, it follows that the E.M.F. round the
circuit must have been proportional to the resistance, i.e.
to the number of turns of the coil.   Hence, since the turns
of the coils were so close together that each enclosed the
same number of tubes of magnetic induction, it follows
that when the rate of change is given the E.M.F. round
the circuit must be proportional to the number of tubes
of magnetic induction passing through it.

Faraday also showed by rotating the same circuit
at different speeds in the same magnetic field that the
E.M.F. round the circuit is proportional to the speed
of rotation, i.e. to the rate of change of the number of
tubes of magnetic induction passing through the circuit.

These investigations of Faraday's determined the
conditions under which induced currents are produced:
F. E. Neumann was however the first to give, in 1845, an
expression by which the *magnitude* of the electromotive
force could be determined.   We may state the law of
induction of currents as follows—*Whenever the number of*
tubes of magnetic induction passing through a circuit is

changing, there is an E.M.F. *acting round the circuit equal to the rate of diminution in the number of tubes of magnetic induction which pass through the circuit.* The positive direction of the E.M.F. and the positive direction in which a tube passes through the circuit are related to each other like rotation and translation in a right-handed screw.

We shall show later on (page 472) that this law can be connected with Ampère's law (Art. 214) by dynamical principles.

Let us apply this law of induction to the case of a circuit exposed to a variable magnetic field. Let the circuit contain a galvanic battery whose electromotive force is E_0, and let the resistance of the circuit, including that of the battery, be R. If P is the number of tubes of magnetic induction passing at any time t through the circuit, there will be an E.M.F. equal to $-dP/dt$ round the circuit due to induction; hence by Ohm's law, we have if i is the current round the circuit,

$$Ri = E_0 - \frac{dP}{dt},$$

or
$$\frac{dP}{dt} + Ri = E_0 \dots\dots\dots\dots\dots(1).$$

Suppose the magnetic field is due to two currents, one circulating round this circuit and the other through a second circuit in its neighbourhood; let j be the current passing round the second circuit. Let L be the coefficient of self-induction of the first circuit, N that of the second, M the coefficient of mutual induction between the two circuits. Then as the magnetic field is due to the two circuits,

$$P = Li + Mj,$$

and equation (1) becomes

$$\frac{d}{dt}(Li + Mj) + Ri = E_0.$$

If S is the resistance of the second circuit and E_0' the electromotive force of any battery there may be in that circuit, then we have similarly,

$$\frac{d}{dt}(Mi + Nj) + Sj = E_0'.$$

230. Let us compare these equations with the equations of motion of a dynamical system having two degrees of freedom, one degree being fixed by the coordinate x, the other by the coordinate y; these coordinates may be regarded as fixing the positions of two moving pieces. Let the first moving piece be acted upon by the external force E_0, the second by the force E_0'. Let the motions of the first and second moving pieces be resisted by resistances proportional to their velocities, and let $R\dot{x}$, $S\dot{y}$ be these resistances respectively. The momenta corresponding to the two moving pieces will be linear functions of the velocities. Let the momentum of the first moving piece be

$$L\dot{x} + M\dot{y},$$

that of the second

$$M\dot{x} + N\dot{y}.$$

Then, if L, M, N are independent of the coordinates x, y, the equations of motion of the two systems will be

$$\frac{d}{dt}(L\dot{x} + M\dot{y}) + R\dot{x} = E_0,$$

$$\frac{d}{dt}(M\dot{x} + N\dot{y}) + S\dot{y} = E_0'.$$

Comparing these equations with those for the two currents we see that they are identical if we make i, j the currents round the two circuits coincide with $\dot{x}, \dot{y}$ the velocities of the two moving pieces. The electrical equations of a system of circuits are thus identical with the dynamical equations of a system of moving bodies, the current flowing round a circuit corresponds to a velocity, the number of tubes of magnetic induction passing through the circuit to the momentum corresponding to that velocity, the electrical resistance corresponds to a viscous resistance, and the electromotive force to a mechanical force.

A further analogy is afforded by the comparison of the Kinetic Energy of the Mechanical System with the energy in the magnetic field due to the system of currents. The Kinetic Energy of the Mechanical System is equal to

$$\tfrac{1}{2}\dot{x}\left(L\dot{x} + M\dot{y}\right) + \tfrac{1}{2}\dot{y}\left(M\dot{x} + N\dot{y}\right).$$

The energy in the magnetic field is by Art. 220 equal to

$$\tfrac{1}{2}i\left(Li + Mj\right) + \tfrac{1}{2}j\left(Mi + Nj\right).$$

This expression becomes identical with the preceding one if we write $\dot{x}$ for i and $\dot{y}$ for j.

Since the terms in the electrical equations which express the induction of currents correspond to terms in the dynamical equations which express the effects of changes in the momentum, and as these latter effects arise from the inertia of the system, we are thus led to regard a system of electrical currents as also possessing inertia. The inertia of the system will be increased by any circumstance which, for given values of the currents, increases the number of tubes of electromagnetic induction passing through the circuits; the inertia of the system may thus be increased by the introduction of soft iron in the neighbourhood of the circuits.

231. We can illustrate by a mechanical model the analogies between the behaviour of electrical circuits and a suitable mechanical system. Models of this kind have been designed by Maxwell and Lord Rayleigh; a simple one which serves the same purpose is represented in Fig. 116.

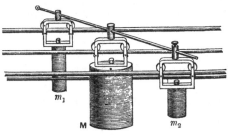

Fig. 116.

It consists of three smooth parallel horizontal steel bars on which masses m_1, M, m_2 slide, the masses being separated from the bars by friction wheels: the three masses are connected together by a light rigid bar, which passes through holes in swivels fixed on to the upper part of the masses; the bar can slide backwards and forwards through these holes, so that the only constraint imposed by the bar is to keep the masses in a straight line.

This system will, if we regard the velocities of m_1, m_2 respectively along their bars as representing currents flowing round two circuits, illustrate the induction of currents. Let us start with the three masses at rest, then suddenly move m_1 forward along its bar, m_2 will then

move backwards, an effect analogous to the production
of the inverse current in the secondary when the current
is started in the primary. If now m_1 is moved uniformly
forward the friction between m_2 and its bar will soon bring
it to rest and it will continue at rest as long as the motion
of m_1 remains uniform : this is analogous to the absence of
current in the secondary when the current in the primary
is uniform. If now we suddenly stop m_1, m_2 will start
off in the direction in which m_1 was moving before being
brought to rest.   This is analogous to the direct current in
the secondary produced by the stoppage of the current
in the primary.   These effects are the more marked the
greater the mass M.

It is instructive to find the quantities in the dynamical
system which correspond to the coefficients of self and
mutual induction.   Let us suppose that the bar on which
M slides is midway between the other two.

Then if $\dot{x}_1$ is the velocity of m_1 along its bar, $\dot{x}_2$ that of
m_2, the velocity of M will be $(\dot{x}_1 + \dot{x}_2)/2$, and T the kinetic
energy of the system is given by the equation

$$T = \tfrac{1}{2}m_1\dot{x}_1{}^2 + \tfrac{1}{2}m_2\dot{x}_2{}^2 + \tfrac{1}{2}M\left(\frac{\dot{x}_1 + \dot{x}_2}{2}\right)^2.$$

The momentum along x_1 is $dT/d\dot{x}_1$ and is therefore
equal to

$$\left(m_1 + \frac{M}{4}\right)\dot{x}_1 + \frac{M}{4}\dot{x}_2.$$

The momentum along x_2 is $dT/d\dot{x}_2$ and is therefore
equal to

$$\frac{M}{4}\dot{x}_1 + \left(m_2 + \frac{M}{4}\right)\dot{x}_2.$$

Thus $m_1 + M/4$, $m_2 + M/4$ correspond to the coefficients of

self-induction of the two circuits, while $M/4$ corresponds
to the coefficient of mutual induction between the circuits.
The effect of increasing the coefficient of mutual induction
between the circuits, such an increase for example as may
be produced by winding the primary and secondary coils
round an iron core, may be illustrated by the effect produced on the model by increasing the mass M relatively
to m_1 and m_2.

The behaviour of the model will illustrate important
electrical phenomena. Thus suppose the mass m_1 is struck
with a given impulse, it will evidently move forward with
greater velocity if m_2 is free to move than if it is fixed,
for if m_2 is free the large mass M will move very slowly
compared with m_1, the connecting bar turning round the
swivel on M almost as if this were fixed: if however m_2 is
fixed, then when m_1 moves forward it has to drag M along
with it, and will therefore move more slowly than in the
preceding case. When m_2 is free to move it moves in
the opposite direction to m_1. Now consider the electrical
analogue, the case when m_2 is free to move corresponds
to the case when there is in the neighbourhood of the
primary circuit a closed circuit round which a current can
circulate: the case when m_2 is fixed corresponds to the
case when this circuit is broken, when it can produce no
electrical effect as no current can circulate round it. The
greater velocity of m_1 when m_2 was free than when it
was fixed shows that when an electrical impulse acts on
a circuit the current produced is greater when there
is another circuit in the neighbourhood than when the
primary circuit was alone in the field; in other words,
the presence of the secondary diminishes the effective
inertia or self-induction of the primary.

232. Effect of a Secondary Circuit. As an example in the use of the equations given in Art. 229 we shall consider the behaviour of a primary and a secondary coil when an electric impulse acts upon the primary. Let us suppose that originally there were no currents in the circuits. Let L, M, N be respectively the coefficients of self-induction of the primary, of mutual induction between the primary and the secondary, and the coefficient of self-induction of the secondary: R, S the resistances of the primary and secondary respectively, x and y the currents through these coils. Then if P' is the external electromotive force acting on the primary, we have by the equations of Art. 229,

$$\frac{d}{dt}(Lx + My) + Rx = P' \quad \ldots\ldots\ldots\ldots(1),$$

$$\frac{d}{dt}(Mx + Ny) + Sy = 0 \quad \ldots\ldots\ldots\ldots(2).$$

The primary is acted on by an impulse, that is the force P' only lasts for a short time, let us call this time τ. Then if x_0, y_0 are the values of x, y due to this impulse we have by integrating equation (1) from $t = 0$ to $t = \tau$

$$Lx_0 + My_0 + R\int_0^\tau x\,dt = \int_0^\tau P'dt.$$

Since τ is indefinitely small and x is finite

$$\int_0^\tau x\,dt = 0;$$

let
$$\int_0^\tau P'dt = P,$$

then we have $\quad\quad Lx_0 + My_0 = P \quad \ldots\ldots\ldots\ldots(3).$

Similarly by integrating (2) we get

$$Mx_0 + Ny_0 = 0 \quad \ldots\ldots\ldots\ldots(4),$$

hence $x_0 = \dfrac{P}{L - \dfrac{M^2}{N}}, \quad y_0 = -\dfrac{PM}{LN - M^2}.$

If the secondary circuit had not been present the current in the primary due to the same impulse would have been P/L: thus the effect of the secondary is to increase the initial current in the primary: it diminishes its effective self-induction from L to $L - M^2/N$. This is an illustration of the effect described in the last article. Equation (4) expresses that the number of tubes of magnetic induction passing through the second circuit is not altered suddenly by the impulse acting on the first circuit.

When the impulse ceases, the circuits are free from external forces, and the equations for x and y are

$$\frac{d}{dt}(Lx + My) + Rx = 0 \dots\dots\dots\dots(5),$$

$$\frac{d}{dt}(Mx + Ny) + Sy = 0 \dots\dots\dots\dots(6).$$

Let us now choose as the origin from which time is measured the instant when the impulse ceases. Integrate these equations from $t = 0$ to $t = \infty$, then since x and y will vanish when $t = \infty$ we have

$$R\int_0^\infty x\,dt = Lx_0 + My_0$$
$$= P \text{ by equation (3),}$$

but $\int_0^\infty x\,dt$ is the total quantity of electricity which passes across any section of the primary circuit, if we denote this by Q we have

$$Q = \frac{P}{R};$$

hence Q is not affected by the presence of a secondary circuit. Thus since the current is greater to begin with when the secondary is present than when it is absent, it must, since Q is the same in the two cases, die away faster on the whole when the secondary is present.

The presence of the secondary increases the rate at which the current dies away just after it is started, but diminishes the rate at which the current ultimately dies away.

Integrating (6) from $t = 0$ to $t = \infty$ we find

$$S \int_0^\infty y\, dt = Mx_0 + Ny_0$$

$$= 0 \text{ by equation (4)};$$

hence the total quantity of electricity passing across any section of the secondary circuit is zero.

To solve equations (5) and (6) put

$$x = A\epsilon^{-\lambda t},$$
$$y = B\epsilon^{-\lambda t};$$

eliminating A and B we find

$$(R - L\lambda)(S - N\lambda) = M^2\lambda^2 \ldots\ldots\ldots(7);$$

hence if λ_1, λ_2 are the roots of this quadratic, we have

$$x = A_1\epsilon^{-\lambda_1 t} + A_2\epsilon^{-\lambda_2 t},$$
$$y = B_1\epsilon^{-\lambda_1 t} + B_2\epsilon^{-\lambda_2 t}.$$

We notice that since $\frac{1}{2}Lx^2 + Mxy + \frac{1}{2}Ny^2$, the expression for the kinetic energy of the currents, must be positive for all values of x and y, $LN - M^2$ must be positive, and therefore λ_1 and λ_2 are positive quantities. If we determine the values of the A's and B's from the values of x and y when $t = 0$, we find after some reductions

$$x = \frac{1}{\lambda_1 - \lambda_2} \frac{PN}{LN - M^2} \left\{ \epsilon^{-\lambda_1 t}\left(\lambda_1 - \frac{S}{N}\right) - \epsilon^{-\lambda_2 t}\left(\lambda_2 - \frac{S}{N}\right) \right\} (8),$$

$$y = -\frac{1}{\lambda_1 - \lambda_2} \frac{PM}{LN - M^2} \left\{ \lambda_1\epsilon^{-\lambda_1 t} - \lambda_2\epsilon^{-\lambda_2 t} \right\} \ldots\ldots\ldots\ldots(9).$$

We see from the quadratic equation (7) that one of its roots is greater than, the other root less than S/N, thus $\lambda_1 - S/N$, $\lambda_2 - S/N$ are of opposite signs, and therefore by (8), x the current through the primary never changes sign; y the current through the secondary begins by being of the opposite sign to x, it changes sign, and finally x and y are of the same sign.

A very important special case of the preceding investigation is when the two circuits are close together, or when the circuits are wound round a core of soft iron which completely fills their apertures; in this case nearly all the lines of magnetic force which pass through one circuit pass through the other also; this is often expressed by saying that there is very little *magnetic leakage* between the circuits. When this condition is fulfilled $L - M^2/N$ is very small compared with L. In the limiting case when this quantity vanishes we see by equation (7) that one of the values of λ, say λ_2, is infinite, while λ_1 is equal to

$$\frac{RS}{LS + NR}.$$

In this case we find from equations (8) and (9) that, except at the very beginning of the motion,

$$x = \frac{P}{L\left\{1 + \dfrac{RN}{LS}\right\}^2} \epsilon^{-\frac{R}{L}\frac{1}{1+\frac{RN}{LS}}t},$$

$$y = \frac{R}{S}\frac{PM}{L^2\left\{1 + \dfrac{RN}{LS}\right\}^2} \epsilon^{-\frac{R}{L}\frac{1}{1+\frac{RN}{LS}}t}.$$

The relation between the currents and the time, when

26—2

$L - M^2/N$ is small, is represented by the curves in Fig. 117; the dotted curve represents the current in the primary when the secondary is absent.

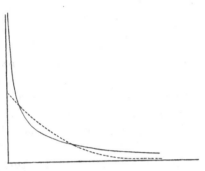

Fig. 117.

233. Currents induced in a mass of metal by an impulse. Let us suppose that the impulse is due to the sudden alteration of a magnetic system. Let N be the number of tubes of magnetic induction due to this system which pass through any circuit; to fix our ideas let us suppose this is the primary circuit in the case considered in Art. 232. Then using the notation of that article

$$P' = - \frac{dN}{dt},$$

by Faraday's law.

Hence $$P = \int_0^\tau P' dt = - (N_\tau - N_0),$$

where N_τ and N_0 represent respectively the number of tubes of magnetic induction passing through the circuit at the times $t = \tau$ and $t = 0$ respectively. We have, however, by equation (3), Art. 232,

$$Lx_0 + My_0 \qquad = P,$$

or $\qquad Lx_0 + My_0 + N_\tau = N_0.$

Now the right-hand side is the number of tubes of magnetic induction which pass through the circuit at the time $t = 0$, i.e. the time when the impulse began to act; the left-hand side represents the number of tubes of magnetic induction, some of them now being due to the currents started in the circuit, which pass through the circuit at the time $t = \tau$ when the impulse ceases to act. The equality of these two expressions shows that the currents generated by the impulse are such as to keep the number of tubes of magnetic induction which pass through the circuit unaltered. The case we have considered is one where there is only one secondary, the reasoning is however quite general, and whenever an impulse acts upon a system of conductors, the currents started in these conductors are such that their electro-magnetic action causes the number of tubes of magnetic induction passing through any of the conducting circuits to be unaltered by the impulse.

Let us apply this result to the case of the currents induced in a mass of metal by the alteration in an external magnetic field.

The number of tubes passing through every circuit that can be drawn in the metal is the same after the impulse as before. Hence we see that the magnetic field in the metal is the same after the impulse as before. This will give an important result as to the distribution of currents inside the metal. For we have seen (Art. 203) that the work done when unit pole is taken round a closed circuit is equal to 4π times the current flowing through that circuit. Now as the magnetic field inside the metal, and therefore the work done when unit pole passes round a closed circuit, is unaltered by the impulse, the current flowing through any such closed curve is also unaltered by the impulse; hence, as there were no currents through it before the impulse acted, there will be none generated by the impulse. In other words, the currents generated in a mass of metal by an electric impulse are entirely on the surface of the metal, and the inside of the conductor is free from currents.

234. The currents will not remain on the surface, they will rapidly diffuse through the metal and die away. We can find the way the currents distribute themselves after the impulse stops by the use of the two fundamental principles of electro-dynamics, (1) that the work done by the magnetic forces when unit pole travels round a closed circuit is equal to 4π times the quantity of current flowing through the circuit, (2) that the total electromotive force round any closed circuit is equal to the rate of diminution of the number of tubes of magnetic induction passing through the circuit.

Let u, v, w be the components of the electric current parallel to the axes of x, y, z at any point; α, β, γ the components of the magnetic force at the same point. The axes are chosen so that if x is drawn to the east, y to the north, z is upwards. Consider a small rectangular circuit $ABCD$, the sides AB, BC being parallel to the axes of z and y respectively. Let $AB=2h$, $BC=2k$. Let α, β, γ be the components of magnetic force at O, the centre of the rectangle; x, y, z the coordinates of O; let the coordinates of P, a point on AB, be x, $y + k$, $z + \zeta$; the z component of the magnetic force at P will be approximately

$$\gamma + \frac{d\gamma}{dz}\zeta + \frac{d\gamma}{dy}k.$$

Let now a unit magnetic pole be taken round the rectangle $ABCD$, the direction of motion round $ABCD$ being related to the positive direction of x like rotation and translation in a right-handed screw. The work done on unit pole as it moves from A to B will be

$$\int_{-h}^{+h}\left(\gamma + \frac{d\gamma}{dz}\zeta + \frac{d\gamma}{dy}k\right)d\zeta,$$

which is equal to $\qquad 2h\gamma + 2hk\dfrac{d\gamma}{dy};$

the work done on the pole as it moves from C to D is

$$-2h\gamma + 2hk\frac{d\gamma}{dy}.$$

We may show similarly that the work done on unit pole as it moves from B to C is equal to

$$-2k\beta - 2hk\frac{d\beta}{dz},$$

and when it moves from D to A, to

$$2k\beta - 2hk\frac{d\beta}{dz}.$$

Adding these expressions we see that the work done on unit pole as it travels round the rectangle $ABCD$ is equal to

$$\left(\frac{d\gamma}{dy} - \frac{d\beta}{dz}\right)4hk.$$

The quantity of current passing through this rectangle is equal to $\qquad 4uhk,$

hence since the work done on unit pole in going round the rectangle is equal to 4π times the current passing through the rectangle, see Art. 203, we have

$$4\pi \times 4uhk = \left(\frac{d\gamma}{dy} - \frac{d\beta}{dz}\right)4hk,$$

or $$4\pi u = \frac{d\gamma}{dy} - \frac{d\beta}{dz} \quad\ldots\ldots\ldots\ldots\ldots(1).$$

By taking rectangles whose sides are parallel to the axes of x and z, and of x, y we get in a similar way

$$4\pi v = \frac{d\alpha}{dz} - \frac{d\gamma}{dx}\ldots\ldots\ldots\ldots\ldots(2),$$

$$4\pi w = \frac{d\beta}{dx} - \frac{d\alpha}{dy} \quad\ldots\ldots\ldots\ldots(3).$$

If X, Y, Z are the components of the electric intensity at O, we can prove by a similar process that the work done on unit charge of electricity in going round the rectangle $ABCD$ is equal to

$$\left(\frac{dZ}{dy} - \frac{dY}{dz}\right)4hk.$$

If a, b, c are the components of magnetic induction at O, the number of tubes of magnetic induction passing through the rectangle is $a \times 4hk$; hence the rate of diminution of the number of unit tubes is equal to

$$-\frac{da}{dt}4hk.$$

But by Faraday's law of Electromagnetic Induction the work done on unit charge in going round the circuit is equal to the rate of diminution in the number of tubes of magnetic induction passing through the circuit, hence

$$-\frac{da}{dt}4hk = \left(\frac{dZ}{dy} - \frac{dY}{dz}\right)4hk,$$

or
$$-\frac{da}{dt} = \frac{dZ}{dy} - \frac{dY}{dz},$$

similarly
$$-\frac{db}{dt} = \frac{dX}{dz} - \frac{dZ}{dx},$$
$$\left.\vphantom{\begin{array}{c}1\\1\\1\end{array}}\right\} \dots\dots\dots\dots(4).$$
$$-\frac{dc}{dt} = \frac{dY}{dx} - \frac{dX}{dy}$$

Let us consider the case when the variable part of magnetization is induced, so that

$$\frac{da}{dt} = \mu \frac{d\alpha}{dt}, \qquad \frac{db}{dt} = \mu \frac{d\beta}{dt}, \qquad \frac{dc}{dt} = \mu \frac{d\gamma}{dt},$$

where μ is the magnetic permeability. If σ is the specific resistance of the metal in which the currents are flowing, and if the currents are entirely conduction currents,

$$\sigma u = X, \qquad \sigma v = Y, \qquad \sigma w = Z.$$

We have by equation (1)

$$4\pi\mu \frac{du}{dt} = \frac{d}{dy}\frac{dc}{dt} - \frac{d}{dz}\frac{db}{dt},$$

hence by putting $Y = \sigma v$, $Z = \sigma w$ in equation (4) we get

$$4\pi\mu \frac{du}{dt} = \sigma \left(\frac{d^2u}{dx^2} + \frac{d^2u}{dy^2} + \frac{d^2u}{dz^2}\right) - \sigma \frac{d}{dx}\left(\frac{du}{dx} + \frac{dv}{dy} + \frac{dw}{dz}\right).$$

We see from equations (1), (2), (3) that

$$\frac{du}{dx} + \frac{dv}{dy} + \frac{dw}{dz} = 0,$$

hence
$$4\pi\mu \frac{du}{dt} = \sigma \left(\frac{d^2u}{dx^2} + \frac{d^2u}{dy^2} + \frac{d^2u}{dz^2}\right);$$

similarly
$$4\pi\mu \frac{dv}{dt} = \sigma \left(\frac{d^2v}{dx^2} + \frac{d^2v}{dy^2} + \frac{d^2v}{dz^2}\right),$$

$$4\pi\mu \frac{dw}{dt} = \sigma \left(\frac{d^2w}{dx^2} + \frac{d^2w}{dy^2} + \frac{d^2w}{dz^2}\right).$$

We can also prove by a similar method that

$$4\pi\mu\frac{da}{dt} = \sigma\left(\frac{d^2a}{dx^2} + \frac{d^2a}{dy^2} + \frac{d^2a}{dz^2}\right),$$

with similar equations for b and c.

These equations are identical in form with those which hold for the conduction of heat, and we see that the currents and magnetic force will diffuse inwards into the metal in the same way as temperature would diffuse if the surface of the metal were heated, and then the heat allowed to diffuse.

235. We may apply the results obtained in the conduction of heat to the analogous problem in the distribution of currents. As a simple example let us take a case in one dimension. Let us suppose that over the infinite face of a plane slab we have initially a uniform distribution of currents, and that these currents are left to themselves. Then from the analogous problem in the conduction of heat we know that after a time t has elapsed the current, at a distance x from the face to which the currents were originally confined, will be proportional to

$$\frac{e^{-\frac{x}{t\sigma/\pi\mu}}}{t^{\frac{1}{2}}}.$$

This expression satisfies the differential equation and vanishes when $t = 0$ except at the face where $x = 0$. The currents at a distance x will attain their maximum value when

$$t = \frac{2x^2}{\sigma/\pi\mu},$$

and the magnitude of the maximum current will be inversely proportional to x.

In the case of copper $\mu = 1$, $\sigma = 1600$, hence the time at which the current is a maximum at a place, one centimetre from the surface, is $2\pi/1600$ seconds, or about $1/250$ of a second, a point ·1 cm. from the surface would receive the maximum current after about $1/25,000$ of a second, while at a point 10 cm. from the surface the current would not reach its maximum for about $4/10$ of a second.

Let us now consider the case of iron: for an average specimen of soft iron we may put $\sigma = 10^4$, $\mu = 10^3$; hence in this case, the time the current, 1 cm. from the surface, will take to reach its maximum value is about $2\pi/10$ seconds, while a place 10 cm. from the face only attains its maximum after 20π seconds. Thus the currents diffuse much more slowly through iron than they do through copper. The diffusion of the currents is regulated by two circumstances, the inertia of the currents which tends to confine them to the outside of the conductor, and the resistance of the metal which tends to make the currents diffuse through the conductor; though the resistance of iron is greater than that of copper, this is far more than counterbalanced by the enormously greater magnetic permeability of the iron which increases the inertia of the currents, and thereby the tendency of the currents to concentrate themselves on the outside of the conductor.

When t is much greater than $x^2/(\sigma/\pi\mu)$, $e^{-\frac{x^2}{t\sigma/\pi\mu}}$ differs little from unity, in this case the currents are almost independent of x and vary inversely as $t^{\frac{3}{2}}$, thus the currents ultimately get nearly uniformly distributed, and gradually fade away.

236. Periodic electromotive forces acting on

a circuit possessing inertia. So far we have confined our attention to the case of impulses; we now proceed to consider the case when electromotive forces act on a circuit for a finite time. If these forces are steady the currents will speedily become steady also, and there will be no effects due to induction; when, however, these forces are periodic, induction will produce very important effects which we shall now proceed to investigate. We shall commence with the case of a single circuit whose co-efficient of self-induction is L and whose resistance is R; we shall suppose that this circuit is acted on by an external electromotive force varying harmonically with the time, the force at the time t being equal to $E \cos pt$; this expression represents a force making $p/2\pi$ complete vibrations a second, it changes its direction p/π times per second. If i is the current through the coil, we have

$$L \frac{di}{dt} + Ri = E \cos pt \dots\dots\dots\dots\dots(1);$$

the solution of this equation is

$$i = \frac{E \cos(pt - \alpha)}{\{R^2 + L^2 p^2\}^{\frac{1}{2}}} \dots\dots\dots\dots(2),$$

where

$$\tan \alpha = \frac{Lp}{R} \dots\dots\dots\dots\dots(3).$$

The maximum value of the electromotive force is E, while the maximum value of the current is

$$E/\{R^2 + L^2 p^2\}^{\frac{1}{2}}:$$

if a steady force E acted on the circuit the current would be E/R. Thus the inertia of the circuit makes the maximum current bear to the maximum electromotive force a smaller ratio than a steady current through the same circuit bears to the steady electromotive force producing it. The ratio of the maximum electromotive force to the maximum current, when the force is periodic, is equal to $\{R^2 + L^2 p^2\}^{\frac{1}{2}}$; this quantity is called the *impedance* of the circuit.

We see from equation (2) that the phase of the current lags behind that of the electromotive force. When the force oscillates so rapidly that Lp is large compared with R, we see from equation (3) that α will be approximately equal to $\pi/2$. In this case the current through the coil will be greatest when the electromotive force acting on the circuit is zero, and will vanish when the electromotive force is greatest.

In this case, since Lp is large compared with R, we have approximately

$$i = \frac{E}{Lp} \sin pt;$$

thus the current through the circuit is approximately independent of the resistance and depends only upon the coefficient of self-induction and on the frequency of the electromotive force. Thus a very rapidly alternating electromotive force will send far more current through a short circuit with a small coefficient of self-induction, even though it is made of a badly conducting material, than through a long circuit with large self-induction, even though this circuit is made of an excellent conductor. For steady electromotive forces on the other hand, the current sent through the second circuit would be enormously greater than that through the first.

The work done by the current per unit time, which appears as heat, is equal to the mean value of either

$E \cos pt \cdot i$ or Ri^2, and is equal to

$$\tfrac{1}{2} \frac{E^2 R}{R^2 + L^2 p^2}.$$

Thus when the electromotive force changes so slowly that Lp is small compared with R, the work done per unit time varies inversely as R; while when the force varies so rapidly that Lp is large compared with R, the work done varies directly as R. If E, p and L are given the work done is a maximum when

$$R = Lp.$$

237. Circuit rotating in the Earth's field. An external electromotive force of the type considered in the last article is produced when a conducting circuit rotates with uniform velocity ω in the earth's magnetic field about a vertical axis. If θ is the angle the plane of the circuit makes with the magnetic meridian, H the horizontal component of the earth's magnetic force, A the area of the circuit, then the number of tubes of magnetic induction passing through the circuit is

$$HA \sin \theta :$$

the rate of diminution of this is

$$- HA \cos \theta \frac{d\theta}{dt}.$$

If the circuit revolves with uniform angular velocity ω, $\theta = \omega t$, and the rate of diminution in the number of tubes of magnetic induction passing through the circuit is

$$- HA\omega \cos \omega t,$$

as this, by Faraday's law, is the electromotive force acting on the circuit. The case is identical with that just considered if we write ω for p and $- HA\omega$ for E; thus if L is the coefficient of self-induction of the circuit, R the resistance, i the current through the circuit,

$$i = -\frac{HA\omega \cos(\omega t - \alpha)}{\{L^2\omega^2 + R^2\}^{\frac{1}{2}}}.$$

The motion of the circuit is resisted by a couple whose moment is, by Art. 214, equal to the current multiplied by the differential coefficient with respect to θ of the number of tubes of magnetic induction due to the earth's field passing through the circuit; thus the moment of the couple is

$$iHA \cos \theta,$$

or

$$\frac{H^2A^2\omega \cos \omega t \cos(\omega t - \alpha)}{\{L^2\omega^2 + R^2\}^{\frac{1}{2}}}.$$

Thus the couple always tends to oppose the rotation of the coil unless θ is between $\dfrac{\pi}{2}$ and $\dfrac{\pi}{2} + \alpha$ or between $\dfrac{3\pi}{2}$ and $\dfrac{3\pi}{2} + \alpha$.

To maintain the motion of the circuit work must be spent; the amount of work spent in any time is equal to the mechanical equivalent of the heat developed in the circuit.

The mean value of the retarding couple is

$$\tfrac{1}{2}\frac{H^2A^2\omega \cos \alpha}{\{L^2\omega^2 + R^2\}^{\frac{1}{2}}} = \tfrac{1}{2}\frac{H^2A^2R\omega}{L^2\omega^2 + R^2};$$

it vanishes when ω is zero or infinite and is greatest when $\omega = R/L$.

If the circuit rotates so rapidly that $L\omega$ is large compared with R, α is approximately equal to $\pi/2$, and we see that

$$i = -\frac{HA \sin \omega t}{L}.$$

Now by definition Li is the number of tubes of magnetic induction due to the currents which pass through the circuit, while $HA \sin \omega t$ is the number passing through the same circuit due to the earth's magnetic field; we see from the preceding expression for i that the sum of these two quantities, which is the total number of tubes of magnetic induction passing through the circuit, remains zero throughout the whole of the time. This is an illustration of the general principle that when the inertia effects are paramount the number of tubes passing through any conducting circuit remains constant.

238. Circuits in parallel. Suppose that two points A and B are connected by two circuits in parallel. Let R be the resistance of the first circuit, S that of the second; let the first circuit contain a coil whose coefficient of self-induction is L, the second one whose coefficient of self-induction is N. Let the coils be so far apart that their coefficient of mutual induction is zero. Then if a difference of potential $E \cos pt$ be maintained between the points A and B we see by the preceding investigations that i and j, the currents in the two circuits, will be given by the equations

$$i = \frac{E \cos(pt - \alpha)}{\{L^2 p^2 + R^2\}^{\frac{1}{2}}},$$

$$j = \frac{E \cos(pt - \beta)}{\{N^2 p^2 + S^2\}^{\frac{1}{2}}},$$

where $\qquad \tan \alpha = \frac{Lp}{R}, \quad \tan \beta = \frac{Np}{S}.$

If the external electromotive force varies so rapidly that

Lp and Np are large compared with R and S respectively, then

$$i = \frac{E \sin pt}{Lp},$$

$$j = \frac{E \sin pt}{Np},$$

or the currents flowing through the two circuits are inversely proportional to their coefficients of self-induction. Thus with very rapidly alternating currents the distribution of the currents is almost independent of their resistances and depends almost entirely on their self-inductions. Thus if one of the coils had a moveable iron core, the current through the coil would be very much increased by removing the iron, as this would greatly diminish the self-induction of the circuit.

239. Transformers. We have hitherto confined our attention to the case when the only circuit present was the one acted upon by the periodic electromotive force. We shall now consider the case when in addition to the circuit acted upon by the external electromotive force, which we shall call the primary circuit, another circuit is present in which currents are induced by the alternating currents in the primary: we shall call this circuit the secondary circuit, and suppose that it is not acted upon by any external electromotive force beyond that due to the alternating current in the primary. A very important example of this is afforded by the 'transformer.' In this instrument a periodic electromotive force acts on the primary, which consists of a large number of turns of wire; in the ordinary use of the transformer for electric lighting this electromotive force is so large that it would be dangerous to lead the primary circuit about a building; the current for lighting is derived from a secondary circuit consisting of a smaller number of turns of wire. The primary and secondary circuits are wound round an iron core as in Fig. 118.

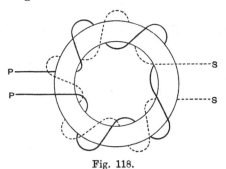

Fig. 118.

The tubes of magnetic induction concentrate in this core, so that most of the tubes which pass through the primary pass also through the secondary.

The current in this secondary is larger than that in the primary, but the electromotive force acting round it is smaller. The current in the secondary bears to that in the primary approximately the same ratio as the electromotive force round the primary bears to that round the secondary.

Let L, M, N be respectively the coefficients of self-induction of the primary, of mutual induction between the primary and the secondary and of self-induction of the secondary, let R and S be the resistances of the primary and secondary respectively, x and y the currents through these coils. Let $E \cos pt$ be the electromotive force acting on the primary. To find x and y we have the following equations:

$$L\frac{dx}{dt} + M\frac{dy}{dt} + Rx = E\cos pt\ldots\ldots\ldots(1),$$

$$M\frac{dx}{dt} + N\frac{dy}{dt} + Sy = 0\ldots\ldots\ldots\ldots(2).$$

The values of x and y are

$$x = A\cos(pt-\alpha)\ldots\ldots\ldots\ldots(3),$$
$$y = B\cos(pt-\beta)\ \ldots\ldots\ldots\ldots(4).$$

By substituting these values in equations (1) and (2), we find

$$B^2 = \frac{M^2p^2}{N^2p^2+S^2}A^2\ \ldots\ldots\ldots(5),$$

$$A^2 = \frac{E^2}{L'^2p^2+R'^2},$$

$$L' = L - \frac{M^2Np^2}{N^2p^2+S^2},$$

$$R' = R + \frac{M^2p^2S}{N^2p^2+S^2},$$

$$\tan\alpha = \frac{L'p}{R'},$$

$$\tan(\beta-\alpha) = -\frac{S}{Np}.$$

From the expressions for A and α in terms of E we see that the effect of the secondary circuit is to make the primary circuit behave like a single circuit whose co-efficient of self-induction is L' and whose resistance is R'. We see from the expressions for L' and R', that L' is less than L, while R' is greater than R. Thus the presence of the secondary circuit diminishes the apparent self-induction of the primary circuit, while it increases its resistance.

When the electromotive force changes so rapidly that Np is large compared with S, we have approximately

$$L' = L - \frac{M^2}{N},$$

$$R' = R + \frac{M^2}{N^2} S,$$

$$B = \frac{M}{N} A,$$

$$\beta - \alpha = \pi.$$

This value of the apparent self-induction is the same as that under an electrical impulse, see Art. 231. In a well-designed transformer $L - M^2/N$ is exceedingly small compared with L. When the secondary circuit is not completed S is infinite; in this case $L' = L$. When the secondary circuit is completed through electric lamps &c., S is in practice small compared with Np, so that $L' = L - M^2/N$. Thus the completion of the circuit causes a great diminution in the value of the apparent self-induction of the primary circuit. The work done per unit time in the transformer is equal to the mean value of $E \cos pt \cdot x$, it is thus equal to

$$\frac{1}{2} \frac{E^2 \cos \alpha}{\{L'^2 p^2 + R'^2\}^{\frac{1}{2}}}$$

$$= \frac{1}{2} \frac{E^2 R'}{L'^2 p^2 + R'^2}.$$

When the secondary circuit is broken S is infinite and therefore $L' = L$, $R' = R$, and the work done on the transformer per unit time or the power spent on it is equal to

$$\frac{1}{2} \frac{E^2 R}{L^2 p^2 + R^2}.$$

When the circuit is completed, and S is small compared

with Np, $L' = L - M^2/N$, $R' = R + M^2S/N^2$, and then the power spent is equal to

$$\frac{1}{2} \frac{E^2 \left(R + \frac{M^2}{N^2} S \right)}{\left(L - \frac{M^2}{N} \right)^2 p^2 + \left(R + \frac{M^2}{N^2} S \right)^2}.$$

This is very much greater than the power spent when the secondary circuit is not completed; this must evidently be the case, as when the secondary circuit is completed lamps are raised to incandescence, the energy required for this must be supplied to the transformer. The power spent when the secondary circuit is not completed is wasted as far as useful effect is concerned, and is spent in heating the transformer. The greater the coefficient of self-induction of the primary, the smaller is the current sent through the primary by a given electromotive force, and the smaller the amount of power wasted when the secondary circuit is broken. When the secondary circuit is closed the self-induction of the primary is diminished from L to L'; since there is less effective self-induction in the primary, the current through it, and consequently the power given to it, is greatly increased.

We see from the expression just given that the power absorbed by the transformer is greatest when

$$R + \frac{M^2}{N^2} S = \left(L - \frac{M^2}{N} \right) p,$$

that is, when

$$S = -\frac{N^2}{M^2} R + \frac{N}{M^2} (LN - M^2) p.$$

When there is no magnetic leakage, i.e. when

$$LN = M^2,$$

the power absorbed continually increases as the resistance in the secondary diminishes; when however LN is
not equal to M^2 the power absorbed does not necessarily
increase as S diminishes, it may on the contrary reach
a maximum value for a particular value of S, and any
diminution of S before this value will be accompanied
by a decrease in the energy absorbed by the transformer.
The greater the frequency of the electromotive force, the
larger will be the resistance of the secondary when the
absorption of power by the transformer is greatest. When
the frequency is very great, such as, for instance, when
a Leyden jar is discharged (see page 436), the critical value
of the resistance in the secondary may be exceedingly
large. In this case the difference between the maximum
absorption of power and that corresponding to $S = 0$ may
be very great. Thus when $S = 0$, the power absorbed
is equal to

$$\frac{1}{2}\frac{E^2R}{L'^2p^2 + R^2},$$

or approximately for very high frequencies

$$\frac{1}{2}\frac{E^2R}{L'^2p^2},$$

while the maximum power absorbed is

$$\frac{1}{4}\frac{E^2}{L'p},$$

which exceeds that when $S = 0$ in the proportion of $L'p$
to $2R$.

The currents x, y in the primary and secondary are
represented by the equations

$$x = A \cos (pt - \alpha),$$
$$y = B \cos (pt - \beta).$$

Thus the ratio of the maximum value of the current in the primary to that in the secondary is B/A: by equation (5), we have

$$\frac{B}{A} = \frac{Mp}{\{N^2p^2 + S^2\}^{\frac{1}{2}}},$$

or, when Np is large compared with S,

$$\frac{A}{B} = \frac{M}{N},$$

$$\beta - \alpha = \pi.$$

If the primary and secondary coils cover the same length of the core, and are wound on a core of great permeability, then M/N is equal to m/n, where m is the number of turns in the primary, and n the number in the secondary.

If we have a lamp whose resistance is s in the secondary the potential difference between its electrodes is sy, i.e.

$$sB \cos (pt - \beta).$$

The maximum value of this expression is sB; substituting the value of B, we find that when Np is large compared with S this value is equal to

$$\frac{s\frac{M}{N}E}{\{L'^2p^2 + R'^2\}^{\frac{1}{2}}}.$$

This is greatest when $L' = 0$, in which case it is equal to

$$\frac{s\frac{M}{N}E}{R'};$$

and this, as S is small compared with Np, is equal to

$$\frac{s\frac{M}{N}E}{R + \frac{M^2}{N^2}S}.$$

If R is small compared with SM^2/N^2 this is approximately

$$\frac{s}{S}\frac{N}{M} E.$$

Thus if for example $M/N = 20$, the maximum current through the secondary is 20 times that through the primary; while the electromotive force between the terminals of the lamp is approximately

$$\frac{s}{20S} E.$$

Now s is always smaller than S, as S is the resistance of the whole secondary circuit, while s is the resistance of only a part of it: the electromotive force between the terminals of any lamp is thus in this case always less than 1/20 of the electromotive force between the terminals of the secondary. In getting this value we have assumed the conditions to be those most favourable to the production of a high electromotive force in the secondary; if there is any magnetic leakage, i.e. if L' is not zero, then at high frequencies the electromotive force in the secondary would be very much less than the value just found, in fact where there is any magnetic leakage, the ratio of the electromotive force in the secondary to that in the primary is indefinitely small when the frequency is infinite.

240. Distribution of rapidly alternating currents. When the frequency of the electromotive force is so great that in the equations of the type

$$L \frac{dx}{dt} + M \frac{dy}{dt} + \dots Rx = \text{external electromotive force},$$

the term Rx depending on the resistance is small compared with the terms Ldx/dt, Mdy/dt depending on induction, which, if the electromotive force is supposed to vary as $\cos pt$, will be the case when Lp, Mp are large compared with R; the equations determining the currents take the form

$$\frac{d}{dt}(Lx + My + \ldots) = \text{external electromotive force,}$$

$$= -\frac{d\mathbf{N}}{dt},$$

where $\mathbf{N}$ is the number of tubes of induction due to the external system passing through the circuit whose co-efficient of self-induction is L.

We see from this that

$$Lx + My + \ldots + \mathbf{N} = \text{constant,}$$

and since x, $y \ldots \mathbf{N}$ all vary harmonically, the constant must be zero. Now $Lx + My + \ldots$ is the number of tubes of magnetic induction which pass through the circuit we are considering due to the currents flowing in this and the neighbouring circuits, while $\mathbf{N}$ is the number of tubes passing through the same circuit due to the external system. Hence the preceding equation expresses that the total number of tubes passing through the circuit is zero. The same result is true for any circuit.

Now consider the case of the currents induced in a mass of metal by a rapidly alternating electromotive force. The number of tubes of magnetic induction which pass through any circuit which can be drawn in the metal is zero, and hence the magnetic induction must vanish throughout the mass of the metal. The magnetic force will consequently also vanish throughout the same region. But since the magnetic force vanishes, the work done when unit pole is taken round any closed curve in the region must also vanish, and therefore by Art. 203 the current flowing through any closed curve in the region must also vanish; this implies that the current vanishes throughout the mass of metal, or in other words, that the currents generated by infinitely rapidly alternating forces are confined to the surface of the metal, and do not penetrate into its interior.

We showed in Art. 235 that the currents generated by an electrical impulse started from the surface of the conductor and then gradually diffused inwards. We may approximate to the condition of a rapidly alternating force by supposing a series of positive and negative impulses to follow one another in rapid succession. The currents started by a positive impulse have thus only time to diffuse a very short distance from the surface before the subsequent negative impulse starts opposite currents from the surface; the effect of these currents at some distance from the surface is to tend to counteract the original currents, and thus the intensity of the current falls off rapidly as the distance from the surface of the conductor increases.

The amount of concentration of the current depends on the frequency of the electromotive force and of the conductivity of the conductor. If the frequency is infinite and the conductivity finite, or the frequency finite and the conductivity infinite, then the current is confined to an indefinitely thin skin near the surface of the conductor. If, however, both the frequency and the conductivity are finite, then the thickness of the skin occupied by the current is finite also, while the magnitude of the current diminishes rapidly as we recede from the surface. Any increase in the frequency or in the conductivity increases the concentration of the current.

The case is analogous to that of a conductor of heat, the temperature of whose surface is made to vary harmonically, the fluctuations of temperature corresponding to the alterations in the surface temperature diminish in intensity as we recede from the surface, and finally cease to be appreciable. The fluctuations, however, with a long period are appreciable at a greater depth than those with a short one. We may for example suppose the temperature of the surface of the earth to be subject to two variations, one following the seasons and having a yearly period, the other depending on the time of day and having a daily period. These fluctuations become less and less apparent as the depth of the place of observation below the surface of the earth increases, and finally they become too small to be measured. The annual variations can, however, be detected at depths at which the diurnal variations are quite inappreciable.

This concentration of the current near the surface of the conductor, which is sometimes called 'the throttling of the current,' increases the resistance of the conductor to the passage of the current. When, for example, a rapidly alternating current is flowing along a wire, the current will flow near to the outside of the wire, and if the frequency is very great the inner part of the wire will be free from current; thus since the centre of the wire is free from current, the current is practically flowing through a tube instead of a solid wire. The area of the cross section of the wire, which is effective in carrying this rapidly alternating current, is thus smaller than the effective area when the current is continuous, as in this case the current distributes itself uniformly over the whole of the cross section of the wire. As the effective area for

the rapidly alternating currents is less than that for continuous currents, the resistance, measured by the heat produced in unit time when the total current is unity, is greater
for the alternating currents than for continuous currents.

**241.   Distribution of an alternating current in
a Conductor.**   The equations given in Art. 234 enable
us to find how an alternating current distributes itself
in a conductor.   We shall consider a case in which the
analysis is simple, but which will serve to illustrate the
laws of the phenomenon we are discussing.   This case
is that of an infinite mass of a conductor bounded by a
plane face.   Take the axis of x at right angles to this
face, and the origin of coordinates in the face; let the
currents be everywhere parallel to the axis of z, and the
same at all points in any plane parallel to the face of the
conductor.   Then if μ is the magnetic permeability and
σ the specific resistance of the conductor, w the current at
the point x, y, z at the time t parallel to the axis of z,
we have by the equations of Art. 234,

$$4\pi\mu \frac{dw}{dt} = \sigma \left(\frac{d^2w}{dx^2} + \frac{d^2w}{dy^2} + \frac{d^2w}{dz^2} \right),$$

or, since w is independent of y and z,

$$4\pi\mu \frac{dw}{dt} = \sigma \frac{d^2w}{dx^2} \quad \ldots\ldots\ldots\ldots\ldots(1).$$

We shall suppose that the currents are periodic, making
$p/2\pi$ complete alternations per second.   We may put,
writing i for $\sqrt{-1}$,

$$w = \epsilon^{ipt}\, \omega,$$

where ω is a function of x, but not of t.   Substituting this
value of w in equation (1) we get

$$4\pi\mu ip\omega = \sigma \frac{d^2\omega}{dx^2},$$

or if $n^2 = 4\pi\mu ip/\sigma$,

$$n^2\omega = \frac{d^2\omega}{dx^2}.$$

The solution of this is

$$\omega = A\epsilon^{-nx} + B\epsilon^{nx},$$

where A and B are constants.

Now
$$n = \left\{\frac{4\pi\mu p}{\sigma}\right\}^{\frac{1}{2}} i^{\frac{1}{2}}$$

$$= \left\{\frac{4\pi\mu p}{\sigma}\right\}^{\frac{1}{2}} \left\{\frac{1}{\sqrt{2}} + \frac{i}{\sqrt{2}}\right\}$$

$$= \left\{\frac{2\pi\mu p}{\sigma}\right\}^{\frac{1}{2}} (1 + i).$$

We shall suppose that the conductor stretches from $x = 0$ to $x = \infty$ and that the cause which induces the currents lies on the side of the conductor for which x is negative. It is evident that in this case the magnitude of the current cannot increase indefinitely as we recede from the face nearest the inducing system; in other words, w cannot be infinite when x is infinite: this condition requires that B should vanish; in this case we have

$$\omega = A\epsilon^{-\left\{\frac{2\pi\mu p}{\sigma}\right\}^{\frac{1}{2}}x}\,\epsilon^{-i\left(\frac{2\pi\mu p}{\sigma}\right)^{\frac{1}{2}}x},$$

and therefore

$$w = A\epsilon^{-\left(\frac{2\pi\mu p}{\sigma}\right)^{\frac{1}{2}}x}\,\epsilon^{i\left\{pt - \left(\frac{2\pi\mu p}{\sigma}\right)^{\frac{1}{2}}x\right\}}.$$

Thus if $w = A\cos pt$ when $x = 0$,

$$w = A\epsilon^{-\left(\frac{2\pi\mu p}{\sigma}\right)^{\frac{1}{2}}x} \cos\left\{pt - \left(\frac{2\pi\mu p}{\sigma}\right)^{\frac{1}{2}}x\right\}$$

at a distance x from the surface.

This result shows that the maximum value of the current at a distance x from the face is proportional to $\epsilon^{-\left(\frac{2\pi\mu p}{\sigma}\right)^{\frac{1}{2}}x}$. Thus the magnitude of the current diminishes in geometrical progression as the distance from the face increases in arithmetical progression.

In the case of a copper conductor exposed to an electromotive force making 100 alternations per second, $\mu = 1$, $\sigma = 1600$, $p = 2\pi \times 100$; hence $\{2\pi\mu p/\sigma\}^{\frac{1}{2}} = \pi/2$, so that the maximum current is proportional to $\epsilon^{-\frac{\pi x}{2}}$. Thus at 1 cm. from the surface the maximum current would only be ·208 times that at the surface, at a distance of 2 cms. only ·043, and at a distance of 4 centimetres less than 1/500 part of the value at the surface.

If the electromotive force makes a million alternations per second $\{2\pi\mu p/\sigma\}^{\frac{1}{2}} = 50\pi$; the maximum current is thus proportional to $\epsilon^{-50\pi x}$, and at the depth of one millimetre is less than one six-millionth part of its surface value.

The concentration of the current in the case of iron is even more remarkable. Consider a sample of iron for which $\mu = 1000$, $\sigma = 10000$, exposed to an electromotive force making 100 alternations per second, so that $p = 2\pi \times 100$. In this case $\{2\pi\mu p/\sigma\}^{\frac{1}{2}} = 20$ approximately, and thus the maximum current at a depth of one millimetre is only ·13 times the surface value, while at 5 millimetres it is less than one twenty-thousandth part of its surface value.

If the electromotive force makes a million alternations per second, then for this specimen of iron $\{2\pi\mu p/\sigma\}^{\frac{1}{2}}$

is approximately 2000, and the maximum current at the distance of one-tenth of a millimetre from the surface is about one five-hundred-millionth part of its surface value.

We see from the preceding expressions for the current that the distance required to diminish the maximum current to a given fraction of its surface value is directly proportional to the square root of the specific resistance, and inversely proportional to the square root of the number of alternations per second.

242. Magnetic Force in the Conductor. The currents in the conductor are all parallel to the axis of z, and are independent of the coordinates y, z.

Now the equations of Art. 234 may be written in the form

$$-\frac{da}{dt} = \sigma \left(\frac{dw}{dy} - \frac{dv}{dz} \right), \quad -\frac{db}{dt} = \sigma \left(\frac{du}{dz} - \frac{dw}{dx} \right),$$
$$-\frac{dc}{dt} = \sigma \left(\frac{dv}{dx} - \frac{du}{dy} \right),$$

where a, b, c are the components of the magnetic induction, u, v, w those of the current. In the case we are considering $u = v = 0$, and w is independent of y and z; hence $a = c = 0$, and the magnetic induction is parallel to the axis of y. Thus the currents in the plate are accompanied by a magnetic force parallel to the surface of the plate and at right angles to the direction of the current.

From the above equations we have

$$\frac{db}{dt} = \sigma \frac{dw}{dx},$$

and by Art. 241 $w = A\epsilon^{-mx}\cos(pt - mx)$,

where $\qquad\qquad m = (2\pi\mu p/\sigma)^{\frac{1}{2}}$.

Hence $\quad b = -\dfrac{\sqrt{2}\sigma m}{p} A\epsilon^{-mx}\cos\left(pt - mx - \dfrac{\pi}{4}\right)$

$\qquad\quad = -\dfrac{2\sqrt{2}\pi\mu}{m} A\epsilon^{-mx}\cos\left(pt - mx - \dfrac{\pi}{4}\right)$.

Thus the magnetic force in the conductor diminishes as we recede from the surface according to the same law as the current.

243. Mechanical Force acting on the Conductor. When a current flows in a magnetic field a mechanical force acts on the conductor carrying the current (see Art. 214). The direction of the force is at right angles to the current and also to the magnetic induction, and the magnitude of the force per unit volume of the conductor is equal to the product of the current and the magnetic induction at right angles to it.

In the case we are considering the magnetic induction and the current are at right angles. If w is the intensity of the current, the current flowing through the area $dx\,dy$ is $w\,dx\,dy$; hence the force on the volume $dx\,dy\,dz$ parallel to x, and in the positive direction of x, is equal to

$$-wb\,dx\,dy\,dz.$$

The total force parallel to x acting on the conductor is

$$-\iiint wb\,dx\,dy\,dz,$$

but since b and w are both independent of y and z, the force acting on the conductor per unit area of its face is

$$-\int_0^\infty wb\,dx.$$

Now if α, β, γ are the components of the magnetic force

$$4\pi w = \frac{d\beta}{dx} - \frac{d\alpha}{dy};$$

hence, since $b = \mu\beta$, we see that the force on the conductor parallel to x is

$$-\frac{\mu}{4\pi} \int_0^\infty \beta \frac{d\beta}{dx} . dx.$$

$$= \frac{\mu}{8\pi} \{\beta_0{}^2 - \beta_\infty{}^2\},$$

where β_0 is the value of β when $x = 0$, i.e. at the surface of the conductor, and β_∞ is the value of β when $x = \infty$. But it follows from the expression for b given in the last article that $\beta_\infty = 0$; hence the force on the conductor parallel to x per unit area of its face due to the action of the magnetic field on the currents is equal to

$$\frac{\mu\beta_0{}^2}{8\pi}.$$

The magnetic force is not uniform in the conductor but diminishes as we recede from the surface; hence, if the conductor is a magnetic substance, there will, in addition to the mechanical force due to the action of the magnetic field on the currents, be a force due to the effort of the magnetic substance to move towards the stronger parts of the field. The magnitude of the force parallel to x per unit volume is by Art. 164 equal to $\dfrac{(\mu - 1)}{8\pi} \dfrac{d\beta^2}{dx}$; thus the force acting per unit area of the face of the slab due to this cause is

$$\int_0^\infty \frac{\mu - 1}{8\pi} \frac{d\beta^2}{dx} dx$$

$$= -\frac{(\mu - 1)}{8\pi} \beta_0{}^2.$$

Adding this to the force $\dfrac{\mu\beta_0^2}{8\pi}$ due to the action of the magnetic field on the currents we find that the total force parallel to x is per unit area of surface of the slab $\beta_0^2/8\pi$, which for equal values of β_0 is the same for magnetic as for non-magnetic substances.

This force is always positive, and hence the conductor tends to move along the positive direction of x; in other words, the conductor is repelled from the system which induces the currents in the conductor. These repulsions have been shown in a very striking way in experiments made by Professor Elihu Thomson and also by Dr Fleming. In these experiments a plate placed above an electromagnet round which a rapidly alternating current was circulating, was thrown up into the air, the repulsion between the plate and the magnet arising from the cause we have just investigated.

The expression $\dfrac{\beta_0^2}{8\pi}$ is the repulsion at any instant, but since β_0 is proportional to $\cos(pt + \epsilon)$ the mean value of β_0^2 is $H^2/2$ if H is the maximum value of β_0. Hence the mean value of the repulsion is equal to

$$\frac{H^2}{16\pi}.$$

244. The screening off of Electromagnetic Induction. We have seen in Art. 242 that the magnetic force diminishes rapidly as we recede from the surface of the conductor, and becomes inappreciable at a finite distance, say d, from the surface. At a point P whose distance from the surface is greater than d we may neglect both the current and the magnetic force. Thus the electromagnetic action of the currents in the sheet of the conductor whose thickness is d just counterbalances at P the electromagnetic action of the original inducing system situated on the other side of the face of the conductor.

Hence the slab of thickness d may be regarded as screening off from P the electromagnetic effect of the original system. In the investigation in Art. 242 we supposed that the conductor was infinitely thick, but since the currents are practically confined to the slab whose thickness is d, it is evident that the screening is done by this layer and that no appreciable advantage is gained by increasing the thickness of the slab beyond d. The thickness d of the slab required to screen off the magnetic force depends upon the frequency of the alternations and on the magnetic permeability and specific resistance of the conductor. By Arts. 241 and 242 the current and magnetic force at a distance x from the surface are proportional to ϵ^{-mx}, where $m = \{2\pi\mu p/\sigma\}^{\frac{1}{2}}$; hence for a thickness d to reduce the magnetic force to an inappreciable fraction of its surface value md must be considerable. If we regard the system as screened off when the magnetic effect is reduced to a definite fraction of its undisturbed value, then d the thickness of the screen is inversely proportional to m. The greater the frequency the thinner the screen. Thus from the examples given in Art. 241 we see that if the system makes a million oscillations a second, a screen of copper less than a millimetre thick will be perfectly efficient, while a screen of iron a very small fraction of a millimetre in thickness will stop practically all induction. If the system only makes 100 alternations a second, the screen if of copper must be several centimetres and if of iron several millimetres thick.

245. Discharge of a Leyden Jar. One of the most interesting applications of the laws of induction of currents is to the case of a Leyden jar, the two coatings of which are connected by a conducting circuit possessing self-induction. Let us consider a jar whose inside A is connected to the outside B by a circuit whose resistance is R and whose coefficient of self-induction is L. Let i be the current flowing through the circuit from A to B; V_A and V_B the potentials of A and B respectively. Then by the laws of the induction of currents

$$L\frac{di}{dt} + Ri = \text{electromotive force tending to increase } i$$

$$= V_A - V_B \dots\dots\dots\dots\dots\dots\dots\dots\dots\dots(1).$$

If Q is the charge on the inside of the jar, and C the capacity of the jar, then

$$C(V_A - V_B) = Q,$$

or $$(V_A - V_B) = \frac{Q}{C}.$$

The alteration in the charge is due to the current flowing through the conductor, and i is the rate at which the charge is diminishing, so that

$$i = -\frac{dQ}{dt}.$$

Substituting this value of i in equation (1), we get

$$L\frac{d^2Q}{dt^2} + R\frac{dQ}{dt} + \frac{Q}{C} = 0 \dots\dots\dots\dots\dots(2).$$

The form of the solution of this equation will depend upon whether the roots of the quadratic equation

$$Lx^2 + Rx + \frac{1}{C} = 0$$

are real or imaginary.

Let us first take the case when they are imaginary, i.e. when

$$R^2 < 4\frac{L}{C}.$$

In this case the solution of (2) takes the form

$$Q = A\epsilon^{-\frac{R}{2L}t}\cos\left\{\left(\frac{1}{LC}-\frac{R^2}{4L^2}\right)^{\frac{1}{2}}t+\alpha\right\}\dots\dots(3),$$

where A and α are arbitrary constants.

We see from this expression that Q is alternately positive and negative and vanishes at times following one another at the interval

$$\pi\bigg/\left\{\frac{1}{LC}-\frac{R^2}{4L^2}\right\}^{\frac{1}{2}}.$$

The charge Q is thus represented by a harmonic function whose amplitude decreases in geometrical progression as the time increases in arithmetical progression.

The discharge of the jar is oscillatory, so that if, for example, to begin with, the inside of the jar is charged positively, the outside negatively; then on connecting by the circuit the inside and the outside of the jar, the positive charge on the inside diminishes; when however it has all disappeared there is a current in the circuit, and the inertia of this current keeps it going, so that positive electricity still continues to flow from the inside of the jar; this loss of positive electricity causes the inside to become charged with negative electricity, while the outside gets positively charged. Thus the jar which had originally positive on the inside, negative on the outside, has now negative on the inside, positive on the outside. The potential difference developed in the jar by these charges tends

to stop the current and finally succeeds in doing so. When
this happens the charges on the inside and outside would
be equal and opposite to the original charges if the resistance of the circuit were negligible; if the resistance
is finite the new charges will be of opposite sign to the
old ones, but smaller. The current now begins to flow
in the opposite direction, and goes on flowing until the
inside is again charged positively, the outside negatively;
if there were no resistance the charges on the inside and
outside would regain their original values, so that the
state of the system would be the same as when the discharge began; if the resistance is finite the charges are
smaller than the original ones. The system goes on then
as before until the charges become too small to be appreciable. The charges in the jar and the currents in the
wire are thus periodic, the charges surging backwards and
forwards between the coatings of the jar.

The oscillatory character of the discharge was sus-
pected by Henry from observations on the magnetization
of needles placed inside a coil in the discharging circuit.
The preceding theory was given by Lord Kelvin in 1853.
The oscillations were detected by Feddersen in 1857.
The method he used consisted of putting an air break
in the wire circuit joining the inside to the outside of
the jar. This air break is luminous when a current passes
through it, shining out brightly when the current passing
through it is great, while it is dark when the current
vanishes. Hence if we observe the image of this air space
formed by reflection at a rotating mirror, it will, if the
discharge is oscillatory, be drawn out into a band with
dark and bright spaces, the interval between two dark
spaces depending on the speed of the mirror and the

frequency of the electrical vibrations. Feddersen observed that the appearance of the image of the air break formed by a rotating mirror was of this character. He showed moreover that the oscillatory character of the discharge was destroyed by putting a large resistance in the circuit, for he found that in this case the image of the air space was a broad band of light gradually fading away in intensity instead of a series of bright and dark bands.

When the discharge is oscillatory the frequency of the discharges is often exceedingly large, a frequency of a million complete oscillations a second being by no means a high value for such cases. We see by the expression (3) that when $R = 0$, the time of vibration is $2\pi\sqrt{LC}$; thus this time is increased when the self-induction or the capacity is increased. By inserting coils with very great self-induction in the circuit, Sir Oliver Lodge has produced such slow electrical vibrations that the sounds generated by the successive discharges form a musical note.

In the preceding investigation we have supposed that R^2 was less than $4L/C$; if however R is greater than this value, the solution of equation (2) changes its character, and we have now

$$Q = A\epsilon^{-\lambda_1 t} + B\epsilon^{-\lambda_2 t},$$

where $-\lambda_1$, $-\lambda_2$ are the roots of the quadratic equation

$$L\lambda^2 + R\lambda + \frac{1}{C} = 0.$$

Hence
$$\lambda_1 = \frac{R}{2L} + \sqrt{\frac{R^2}{4L^2} - \frac{1}{CL}},$$

$$\lambda_2 = \frac{R}{2L} - \sqrt{\frac{R^2}{4L^2} - \frac{1}{CL}}.$$

If we take $t = 0$ when the circuit is closed, then dQ/dt vanishes when $t = 0$ and we get, if Q_0 is the value of Q when $t = 0$,

$$Q = \frac{Q_0}{\lambda_1 - \lambda_2} (\lambda_1 \epsilon^{-\lambda_2 t} - \lambda_2 \epsilon^{-\lambda_1 t}),$$

$$\frac{dQ}{dt} = \frac{\lambda_1 \lambda_2}{\lambda_1 - \lambda_2} Q_0 (\epsilon^{-\lambda_1 t} - \epsilon^{-\lambda_2 t}).$$

Hence dQ/dt never vanishes except when $t = 0$ and when $t = \infty$. Thus Q which is zero when $t = \infty$ never changes sign. The charge in this case instead of becoming positive and negative never changes sign but continually diminishes, and ultimately becomes too small to be observed. This result is confirmed by Feddersen's observations with the rotating mirror.

The behaviour of the Leyden jar is analogous to that of a mass attached to a spring whose motion is resisted by a force proportional to the velocity. If M is the mass attached to the spring, x the extension of the spring, nx the pull of the spring when the extension is x, $r dx/dt$ the frictional resistance, then the equation of motion of the spring is

$$M \frac{d^2 x}{dt^2} + r \frac{dx}{dt} + nx = 0.$$

Comparing this with the equation for Q we see that if we compare the extension of the spring to the charge on the jar, then the coefficient of self-induction of the circuit will correspond to the mass attached to the spring, the electrical resistance of the circuit to the frictional resistance of the mechanical system, and the reciprocal of the capacity of the condenser to n, the stiffness of the spring.

The pulling out of the spring corresponds to the charging of the jar, the release of the spring to the completion of the circuit between the inside and the outside of the jar; when the spring is released it will if the friction is small oscillate about its position of equilibrium, the spring being alternately extended and compressed, and the oscillations will gradually die away in consequence of the resistance; this corresponds to the oscillatory discharge of the jar. If however the resistance to the motion of the spring is very great, if for example it is placed in a very viscous liquid like treacle, then when it is released it will move slowly towards its position of equilibrium but will never go through it. This case corresponds to the non-oscillatory discharge of the jar when there is great resistance in the circuit.

We have seen that the resistance of a conductor to a variable current is not the same as to a steady one, and thus since the currents which are produced by the discharge of a condenser are not steady, R, which appears in the expression (2), is not the resistance of the circuit to steady currents. Now R the resistance depends upon the frequency of the currents, while as the expression (3) shows, the frequency of the electrical vibrations depends to some extent on the resistance; hence the preceding solution is not quite definite, it represents however the main features of the case. For a complete solution we may refer the reader to *Recent Researches in Electricity and Magnetism*, J. J. Thomson, Art. 294.

246. Periodic Electromotive Force acting on a circuit containing a condenser.

Let an external electromotive force equal to $E \cos pt$ act on the circuit which connects the coatings of the jar, let C be the capacity of the jar, L the coefficient of self-induction, and R the resistance of the circuit connecting its coatings. Then if x is the charge on one of the coatings of the jar (which of the coatings is to be taken is determined by the condition that an increase in x corresponds to a current in the direction of the external electromotive force), we can prove in the same way as we proved equation (2) Art. 245, that

$$L\frac{d^2x}{dt^2} + R\frac{dx}{dt} + \frac{x}{C} = E\cos pt \ldots\ldots\ldots(1).$$

The solution of this equation is

$$x = \frac{E\sin(pt - \alpha)}{p\left\{\left(L - \frac{1}{Cp^2}\right)^2 p^2 + R^2\right\}^{\frac{1}{2}}} \ldots\ldots\ldots(2),$$

and thus

$$\frac{dx}{dt} = \frac{E\cos(pt - \alpha)}{\left\{\left(L - \frac{1}{Cp^2}\right)^2 p^2 + R^2\right\}^{\frac{1}{2}}} \ldots\ldots\ldots\ldots(3),$$

where

$$\tan \alpha = \frac{\left(L - \frac{1}{Cp^2}\right)p}{R}.$$

Comparing these equations with those of Art. 234 we see that the circuit behaves as if the jar were done away with and the self-induction changed from L to $L - 1/Cp^2$. We also see from (3) that if Cp^2 is greater than $1/2L$, the current produced by the electromotive force in the circuit broken by the jar (whose resistance is infinite) is actually greater than the current which would flow if the jar were replaced by a conductor of infinite conductivity. If $Cp^2 = 1/L$ the apparent self-induction of the circuit is zero, and the circuit behaves like an induction-less closed circuit of resistance R. Thus by cutting the circuit and connecting the ends to a condenser of suitable capacity we can increase enormously the current passing through the circuit. We can perhaps see the reason for this more clearly if we consider the behaviour of the mechanical system, which we have used to illustrate the oscillatory discharge of a Leyden jar, viz. the rectilinear motion of a mass attached to a spring and resisted by a frictional force proportional to the velocity. Suppose that X, an external force, acts on this system; then at any instant X must be in equilibrium with (1) the resultant of the rate of diminution of the momentum of the mass, (2) the force due to the compression or extension of the spring, (3) the resistance. If the frequency of X is very great, then for a given momentum (1) will be very large, so that unless (1) is counterbalanced by (2) a finite force of very great frequency will produce an exceedingly small momentum. Suppose however the frequency of the external force is the same as that of the free vibrations of the system when the friction is zero, then when the mass vibrates with this frequency, (1) and (2) will balance each other, so that all the external force has to do is to balance the resistance; the system will therefore behave like one without either mass or stiffness resisted by a frictional force.

247. A circuit containing a condenser is parallel with one possessing self-induction.

Let ABC, AEC, Fig. 119, be two circuits. Let L be the coefficient of self-induction of ABC, R the resistance of this circuit, C the capacity of the condenser in AEC, r

the resistance of wires leading from A and C to the plates. Then if i is the current through ABC, x the charge on the

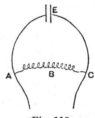

Fig. 119.

plate nearest to A, we have, neglecting the self-induction of the circuit AEC,

$$L\frac{di}{dt} + Ri = r\frac{dx}{dt} + \frac{x}{C},$$

since each of these quantities is equal to the electromotive force between A and C.

If $\qquad\qquad i = \cos pt,$

then $\qquad x = \dfrac{(L^2p^2 + R^2)^{\frac{1}{2}}}{\left\{\dfrac{1}{C^2} + r^2p^2\right\}^{\frac{1}{2}}}\sin(pt + \alpha),$

where $\qquad \alpha = \tan^{-1}\dfrac{Lp}{R} + \tan^{-1}\dfrac{1}{rpC}.$

Hence $\qquad \dfrac{dx}{dt} = \sqrt{\dfrac{L^2p^2 + R^2}{\dfrac{1}{C^2p^2} + r^2}}\cos(pt + \alpha).$

Thus the maximum current along AEC is to that along ABC as $\sqrt{L^2p^2 + R^2}$ is to $\sqrt{\dfrac{1}{C^2p^2} + r^2}$, or, if we can neglect the resistances of the wires to the condenser, as $\sqrt{L^2p^2 + R^2} : 1/Cp$. We see that for very high frequencies practically all the current will go along the condenser circuit.

Thus when the frequency is very high a piece of a circuit with a little electrostatic capacity will be as efficacious in robbing neighbouring circuits of current as if the places where the electricity accumulates were short-circuited by a conductor.

248. Lenz's Law. When a circuit is moved in a magnetic field in such a way that a change takes place in the number of tubes of magnetic induction passing through the circuit, a current is induced in the circuit; the circuit conveying this current being in a magnetic field will be acted upon by a mechanical force. Lenz's Law states that the direction of this mechanical force is such that the force tends to stop the motion which gave rise to the current. The result follows at once from the laws of the induction of currents. For suppose Fig. 120

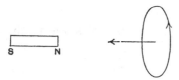

Fig. 120.

represents a circuit which, as it moves from right to left, encloses a larger number of tubes of induction passing through it from left to right. The current induced will tend to keep the number of tubes of induction unaltered, so that since the number of tubes of magnetic induction due to the external magnetic field which pass through the circuit from left to right increases as the circuit moves towards the left, the tubes due to the induced

current will pass through the circuit from right to left.
Thus the magnetic shell equivalent to the induced current
has the positive side on the left, the negative on the
right. Since the number of tubes of induction due to
the external field which pass through this shell in the
negative direction, i.e. which enter at the positive and
leave at the negative side, increases as the shell is moved
to the left, the force acting on the shell is, by Art. 214,
from left to right, which is opposite to the direction of
motion of the circuit.

There is a simple relation between the mechanical
and electromotive forces acting on the circuit. Let P be
the electromotive force, X the mechanical force parallel
to the axis of x, i the current flowing round the circuit,
u the velocity with which the circuit is moving parallel
to x, N the number of unit tubes of magnetic induction
passing through the circuit. Then

$$P = -\frac{dN}{dt},$$

and if the induced current is due to the motion of the
circuit

$$\frac{dN}{dt} = \frac{dN}{dx} \cdot u;$$

hence

$$P = -u\frac{dN}{dx}.$$

Again, by Art. 214, we have

$$X = i\frac{dN}{dx},$$

so that

$$Xu = -Pi.$$

If we wish merely to find the direction of the current
induced in a circuit moving in a magnetic field, Lenz's law
is in many cases the most convenient method to use.

An example of this law is afforded by the coil revolving in a magnetic field (Art. 237); the action of the magnetic field on the currents induced in the coil produces a couple which tends to stop the rotation of the coil. The magnets of galvanometers are sometimes surrounded by a copper box, the motion of the magnet induces currents in the copper, and the action of these currents on the magnets by Lenz's law tends to stop the magnet, and thus brings it to rest more quickly than if the copper box were absent. The quickness with which the oscillations of the moving coil in the Desprez-D'Arsonval Galvanometer (Art. 224) subside is another example of the same effect; when the coil moves in the magnetic field currents are induced in it, and the action of the magnetic field on these currents stops the coil. Again, if a magnet is suspended over a copper disc, and the disc is rotated, the movement of the disc in the magnetic field induces currents in the disc; the action of the magnet on these currents tends to stop the disc, and there is thus a couple acting on the disc in the direction opposite to its rotation. There must, however, be an equal and opposite couple acting on the magnet, i.e. there must be a couple on the magnet in the direction of rotation of the disc; this couple, if the magnet is free to move, will set it rotating in the direction of rotation of the disc, so that the magnet and the disc will rotate in the same direction. This is a well-known experiment; the disc with the magnet freely suspended above it is known as Arago's disc. Another striking experiment illustrating Lenz's law is to rotate a metal disc between the poles of an electro-magnet, the plane of the disc being at right angles to the lines of magnetic force; it is found that the work required to turn the disc when the magnet is 'on' is much greater than when it is 'off.' The extra work is accounted for by the heat produced by the currents induced in the disc.

249. Methods of determining the coefficients of self and mutual induction of coils. When the coils are circles, or solenoids, the coefficients of induction can be calculated. When, however, the coils are not of these simple shapes the calculation of the coefficients would be difficult or impossible ; they may, however, be determined by experiment by means of the following methods.

250. Determination of the coefficient of self-induction of a coil. Place the coil in *BD*, one of the

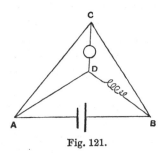

Fig. 121.

arms of a Wheatstone's Bridge, and balance the bridge for steady currents, insert in *CD* a ballistic galvanometer, and place a key in the battery circuit. When this key is pressed down so as to complete the circuit, although there will be no current through the galvanometer when the currents get steady, yet a transient current will flow through the galvanometer, in consequence of the electromotive forces which exist in *BD* arising from the self-induction of the coil. This current though only transient is very intense while it lasts and causes a finite quantity of electricity to pass through the galvanometer, producing a finite kick. We can calculate this quantity as follows: an electromotive force E in BD will produce a current through the galvanometer proportional to E, let this current be kE. In consequence of the self-induction of the coil there will be an electromotive force in BD equal to

$$-\frac{d}{dt}(Li),$$

where L is the coefficient of self-induction of the coil and i the current passing through the coil. This electromotive force will produce a current q through the galvanometer where q is given by the equation

$$q = -k\frac{d}{dt}(Li).$$

If Q is the total quantity of electricity which passes through the galvanometer

$$Q = \int q\, dt$$
$$= -k\int \frac{d}{dt}(Li)\, dt,$$

the integration extending from before the circuit is completed until after the currents have become steady. The right-hand side of this equation is equal to

$$-kLi_0,$$

where i_0 is the value of i when the currents are steady. By the theory of the ballistic galvanometer, given in Art. 225, we see that if θ is the kick of the galvanometer

$$Q = \sin \tfrac{1}{2}\theta \cdot \frac{HT}{\pi G},$$

T. E.   29 where T is the time of swing of the galvanometer needle, G the galvanometer constant, and H the horizontal component of the earth's magnetic force.

Hence we have

$$kLi_0 = \sin \tfrac{1}{2}\theta \cdot \frac{HT}{\pi G} \quad\dots\dots\dots\dots\dots(1).$$

Let us now destroy the balance of the Wheatstone's Bridge by inserting a small additional resistance r in BD, this will send a current p through the galvanometer. To calculate p we notice that the new resistance has approximately the current i_0 running through it, and the effect of its introduction is the same as if an electromotive force ri_0 were introduced into DB, this as we have seen produces a current kri_0 through the galvanometer; hence

$$p = kri_0.$$

This current will produce a permanent deflection ϕ of the galvanometer, and by Art. 222

$$p = \tan \phi \frac{H}{G},$$

or

$$kri_0 = \tan \phi \frac{H}{G} \quad\dots\dots\dots\dots\dots(2).$$

Hence from equations (1) and (2), we get

$$L = r \frac{\sin \tfrac{1}{2}\theta}{\tan \phi} \frac{T}{\pi}.$$

251. Determination of the coefficient of mutual induction of a pair of coils. Let A and B, Fig. 122, represent the pair of coils of which A is placed in series with a galvanometer, and B in series with a battery; this second circuit being provided with a key for breaking or closing the circuit.

Let R be the resistance of the circuit containing A. Suppose that originally the circuit containing B is broken and that the key is then pressed down, and that after the current becomes steady the current i flows through this circuit. Then before the key is pressed down no

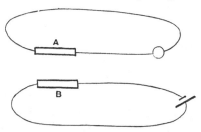

Fig. 122.

tubes of magnetic induction pass through the coil A, while when the current i flows through B the number of such unit tubes is Mi, where M is the coefficient of mutual induction between A and B. Thus the circuit containing A has received an electrical impulse equal to Mi, so that Q, the quantity of electricity flowing through the galvanometer, will be Mi/R, and if θ is the kick of the galvanometer, we have

$$\frac{Mi}{R} = \sin \tfrac{1}{2}\theta \, \frac{HT}{\pi G} \quad \dots\dots\dots\dots(1),$$

using the same notation as before. We can eliminate a good many of the quantities by a method somewhat similar to that used in the last case. Cut the circuit containing the coil A and connect its ends to two points on the circuit B separated by a small resistance S; then if R is

29—2

very large compared with S this will not alter appreciably the current flowing round B; on this supposition the current flowing round the galvanometer circuit will be

$$\frac{S}{R+S}\,i,$$

and if ϕ is the corresponding deflection of the galvanometer

$$\frac{S}{R+S}\,i = \tan\phi \cdot \frac{H}{G} \quad \ldots\ldots\ldots\ldots(2).$$

Hence from equations (1) and (2), we get

$$M = \frac{RS}{R+S}\,\frac{\sin\frac{1}{2}\theta}{\tan\phi}\,\frac{T}{\pi}.$$

252. Comparison of the coefficients of mutual induction of two pairs of coils. Let A, a be one pair of coils, B, b the other. Connect a and b in one circuit with the battery, and connect the points P and Q (Fig. 123)

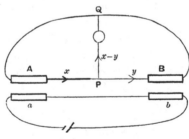

Fig. 123.

to the two electrodes of a ballistic galvanometer. Insert resistances in PAQ and PBQ until there is no kick of the galvanometer when the circuit through a and b is made or broken. Let R be the resistance then in PAQ,

S that in PBQ, and let M_1, M_2 be the coefficients of mutual induction between the coils Aa, Bb respectively, then

$$\frac{M_1}{R} = \frac{M_2}{S}.$$

To prove this we notice that, by Art. 190, if we have any closed circuit consisting of various parts, the sum of the products obtained by multiplying the resistance of each part by the current passing through it is equal to the electromotive force acting round the circuit. In the case when the electromotive forces are transient, we get by integrating this result, that the sum of the products got by multiplying the resistance of each part of the circuit by the quantity of electricity which has passed through it is equal to the electromotive impulse acting round the circuit. Let us apply this to our case: if i is the steady current flowing through the coils a and b, the electromotive impulse acting on A due to the closing of the circuit is $M_1 i$, while that on B is $M_2 i$. If x is the quantity of electricity which passes through A when the circuit through a, b is closed, y that through B, $x - y$ will be the quantity which passes through the galvanometer; hence applying the above rule to the circuit APQ, we have if K is the resistance of the galvanometer circuit

$$Rx + K(x - y) = M_1 i.$$

Applying the same rule to the circuit BPQ, we get

$$Sy + K(y - x) = M_2 i.$$

But if the total quantity which passes through the galvanometer is zero, we have $x = y$, and therefore

$$\frac{M_1}{R} = \frac{M_2}{S}.$$

253. Comparison of the coefficients of self-induction of two coils. Place the two coils whose coefficients of self-induction are L and N respectively in the arms AB, BD of a Wheatstone's Bridge, Fig. 121, balanced for steady currents, then adjust the resistances in AD, BD so that no kick of the galvanometer occurs when the battery circuit is made; these alterations in the resistances of AD and BD will entail proportional alterations in those of AC and BC in order to keep the bridge balanced for steady currents. Then when there is no kick of the galvanometer when the circuit is made, and no steady deflection when it is kept flowing, we have

$$\frac{L}{N} = \frac{P}{Q} = \frac{R}{S},$$

where P, Q, R, S are the resistances of the arms AD, BD, AC, BC respectively.

We can see this as follows: suppose we have a balanced Wheatstone's Bridge with the resistances in as above, then for steady currents the balance will be undisturbed if P and Q are altered in such a way that their ratio remains unchanged; but the alteration of P and Q in this way is equivalent to the introduction into AD and BD of electromotive forces proportional to P and Q. For since no current flows through the galvanometer the same current flows through AD as through BD, and the preceding statement follows by Ohm's Law. Hence we see that the introduction into the arms AD and BD of electromotive forces proportional to P and Q, will not alter the balance of the bridge, and, conversely, that if this balance is not altered by the introduction of an electromotive force A into the arm AD, and another, B, into the arm BD, then A/B must be equal to P/Q.

Now if we have coils in AD and BD whose coefficients of self-induction are L, N, then since after the current gets steady, the same current, i say, flows through each of these coils, there must be, whilst the current is getting steady, an impulse Li in AD, and another equal to Ni in BD. Since these impulses do not send any electricity through the galvanometer they must, by the preceding reasoning, be proportional to P and Q, hence

$$\frac{L}{N} = \frac{P}{Q}.$$

254. Heat developed by the hysteresis of iron. We can, as Dr John Hopkinson showed, deduce from the law of Electromagnetic Induction the expression given on p. 261 for the heat produced in iron per unit volume when the magnetic force undergoes a cyclical change. Take the case of a solenoid filled with iron and carrying a current whose value i is changing cyclically; let l be the length of the solenoid, n the number of turns of wire per unit length, a the area of cross section of the core and B the magnetic induction. The electromotive force in the solenoid due to induction is $-nla\dfrac{dB}{dt}$, hence the work spent by the current in time T in consequence of the presence of the iron is

$$\int_0^T inla\frac{dB}{dt} . dt.$$

But if H is the magnetic force

$$H = 4\pi ni,$$

so that the work spent by the current, appearing as heat in the iron, is equal to

$$\frac{1}{4\pi}\int_0^T laH\frac{dB}{dt}\,dt.$$

Since the volume of the iron is la, the heat produced per unit volume is

$$\frac{1}{4\pi}\int H\frac{dB}{dt}\,dt$$

$$=\frac{1}{4\pi}\int H dB.$$

This is the value already obtained on p. 261.

CHAPTER XII

ELECTRICAL UNITS:
DIMENSIONS OF ELECTRICAL QUANTITIES

255. In Art. 9 we defined the unit charge of electricity, as the charge which repelled an equal charge with unit mechanical force when the two charges were at unit distance apart and surrounded by air at standard temperature and pressure. When we know the unit charge the various other electrical units easily follow. Thus the unit current is the one that conveys unit charge in unit time; unit electric intensity is that which acts on unit charge with unit mechanical force; unit difference of potential is the potential between two points when unit work is done by the passage of unit charge from one point to the other. Unit resistance is the resistance between two points of a conductor between which the potential difference is unity when the conductor is traversed by unit current.

The step from the electrical to the magnetic quantities is made by means of the law that the work done when unit magnetic pole is taken round a closed circuit is equal to 4π times the current flowing through the circuit. This law is to some extent a matter of definition. All that is shown by experiment is that the work done when

unit pole is taken round the circuit is proportional to the
current flowing through the circuit, and, as long as the
current remains the same, is independent of the nature
of the substances passed through by the pole in its tour
round the circuit.   If we said that p times the work done
was equal to 4π times the current, these conditions would
still be fulfilled provided p was independent of the current,
the magnetic force and the nature of the substances in
the field.   Though, as we shall see later, it would be
possible to get a somewhat more symmetrical system of
units by a proper choice of p, yet in practice, to avoid
the introduction of an unnecessary constant, p is always
taken as unity.   When $p = 1$, it follows from Art. 210 that
the magnetic force at the centre of a circle of radius a
traversed by a current i is $2\pi i/a$; thus unit magnetic
force will be the force at the centre of a circle of radius
2π traversed by unit current.   Thus knowing the unit
current we can at once determine the unit magnetic force.
Having got the unit magnetic force, the unit magnetic
pole follows at once, since it is the pole which is
acted on by unit magnetic force with the unit mechanical force.   From these units we can go on and deduce
without ambiguity the units of the other magnetic quantities.   The System of units arrived at in this way is
called the Electrostatic System of Units.

Starting from the unit charge as defined in Art. 9,
we thus arrive at a unit magnetic pole.   In Art. 114,
however, we gave another definition of unit magnetic pole
deduced from the repulsion between two similar poles.
The unit magnetic pole as defined in Art. 114 does not
coincide with the unit pole at which we arrive, starting,
as we have just done, from the unit charge of electricity.
The numerical relation between the two units depends

upon what units of length and time we employ; if these are the centimetre and second, then the unit magnetic pole on the electrostatic system of units is about 3×10^{10} times as great as the unit pole defined in Art. 114.

Instead of starting with unit charge of electricity we may start with unit magnetic pole as defined in Art. 114. The units of the other magnetic quantities would at once follow from considerations similar to those by which we deduced the unit electrical quantities from the unit electrical charge. The electrical units would follow from the magnetic ones, by the principle that the magnetic force at the centre of a circular current of radius a is $2\pi i/a$, where i is the strength of the current; thus the unit current is that which produces unit magnetic force at the centre of a circle whose radius is 2π. In this way we can get the unit current, and from this the units of the other electrical quantities follow without difficulty. The System of units got in this way is called the Electromagnetic System of Units.

The electromagnetic system of units does not coincide with the electrostatic system. The electromagnetic unit charge of electricity bears to the electrostatic unit charge a ratio which depends on the units of length and time; if these are the centimetre and second the electromagnetic unit of electricity is found to be about 3×10^{10} times the electrostatic unit. The ratio of the electromagnetic unit of charge to the electrostatic unit is equal to the ratio of the electrostatic unit pole to the electromagnetic unit.

In the following table the relations between the electrostatic and electromagnetic units of various electric and magnetic quantities are given. Here v is the ratio of the electromagnetic unit charge of electricity to the electrostatic unit.

Quantity	Symbol	Electrostatic unit in terms of Electromagnetic
Quantity of Electricity	e	$1/v$
Electric intensity	F	v
Potential difference	V	v
Current	i	$1/v$
Resistance of a conductor	R	v^2
Electric Polarization	D	$1/v$
Capacity of a condenser	C	$1/v^2$
Strength of Magnetic Pole	m	v
Magnetic force	H	$1/v$
Magnetic induction	B	v
Magnetic permeability	μ	v^2
Coefficient of Self-Induction	L	v^2

Certain combinations of these quantities are equal to purely geometrical or dynamical quantities, such as length, force, energy. The numerical expression of such combinations must evidently be the same whatever system of units we employ; thus, for example, the mechanical force on a charge e placed in a field of electric intensity is Fe, but this force is a definite number of dynes, quite independent of any arbitrary system of measuring electric quantities, thus $F \times e$ must be the same whatever system of electrical units we employ.

The following are examples of such combinations.

$$\text{Time} = \frac{e}{i}.$$

$$\text{Length} = \frac{V}{F}.$$

$$\text{Force} = Fe; \ mH.$$

$$\text{Energy} = \tfrac{1}{2}Ve; \ \tfrac{1}{2}\frac{e^2}{C}; \ Ri^2t; \ \tfrac{1}{2}Li^2.$$

Energy per unit volume $= FD/8\pi; \ \mu H^2/8\pi.$

Thus since Fe is independent of the electrical units chosen, if we adopt a new system in which the unit of e is v times the old unit, the new unit of F must be $1/v$ times the old unit. Again, Ri^2 is another quantity unaltered by the change of units, so that if the new unit of i is v times the old, the new unit of R must be $1/v^2$ times the old unit.

Dimensions of Electrical Quantities

256. For the general theory of Dimensions we shall refer the reader to Maxwell's *Theory of Heat,* Chap. IV.; we shall in this chapter confine our attention to the dimensions of electrical quantities.

It may be well to state at the outset that the 'dimensions' of electrical quantities are a matter of definition and depend entirely upon the system of units we adopt. Thus we shall find that on the electromagnetic system of units a resistance has the same dimensions as a velocity, while on the electrostatic system of units it has the same dimensions as the reciprocal of a velocity. In fact we might choose a system of units so as to make any one electrical quantity of any assigned dimensions; when the dimensions of this are fixed that of the others becomes quite determinate.

A symbol representing an electrical quantity merely tells us how much of the quantity there is, and does not tell us anything about the nature of the quantity; this would require a dynamical theory of electricity. A theory of dimensions cannot tell us what electricity is; its object is merely to enable us to find the change in the numerical measure of a given charge of electricity or any other electrical quantity when the units of length, mass and time are changed in any determinate way.

We have to fix the electrical quantities by one or other of their properties. Thus, to take an example, we may fix a charge of electricity by the repulsion it exerts on an equal charge, as is done in the electrostatic system of units, or by the force experienced by a magnetic pole when the charge is being transferred from one place to another by a current, as is done in the electromagnetic system; these two measures are of different dimensions. To take a simpler case we might fix a quantity of water by the number of hydrogen atoms it contains, by its mass, or by its volume at a definite temperature; all these measures would be of different dimensions.

On the electrostatic system of units the force between two equal charges e, separated by a distance L in a medium whose specific inductive capacity is K, is e^2/KL^2, and since this is of the dimensions of a force we have the dimensional equation

$$\frac{e^2}{KL^2} = \frac{ML}{T^2} \quad \ldots\ldots\ldots\ldots\ldots\ldots(1),$$

M, L, T representing mass, length and time.

This result, with the meaning assigned to K in Art. 68, is only true on the electrostatic system of units. We may, however, generalize the meaning of K and say that whatever be the system of units, the repulsion between the charges is e^2/KL^2, where K is defined as the 'specific inductive capacity of the medium on the new system of units.' We may regard this as the definition of K on this system. The ratio of the K's for two substances on this system is of course the same as the ratio of the K's on the electrostatic system. We shall regard the dimensions of K as indeterminate and keep them in the expression for the dimensions of the electrical quantities[1]. From equation (1) we have the dimensional equation

$$e = M^{\frac{1}{2}} L^{\frac{3}{2}} T^{-1} K^{\frac{1}{2}}.$$

Similarly on the electromagnetic system of units the repulsion between two poles of strength m separated by a distance L in a medium whose magnetic permeability is μ is $m^2/\mu L^2$, μ for this system of units being a quantity of no dimensions. We shall suppose that whatever be the system of units the force between the poles is equal to $m^2/\mu L^2$: where μ thus determined is defined as the magnetic permeability of the medium on this system of units. Thus, for example, if m is the measure, on the electrostatic system of units, of the strength of a pole, the force between two equal poles separated by unit distance in air is not m^2 but $9 \times 10^{20} m^2$. Hence we say the magnetic permeability of air on the electrostatic system of units is $1/9 \times 10^{20}$. We shall regard the dimensions of μ as being left undetermined and retain μ in the expressions for the dimensions of the electric quantities. Since $m^2/\mu L^2$ is of the dimensions of a force we have the dimensional equation

$$m = M^{\frac{1}{2}} L^{\frac{3}{2}} T^{-1} \mu^{\frac{1}{2}}.$$

We shall find it instructive to suppose that the electric and magnetic units are connected together by the relation that p times the work done by unit pole in traversing a closed circuit is equal to 4π times the current flowing through the circuit: the convention made on both the electrostatic and magnetic systems is that p is a quantity

---

[1] Rücker, *Phil. Mag.* vol. 27, p. 104.

of no dimensions and always equal to unity. We shall for the present leave the dimensions of p undecided.

The dimensional equation connecting the electric and magnetic quantities is therefore

$$p \times H \times L = i,$$

where H is magnetic force, L a length and i a current.

Taking this relation and starting with the electric charge, we can get by the equations given in Art. 255 the dimensions of all the electrical and magnetic quantities in terms of M, L, T, p, K: or starting with the magnetic pole we can get them in terms of M, L, T, p, μ. The results for some of the more important electrical quantities are given in the following table.

Quantity	Symbol	Dimensions in terms of K and p	Dimensions in terms of μ and p
Charge	e	$K^{\frac12}M^{\frac12}L^{\frac32}T^{-1}$	$p\mu^{-\frac12}M^{\frac12}L^{\frac12}$
Electric intensity	F	$K^{-\frac12}M^{\frac12}L^{-\frac12}T^{-1}$	$p^{-1}\mu^{\frac12}M^{\frac12}L^{\frac12}T^{-2}$
Potential difference	V	$K^{-\frac12}M^{\frac12}L^{\frac12}T^{-1}$	$p^{-1}\mu^{\frac12}M^{\frac12}L^{\frac32}T^{-2}$
Current	i	$K^{\frac12}M^{\frac12}L^{\frac32}T^{-2}$	$p\mu^{-\frac12}M^{\frac12}L^{\frac12}T^{-1}$
Resistance	R	$K^{-1}L^{-1}T$	$p^{-2}\mu LT^{-1}$
Electric polarization	D	$K^{\frac12}M^{\frac12}L^{-\frac12}T^{-1}$	$p\mu^{-\frac12}M^{\frac12}L^{-\frac32}$
Capacity	C	KL	$p^2\mu^{-1}L^{-1}T^2$
Specific inductive capacity	K	K	$p^2\mu^{-1}L^{-2}T^2$
Strength of Magnetic pole	m	$pK^{-\frac12}M^{\frac12}L^{\frac12}$	$\mu^{\frac12}M^{\frac12}L^{\frac32}T^{-1}$
Magnetic force	H	$p^{-1}K^{\frac12}M^{\frac12}L^{\frac12}T^{-2}$	$\mu^{-\frac12}M^{\frac12}L^{-\frac12}T^{-1}$
Magnetic induction	B	$pK^{-\frac12}M^{\frac12}L^{-\frac32}$	$\mu^{\frac12}M^{\frac12}L^{-\frac12}T^{-1}$
Magnetic permeability	μ	$p^2K^{-1}L^{-2}T^2$	μ

We see from this table that the dimensions of K, μ, and p must on all systems of measurement be connected by the relation

$$\frac{p^2}{\mu K} = \frac{L^2}{T^2} = (\text{velocity})^2.$$

On Maxwell's theory of the electric field $p/\sqrt{\mu K}$ is equal to the velocity with which electric disturbances travel through a medium whose magnetic permeability is μ and specific inductive capacity K.

On the electrostatic system of units K is of no dimensions, as the specific inductive capacity of air is taken as unity whatever may be the units of mass, length and time. Also on this system p is by hypothesis of no dimensions, being always equal to unity. Hence the dimensions of the electrical quantities on this system of units are got by omitting p and K in the third column of the table.

On the electromagnetic system of units μ is of no dimensions, the magnetic permeability of air being taken as unity whatever the units of mass, length and time; p is also of no dimensions on this system. Hence the dimensions of the electrical quantities on this system of units are got by omitting μ and p from the fourth column in the table.

Another system of units could be got by taking μ and K as of no dimensions and p a velocity. If this velocity were taken equal to the ratio of the electromagnetic unit charge to the electrostatic unit, then the unit of electric charge on this system would be the ordinary electrostatic unit of that quantity, while the unit magnetic pole would be the unit as defined on the electromagnetic system. This system would thus have the advantage that the electric quantities would be as defined in the electrostatic system, while the magnetic quantities would be as defined in the magnetic system, and we should not have to introduce any new definitions: whereas if we use the electrostatic system we have to define all the magnetic quantities afresh, and if we use the electromagnetic system we have to re-define all the electrical ones[1].

This system is however never used in practice; the electromagnetic system or one founded upon it is universally used in Electrical Engineering, and the electrostatic system is used for special classes of investigations.

257. The units of resistance, of electromotive force, of capacity on the electromagnetic system are either too large or too small to be practically convenient: hence new units which are definite multiples or submultiples of the electromagnetic units are employed. These units and their relation to the electromagnetic system of units (when the units of length, mass and time are the centimetre, gramme and second) are given in the following table.

The unit of resistance is called the Ohm and is equal to 10^9 electromagnetic units.

---

[1] It should be noticed that it is only when the electromagnetic system of units is used that 'magnetic induction' has the meaning assigned to it in Art. 153. If we use any system of units in which we start from electrical quantities, the 'magnetic induction through unit area' appears as the quantity whose rate of variation is equal to p times the electromotive force round the boundary of the area. The magnetic induction defined in this way is always proportional to the magnetic induction as defined in Art. 153. The two are however only identical on the electromagnetic system of units. With the definition of Art. 153 the magnetic induction is of the same dimensions as magnetic force, since they are both the mechanical force on a unit pole when placed in cavities of different shapes.

The unit of electromotive force is called the Volt and is equal to 10^8 electromagnetic units.

The unit of current is called the Ampère and is equal to 10^{-1} electromagnetic units.

The unit of charge is called the Coulomb and is equal to 10^{-1} electromagnetic units.

The unit of capacity is called the Farad and is equal to 10^{-9} electromagnetic units.

The Microfarad is equal to 10^{-15} electromagnetic units.

The Ampère is the current produced by a Volt through an Ohm.

We shall now proceed to explain the methods by which the various electrical quantities can be measured in terms of these units: when the quantity is so measured it is said to be determined in *absolute measure*.

258. Determination of a Resistance in Absolute Measure. The method given in Art. 226 enables us to compare two resistances, and thus to find the ratio of any resistance to that of an arbitrary standard such as the resistance of a column of mercury of given length and cross section when at a given temperature. In order to make use of the electromagnetic system of units we must find the number of electromagnetic units in our standard resistance, or what amounts to the same thing we must be able to specify a conductor whose resistance is the electromagnetic unit of resistance.

The first method we shall describe, that of the revolving coil, was suggested by Lord Kelvin, and carried out by a committee of the British Association, who were the first to measure a resistance in absolute measure. The method was also one of those used by Lord Rayleigh and Mrs Sidgwick in their determination of the Ohm.

When a coil of wire spins about a vertical axis in the earth's magnetic field, currents are generated in the coil; these currents produce a magnetic force at the centre of the coil. If a magnet is placed at the centre of the coil, this magnetic force gives rise to a couple on the magnet tending to twist the magnet in the direction in which the coil is rotating. The resistance of the coil may be deduced from the deflection of the magnet as follows.

Let H be the horizontal component of the earth's magnetic force, A the area enclosed by one turn of the coil, n the number of turns, θ the angle the plane of the coil makes with the magnetic meridian; let the coil revolve with uniform velocity ω, so that we may put

$$\theta = \omega t.$$

The number of tubes of magnetic induction passing through the coil is equal to

$$nAH \sin \theta,$$

and the rate of diminution of this is

$$- nAH\omega \cos \omega t.$$

Hence, if L is the coefficient of self-induction of the coil, R its resistance, and i the current flowing through the coil, the current being taken as positive when the lines of magnetic force due to the current and those due to the earth pass through the circuit in the same direction, we have

$$L \frac{di}{dt} + Ri = - nAH\omega \cos \omega t.$$

Hence, as in Art. 237, we have

$$i = - \frac{nAH\omega}{R^2 + \omega^2 L^2} \{R \cos \omega t + L\omega \sin \omega t\}.$$

Now if unit current through the coil produces a magnetic force G at the centre, the current i through the coil will produce a magnetic force $Gi \cos \omega t$ at right angles to the magnetic meridian, and a force $Gi \sin \omega t$ along the magnetic meridian, since $\theta = \omega t$. Hence the magnetic force due to the currents in the coil has a component

$$-\frac{nAHG\omega R}{2(R^2+\omega^2 L^2)} - \frac{nAHG\omega}{2(R^2+\omega^2 L^2)}\{R\cos 2\omega t + L\omega \sin 2\omega t\},$$

at right angles to the magnetic meridian; and a component

$$-\frac{nAHGL\omega^2}{2(R^2+\omega^2 L^2)} - \frac{nAHG\omega}{2(R^2+\omega^2 L^2)}\{R\sin 2\omega t - L\omega \cos 2\omega t\},$$

along the magnetic meridian.

Now suppose we have a magnet at the centre of the coil, and let the moment of inertia of this magnet be so great that the time of swing is very large compared with the time of revolution of the coil. The magnetic force acting on the magnet due to the current induced in the coil consists, as we see, of two parts, one constant, the other periodic, the frequency being twice that of the revolution of the coil. By making the moment of inertia of the magnet great enough we may make the effect of the periodic terms as small as we please; we shall suppose that the magnet is heavy enough to allow us to neglect the effect of the periodic terms; when this is done the magnetic force at the centre has a component equal to

$$-\frac{nAHG\omega R}{2(R^2+\omega^2 L^2)}$$

at right angles to the magnetic meridian, and one equal to

$$H - \frac{nAHGL\omega^2}{2(R^2+\omega^2 L^2)}$$

along it.

Hence if ϕ is the angle the axis of the magnet at the centre of the coil makes with the magnetic meridian,

$$\tan \phi = \frac{\dfrac{1}{2}\dfrac{nAHG\omega R}{R^2 + \omega^2 L^2}}{H - \dfrac{1}{2}\dfrac{nAHGL\omega^2}{R^2 + \omega^2 L^2}},$$

or

$$\tan \phi = \frac{\dfrac{1}{2}\dfrac{nAG\omega R}{R^2 + \omega^2 L^2}}{1 - \dfrac{1}{2}\dfrac{nAGL\omega^2}{R^2 + \omega^2 L^2}}.$$

This equation enables us to find R, as A, G, L can be calculated from the dimensions of the rotating coil. When $L\omega$ is small compared with R the equation reduces to the simple form

$$\tan \phi = \frac{1}{2}\frac{nAG\omega}{R}.$$

When the coil consists of a single ring of wire of radius a, $n = 1$, $A = \pi a^2$, $G = 2\pi/a$; hence

$$\tan \phi = \frac{\pi^2 a\omega}{R}.$$

Thus by this method we compare R, which, by Art. 256, is of the dimensions of a velocity, with the velocity of a point on the spinning coil.

The preceding investigation is only approximate as we have neglected the magnetic field due to the magnet placed at the centre of the ring.

259. Lorenz's Method. This was also one of the methods used by Lord Rayleigh and Mrs Sidgwick in their determination of the Ohm. It depends upon the principle that if a conducting disc spins in a magnetic field which is symmetrical about the axis of rotation, and if a circuit is formed by a wire, one end of which is connected to the axis of rotation while the other end presses against the rim of the disc, an electromotive force proportional to the angular velocity will act round the circuit.

We can determine this electromotive force by finding the couple acting on the disc when a current flows round this circuit.

Let I be the current flowing through the wire. When this current enters the disc at its centre it will spread out; let q be the radial current crossing unit length of the circumference of a circle of radius r at the point defined by θ. Let $r\,dr\,d\theta$ be an element of the area of the disc. The radial current flowing through this area is equal to $qr\,d\theta$. Hence by Art. 214, if H is the magnetic force normal to the disc at this area, the tangential mechanical force acting on the area is equal to $Hqr\,dr\,d\theta$. The moment of this force about the axis of the disc is equal to

$$Hqr^2\,dr\,d\theta\,;$$

hence the couple acting on the disc is equal to

$$\iint Hqr^2\,dr\,d\theta,$$

the integration being extended over the area of the disc.

Since the current flowing across a circle drawn on the disc, with its centre at the centre of the disc, must equal the current I flowing into the disc, we have

$$\int qr\,d\theta = I.$$

Since the magnetic field is symmetrical about the axis of rotation, H is independent of θ, hence the couple acting on the disc is equal to

$$I\int Hr\,dr.$$

If N be the number of tubes of magnetic induction passing through the disc

$$N = \int H 2\pi r\, dr,$$

and thus the couple acting on the disc is equal to

$$\frac{1}{2\pi} IN.$$

Now suppose there is a battery whose electromotive force is E in the circuit, then in the time δt the work done by the battery is $EI\delta t$; this work is spent in heating the circuit and in driving the disc. The angle turned through by the disc in this time is $\omega \delta t$, if ω is the angular velocity of the disc; hence the mechanical work done is equal to

$$\frac{1}{2\pi} IN\omega \delta t.$$

By Joule's law the mechanical equivalent of the heat produced in the circuit is equal to

$$RI^2\delta t,$$

where R is the resistance of the circuit. Hence we have by the Conservation of Energy

$$EI\delta t = RI^2\delta t + \frac{1}{2\pi} IN\omega \delta t,$$

or
$$I = \frac{E - \dfrac{1}{2\pi} N\omega}{R};$$

hence there is a counter-electromotive force in the circuit equal to

$$\frac{1}{2\pi} N\omega.$$

This case illustrates the remark made on page 394, since from Ampère's law of the mechanical force acting on currents on a magnetic field we have deduced, by the aid of the principle of the Conservation of Energy, the expression for the electromotive force due to induction, and have thus proved by dynamical principles that the induction of currents is a consequence of the mechanical force exerted by a magnet on a circuit conveying a current.

In Lord Rayleigh's experiments, the disc was placed between two coils through which a current passed, and the axis of the disc and of the two coils were coincident. The magnetic field acting on the disc may be considered as approximately that due to the current through the coils, as this field is very much more intense than that due to the earth. Hence if i is the current through the coils, M the coefficient of mutual induction between the coils and a circuit coinciding with the rim of the disc,

$$N = Mi.$$

So that the electromotive force due to the rotation of the disc is

$$\frac{Mi\omega}{2\pi}.$$

The experiment was arranged as in the diagram, Fig. 124; a galvanometer was placed in the circuit connecting the centre of the disc and the rim, and this circuit was connected to two points P, Q in the circuit in series with the coils, and the resistance between P and Q was adjusted until no current passed through the galvanometer. If R is the resistance between P and Q, and if a current i flows through PQ the E.M.F. between P and Q will be Ri, but, since there is no current through the galvanometer, this balances the electromotive force due to the rotation of the disc; hence

$$Ri = \frac{Mi\omega}{2\pi},$$

or

$$R = \frac{M\omega}{2\pi}.$$

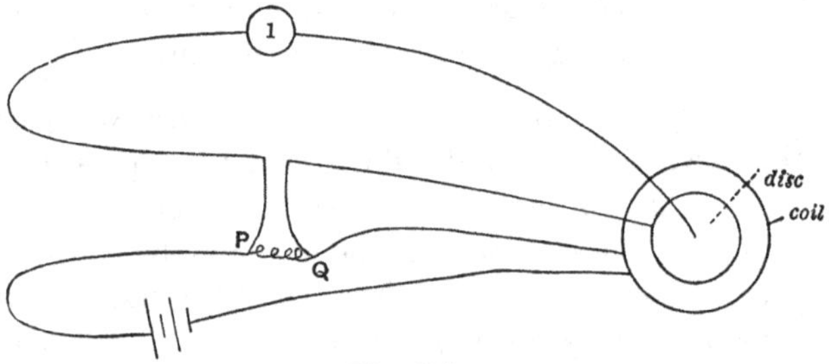

Fig. 124.

Since M can be calculated from the dimensions of the coil and the disc, this formula gives us R in absolute measure.

260. The method given in Art. 251 for determining a coefficient of mutual induction in terms of a resistance may be used to determine a resistance in absolute measure. If we use a pair of coils whose coefficient of mutual induction can be determined by calculation, then equation (2) of Art. 251 will give the absolute measure of a resistance. This method has been employed by Mr Glazebrook.

The result of a large number of experiments made by the preceding methods is that the Ohm is the resistance at 0° C. of a column of mercury 106·3 cm. long and 1 sq. millimetre in cross section.

For a comparison of the relative advantages of the preceding methods the student is referred to a paper by

Lord Rayleigh in the *Philosophical Magazine* for November, 1882.

261. Absolute Measurement of a Current. A current may be determined by measuring the attraction between two coils placed in series with each other and with their planes parallel and at right angles to the line joining their centres. If i is the current through the coils, M the coefficient of mutual induction between the coils, x the distance between their centres, the attraction between the coils is equal to

$$-\frac{dM}{dx}\,i^2.$$

By attaching one of the coils to the scale-pan of a balance and keeping the other fixed we can measure this force, and hence if we calculate dM/dx from the dimensions of the coils we can determine i in absolute measure.

The unit current is very conveniently specified by the amount of silver deposited from a solution of silver nitrate through which this current has been flowing for a given time.

Lord Rayleigh found that the Ampère is the current which flowing uniformly for one second would cause the deposition of ·001118 gramme of silver.

262. The unit electromotive force is that acting on a conductor of unit resistance when conveying unit current. A practical standard of electromotive force is the Clark cell (Art. 183), whose electromotive force at $t°$ Centigrade is equal to

$$1·434\,\{1 - ·00077\,(t - 15)\}\text{ volts.}$$

263. Ratio of Electrostatic and Electromagnetic Units. The table given on page 460 shows that the ratio of the measure of any electrical quantity on the electrostatic system of measurement to the measure of the same quantity on the electromagnetic system, is always some power of a certain quantity which we denoted by "v," and which is the ratio of the electromagnetic unit of electric charge to the electrostatic unit.

The measurement of the same electrical quantity on the two systems of units will enable us to find "v." The quantity which has most frequently been measured with this object is the capacity of a condenser. The electrostatic measure of the capacity can be calculated from the dimensions of the condenser; thus the electrostatic measure of the capacity of a sphere is equal to its radius; the capacity of two concentric spheres of radii a and b is $ab/(b-a)$; the capacity of two coaxial cylinders of length l, radii a and b, is $\frac{1}{2}l/\log b/a$. Thus if we choose a 'condenser of suitable shape the electrostatic measure can be calculated from its dimensions.

The electromagnetic measure can be determined by the following method due to Maxwell. One of the arms AC of a Wheatstone's Bridge is cut at P and Q (Fig. 125), one plate of the condenser is connected to P, the other to a vibrating piece R which oscillates backwards and forwards between P and Q; when R comes into contact with Q the condenser gets charged, when into contact with P it gets discharged. The current through the galvanometer may be divided into two parts. There is first a steady current which flows through AD when no electricity is flowing into the condenser, this we shall denote by $\dot{y}$. Besides this there is at times a transient current which flows while the condenser is being charged. We shall suppose that each time the condenser is being charged a quantity of electricity equal to Y flows through DA in the opposite direction to $\dot{y}$. Then if the condenser is charged n times a second the amount which flows through the galvanometer owing to the charging of the condenser is nY. If the time of swing

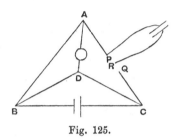

Fig. 125.

of the galvanometer needle is very long compared with $1/n$ of a second this will produce the same effect on the galvanometer as a steady current whose intensity is nY flowing from D to A. Thus if $nY = \dot{y}$, the current due to the repeated charging of the condenser will just balance the steady current and there will be no deflection of the galvanometer.

We now proceed to find Y. This is evidently equal to the quantity of electricity which would flow from A to D if there were no electromotive force in the wire BC and the plates of the condenser with the greatest charge they acquire in the experiment were connected to P and Q respectively.

Let $\dot{Z}$ be the current from the condenser along PA during the discharge, $\dot{Y}$ the current along AD, $\dot{W}$ the current along BD. Let the resistances of AB, BC, CD, DB, DA be c, a, γ, β, α respectively. Let the coefficients of self-induction of these circuits be L_1, L_2, L_3, L_4, L_5 respectively. Then from the circuit ABD, we have

$$L_5 \frac{d^2 Y}{dt^2} + L_1 \frac{d^2 (Y - Z)}{dt^2} - L_4 \frac{d^2 W}{dt^2} + \alpha \frac{dY}{dt}$$
$$+ c \left\{ \frac{dY}{dt} - \frac{dZ}{dt} \right\} - \beta \frac{dW}{dt} = 0.$$

Integrating from just before discharging until after the condenser is completely discharged, and remembering that both initially and finally $\dot{Y}$, $\dot{Z}$, $\dot{W}$ vanish, we have

$$\alpha Y + c (Y - Z) - \beta W = 0 \quad \dots\dots\dots\dots(1),$$

where Y, Z, W are the quantities of electricity which have passed during the discharge through AD, PA, and BD respectively.

Similarly from the circuit DBC, we have

$$(\beta + \gamma + a) W + (\gamma + a) Y - aZ = 0 \quad \dots\dots(2).$$

We find from equations (1) and (2)

$$Y = \frac{c (\beta + \gamma + a) + a\beta}{\beta (\gamma + a) + (\alpha + c) (\beta + \gamma + a)} Z \dots\dots(3).$$

Now Z is the maximum charge in the condenser; hence if C is capacity of the condenser, and **A** and **C** the potentials of A and C respectively when the charge is a maximum, i.e. when no current is flowing into the condenser,

$$Z = C \{ \mathbf{A} - \mathbf{C} \}.$$

If $\dot{y}$ is the current flowing through AD when no current is flowing in the condenser, and **D** denotes the potential of D,

$$\mathbf{A} - \mathbf{D} = \alpha \dot{y},$$
$$\mathbf{D} - \mathbf{C} = \gamma \left\{ 1 + \frac{\alpha + c}{\beta} \right\} \dot{y},$$
$$\therefore \mathbf{A} - \mathbf{C} = \left\{ \alpha + \gamma + \frac{\gamma}{\beta} (\alpha + c) \right\} \dot{y}.$$

263]   DIMENSIONS OF ELECTRICAL QUANTITIES   479

Hence by equation (3)

$$Y = \frac{c(\beta + \gamma + a) + a\beta}{\beta(\gamma + a) + (\alpha + c)(\beta + \gamma + a)} \left\{ \alpha + \gamma + \frac{\gamma}{\beta}(\alpha + c) \right\} C\dot{y}.$$

But when there is no deflection of the galvanometer

$$nY = \dot{y};$$

hence

$$\frac{1}{nC} = \frac{c(\beta + \gamma + a) + a\beta}{\beta(\gamma + a) + (\alpha + c)(\beta + \gamma + a)} \left\{ \alpha + \gamma + \frac{\gamma}{\beta}(\alpha + c) \right\}.$$

If we know the resistances and n, we can deduce from this equation the value of C in electromagnetic measure. In practice the resistance of the battery a is very small compared with the other resistances, hence putting $a = 0$, we find that approximately

$$\frac{1}{nC} = \frac{c\gamma}{\beta} \frac{\left\{ 1 + \dfrac{\alpha\beta}{\gamma(\alpha + c + \beta))} \right\}}{1 - \dfrac{\beta^2}{(\alpha + c + \beta)(\beta + \gamma)}}.$$

By this method we find the electromagnetic measure of the capacity of a condenser; the electrostatic measure can be found from its dimensions.

Now by Art. 255

$$v^2 = \frac{\text{electrostatic measure of a condenser}}{\text{electromagnetic measure of the same condenser}}.$$

Experiments made by this method show that

$$v = 3 \times 10^{10} \text{ cm./sec. very nearly.}$$

CHAPTER XIII

DIELECTRIC CURRENTS AND THE ELECTROMAGNETIC THEORY OF LIGHT

264. The Motion of Faraday Tubes. Dielectric Currents. In Chapter XI. we considered the relation between the currents in the primary and secondary circuits when an alternating current passes through the primary circuit, we did not however discuss the phenomena occurring in the dielectric between the circuits. As we regard the dielectric as the seat of the energy due to the distribution of the currents, the study of the effects in the dielectric is of primary importance. We owe to Maxwell a theory, now in its main features universally accepted, by which we are able to completely determine the electrical conditions, not merely in the conductors but also in every part of the field. We shall also see that Maxwell's views lead to a comprehensive theory of optical as well as of electrical phenomena, and enable us by means of electrical principles to explain the fundamental laws of Optics.

Before specifying in detail the principles of Maxwell's theory, we shall endeavour to show by the consideration of some simple cases that in considering the relation between the work done in taking unit magnetic pole round a closed circuit and the current flowing through that circuit (see Art. 203), we must include under the term *current*, effects other than the passage of electricity through conducting media, if we are to retain the conception that the dielectric is the seat of the energy in electric and magnetic phenomena.

Let us consider the case of a long, straight, cylindrical conductor carrying an alternating electric current. In the dielectric around this wire there is a magnetic field, and, according to the views enunciated in Art. 163, there is in a unit volume of the dielectric at a place where the magnetic force is H an amount of energy equal to $\mu H^2/8\pi$. As the alternating current changes in intensity, the energy in the surrounding field changes, and this change in the energy must be due to the motion of energy from one part of the field to another, the energy moving radially towards or away from the wire conveying the current. If the dielectric medium possesses inertia, and if its properties in any way resemble those of any kind of matter with which we are acquainted, the energy cannot travel from one place to another with an infinite velocity.

As the alternating current changes, the energy in the field will change also; when the current is passing through its zero value, it is evident that the magnetic energy cannot now vanish throughout the field, for we assume that the energy travels at a finite rate, and it is only a finite time since the current was finite. If the magnetic energy did vanish it would imply that the energy could travel over a distance, however great, in a finite time. If, however, the magnetic energy does not vanish simultaneously all over the field, there must be places where the magnetic force does not vanish. But the current through the conductor vanishes and there are no magnetic substances in the field. Hence we conclude that unless we assume that the energy in the magnetic field càn travel from one place to another with an infinite velocity, we must admit that in a variable field magnetic forces can arise apart from magnets or electric currents through conductors.

265. Let us now see if we can find any clue as to what produces the magnetic field under these circumstances. Let us consider the following simple case. Let A, B (Fig. 126) be two vertical metal plates forming a parallel

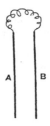

Fig. 126.

plate condenser, and let the upper ends of these plates be connected by a wire of high resistance. Suppose that initially the plate A is charged with a uniform distribution of positive electricity while B is charged with an equal distribution of negative electricity. If the plates are disconnected, horizontal Faraday tubes at rest will stretch from one plate to the other. When the plates are connected by the wire the horizontal Faraday tubes will move vertically upwards towards the wire. Let v be the velocity of these tubes, and σ the surface density of the electricity on the plates, then the upward current passing across unit length in the plate A and the downward current in B are equal to $v\sigma$. By Art. 209 these currents will produce a uniform magnetic field between the plates, the magnetic force being at right angles to the plane of the paper and its magnitude equal to $4\pi v\sigma$. If N is the number of Faraday tubes passing through unit area of a plane in the dielectric parallel to the plates of the condenser $N = \sigma$. Thus the magnetic force between the planes is equal to $4\pi Nv$. The condition of things between the plates is such that we have the Faraday tubes moving at right angles to themselves, and that we have also a magnetic force at right angles both to the Faraday tubes and to the direction in which they are moving; while the intensity of this force is equal to 4π times the product of the number of tubes passing through unit area and the velocity of these tubes.

Let us now see what are the consequences of generalizing this result, and of supposing that the relation between the magnetic force and the Faraday tubes which exists in this simple case is generally applicable to all magnetic fields. Suppose then that whenever we have movements of the Faraday tubes we have magnetic force and conversely, and that the relation between the magnetic force and the Faraday tubes is that the magnetic force is equal to 4π times the product of the 'polarization' (Art. 70) and the velocity of the Faraday tubes at right angles to the direction of polarization; and that the direction of the magnetic force is at right angles to both the direction of polarization and the direction in which the Faraday tubes are moving.

31—2

We shall begin by considering what on this view is the physical meaning of $H' \times OO'$, where OO' is a line so short that the magnetic force may be regarded as constant along its length, and H' is the component of the magnetic force along OO'.

Let OA (Fig. 127) represent in magnitude and direction the velocity of the Faraday tubes, and OP the polarization;

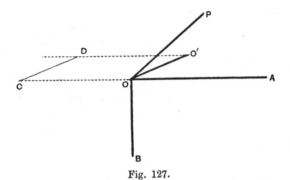

Fig. 127.

then if OB represents the magnetic force, OB will be at right angles to OA and OP and equal to

$$4\pi \cdot OA \cdot OP \sin\phi,$$

where ϕ is the angle POA. The component H' of the magnetic force along OO' will be

$$4\pi \cdot OA \cdot OP \sin\phi \cos\theta,$$

where θ is the angle BOO'. Thus we have

$$H' \times OO' = 4\pi \cdot OA \cdot OP \cdot OO' \sin\phi \cos\theta$$

$$= 24\pi\Delta \quad \dots\dots\dots\dots\dots\dots\dots\dots\dots(1),$$

where Δ is the volume of the tetrahedron three of whose sides are OA, OP, OO'.

Let us now find the number of Faraday tubes which cross OO' in unit time. To do this, draw OC and $O'D$ equal and parallel to AO, OA being the velocity of the Faraday tubes. Then the number of tubes which cross OO' in unit time is the number of tubes passing through the area $OCDO'$.

The area of the parallelogram $OCDO'$ is equal to

$$OA \times OO' \sin AOO'.$$

The number of tubes passing through it is therefore

$$OP \times \sin \theta' \times OA \times OO' \sin AOO'\ldots\ldots\ldots(2),$$

where θ' is the angle between OP and the plane of the parallelogram $OCDO'$; this is the same as the angle between OP and the plane AOO'. But

$$6\Delta = OP \times \sin \theta' \times OA \times OO' \sin AOO',$$

where Δ as before is the volume of the tetrahedron $POO'A$. Hence from (1) and (2) we see that

$$H' \times OO' = 4\pi \text{ (number of Faraday tubes crossing } OO' \text{ in}$$
$$\text{unit time).}$$

Thus $\int H'ds$ where the integral is taken round a closed curve is equal to 4π times the number of tubes which pass inwards across the curve in unit time.

In Art. 203 $\int H'ds$ was taken as equal to 4π times the currents flowing through the space enclosed by the curve, and the only currents discussed in that article were currents flowing through conductors: we shall now consider what interpretation we must attach to the new expression we have just found for $\int H'ds$.

In the first place, any tube which in unit time passes
inwards across one part of the curve and outwards across
another part, will not contribute anything to the total
number of tubes passing across the closed curve, for its
contribution when it passes inwards is equal and opposite
to its contribution when it passes outwards.   Hence all
the tubes we need consider are those which only cross
the curve once, which pass inwards across the curve and
do not leave it within unit time.   These tubes may be

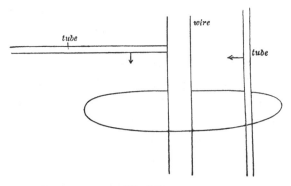

Fig. 128.

divided into two classes, (1) those which remain within
the curve, (2) those which manage to disappear without
again crossing the boundary. The first set will increase the
total polarization over any closed surface bounded by the
curve, and the number of those which cross the boundary
in unit time is equal to the rate of increase in this total
polarization. The existence of the second class of tubes
depends upon the passage of conductors, or of moving
charged bodies, through the area bounded by the curve.

Thus suppose we have a metal wire passing through the circuit, then the tubes which cross the boundary may run into this wire and be annulled, the disappearance of each unit tube corresponding to the passage of unit electricity along the wire; or a tube might have one end on the wire and cross the circuit, its end running along the wire; the passage of such a tube across the boundary means the passage of a unit of electricity along the wire; again, one end of a tube might be on a charged body which moves through the circuit. Thus the number of tubes of class (2) which cross the circuit in unit time is equal to the number of units of electricity which pass in that time along conductors or on charged bodies passing through the circuit, i.e. it is equal to the sum of the conduction and convection currents flowing through the circuit.

Hence the work done when unit pole is taken round a closed circuit is equal to 4π times the sum of the conduction and convection currents flowing through that circuit plus the rate of increase of the total polarization through the circuit. From this we see that a change in the polarization through the circuit produces the same magnetic effect as a conduction current whose intensity is equal to the rate of increase of the polarization. We shall call the rate of increase in the polarization the *dielectric current*. The recognition of the magnetic effects due to these dielectric currents is the fundamental feature of Maxwell's Theory of the Electric Field. We have given a method of regarding the magnetic field which leads us to expect the magnetic effects of dielectric currents. It must be remembered, however, Maxwell's Theory consists in the expression of this result and is not limited to any particular method of explaining it.

266. Propagation of Electromagnetic Disturbances. We shall now proceed to show that Maxwell's Theory leads to the conclusion that an electric disturbance is propagated through air with the velocity of light.

We can employ the equations we deduced in Art. 234, if we regard u, v, w the components of the current, as the components of the sum of the dielectric, convection, and conduction currents. If X, Y, Z are the components of the electric intensity, and K its specific inductive capacity, then the x, y, z components of the polarization are respectively

$$\frac{K}{4\pi} X, \quad \frac{K}{4\pi} Y, \quad \frac{K}{4\pi} Z,$$

the components of the dielectric currents are therefore

$$\frac{K}{4\pi} \frac{dX}{dt}, \quad \frac{K}{4\pi} \frac{dY}{dt}, \quad \frac{K}{4\pi} \frac{dZ}{dt}.$$

If σ is the specific resistance of the medium, the components of the conduction current are

$$\frac{X}{\sigma}, \quad \frac{Y}{\sigma}, \quad \frac{Z}{\sigma}.$$

Hence u, v, w the components of the total effective current are given by the equations

$$u = \frac{K}{4\pi} \frac{dX}{dt} + \frac{X}{\sigma},$$

$$v = \frac{K}{4\pi} \frac{dY}{dt} + \frac{Y}{\sigma},$$

$$w = \frac{K}{4\pi} \frac{dZ}{dt} + \frac{Z}{\sigma}.$$

Hence substituting these values of u, v, w in the equations of Art. 234 we get, using the notation of that Article,

the following equations as the expression of Maxwell's
Theory,

$$4\pi\left\{\frac{K}{4\pi}\frac{dX}{dt}+\frac{X}{\sigma}\right\}=\frac{d\gamma}{dy}-\frac{d\beta}{dz},$$

$$4\pi\left\{\frac{K}{4\pi}\frac{dY}{dt}+\frac{Y}{\sigma}\right\}=\frac{d\alpha}{dz}-\frac{d\gamma}{dx},$$

$$4\pi\left\{\frac{K}{4\pi}\frac{dZ}{dt}+\frac{Z}{\sigma}\right\}=\frac{d\beta}{dx}-\frac{d\alpha}{dy},$$

$$-\frac{da}{dt}=\frac{dZ}{dy}-\frac{dY}{dz},$$

$$-\frac{db}{dt}=\frac{dX}{dz}-\frac{dZ}{dx},$$

$$-\frac{dc}{dt}=\frac{dY}{dx}-\frac{dX}{dy}.$$

Let us now consider the case of a dielectric for which
σ is infinite, so that all the currents are dielectric currents;
putting σ infinite in the preceding equations, and $a=\mu\alpha$,
$b=\mu\beta$, $c=\mu\gamma$, we get

$$\left.\begin{aligned}K\frac{dX}{dt}&=\frac{d\gamma}{dy}-\frac{d\beta}{dz}\\K\frac{dY}{dt}&=\frac{d\alpha}{dz}-\frac{d\gamma}{dx}\\K\frac{dZ}{dt}&=\frac{d\beta}{dx}-\frac{d\alpha}{dy}\end{aligned}\right\}\dots\dots\dots(1),$$

$$\left.\begin{aligned}-\mu\frac{d\alpha}{dt}&=\frac{dZ}{dy}-\frac{dY}{dz}\\-\mu\frac{d\beta}{dt}&=\frac{dX}{dz}-\frac{dZ}{dx}\\-\mu\frac{d\gamma}{dt}&=\frac{dY}{dx}-\frac{dX}{dy}\end{aligned}\right\}\dots\dots\dots(2).$$

Differentiating the first equation in (1) with respect to t, we get

$$K \frac{d^2 X}{dt^2} = \frac{d}{dy}\frac{d\gamma}{dt} - \frac{d}{dz}\frac{d\beta}{dt}.$$

Substituting the values of $d\gamma/dt$, $d\beta/dt$, and noticing that by (1)

$$\frac{dX}{dx} + \frac{dY}{dy} + \frac{dZ}{dz}$$

is independent of the time, we get

$$\mu K \frac{d^2 X}{dt^2} = \frac{d^2 X}{dx^2} + \frac{d^2 X}{dy^2} + \frac{d^2 X}{dz^2} \dots\dots\dots(3).$$

We may by a similar process get equations of the same form for Y, Z, a, b, c.

To interpret these equations let us take the simple case when the quantities are independent of the coordinates x, y. Equation (3) then takes the form

$$\mu K \frac{d^2 X}{dt^2} = \frac{d^2 X}{dz^2} \dots\dots\dots\dots(4).$$

If we put

$$\xi = z - \frac{t}{\sqrt{\mu K}},$$

$$\eta = z + \frac{t}{\sqrt{\mu K}},$$

and change the variables from z and t to ξ and η, we get

$$\frac{d^2 X}{d\xi d\eta} = 0.$$

The solution of which is

$$X = F(\xi) + f(\eta)$$

$$= F\left(z - \frac{t}{\sqrt{\mu K}}\right) + f\left(z + \frac{t}{\sqrt{\mu K}}\right)\dots\dots(5),$$

where F and f denote any arbitrary functions.

Since $F(z - t/\sqrt{\mu K})$ remains constant as long as $z - t/\sqrt{\mu K}$ is constant, we see that if a point travels along the axis of z in the positive direction with the velocity $1/\sqrt{\mu K}$, the value of $F(z - t/\sqrt{\mu K})$ will be constant at this point. Hence the first term in equation (5) represents a value of X travelling in the positive direction of the axis of z with the velocity $1/\sqrt{\mu K}$. Similarly the second term in (5) represents a value of X travelling in the negative direction along the axis of z with the velocity $1/\sqrt{\mu K}$. For example, suppose that when $t = 0$, X is zero except between $z = +\epsilon$, $z = -\epsilon$ where it is equal to unity, and suppose further that dX/dt is everywhere zero when $t = 0$. Then equation (5) shows that after a time t

$$X = \frac{1}{2} \text{ between } z = \frac{t}{\sqrt{\mu K}} - \epsilon, \text{ and } z = \frac{t}{\sqrt{\mu K}} + \epsilon,$$

and between $z = -\dfrac{t}{\sqrt{\mu K}} - \epsilon$, and $z = -\dfrac{t}{\sqrt{\mu K}} + \epsilon$,

and is zero everywhere else. Thus the quantity represented by X travels through the dielectric with the velocity $1/\sqrt{\mu K}$.

It is shown in treatises on Differential Equations that equation (3), the general form of the equation (4), represents a disturbance travelling with the velocity $1/\sqrt{\mu K}$.

Thus Maxwell's Theory leads to the result that electric and magnetic effects are propagated through the dielectric with the velocity $1/\sqrt{\mu K}$.

Let us see what this velocity is when the dielectric is air. Using the electromagnetic system of units we have for air $\mu = 1$, $K = \dfrac{1}{v^2}$, where v is the ratio of the electro-magnetic unit of electricity to the electrostatic unit (Art. 255). Hence on Maxwell's Theory electric and magnetic effects are propagated through air with the velocity "v." Now experiments made by the method described in Art. 263 lead to the result that, within the errors of experiment, v is equal to the velocity of light through air. Hence we conclude that electromagnetic effects are propagated through air with the velocity of light. This result led Maxwell to the view that since light travels with the same velocity as an electromagnetic disturbance, it is itself an electromagnetic phenomenon; a wave of light being a wave of electric and magnetic disturbances.

267. Plane Electromagnetic Waves. Let us consider more in detail the theory of a plane electric wave. If f, g, h are the components of the electric polarization in such a wave, l, m, n the direction cosines of the normal to the wave front, and λ the wave length, then we may put

$$f = f_0 \cos \frac{2\pi}{\lambda} (lx + my + nz - Vt),$$

$$g = g_0 \cos \frac{2\pi}{\lambda} (lx + my + nz - Vt),$$

$$h = h_0 \cos \frac{2\pi}{\lambda} (lx + my + nz - Vt);$$

where V is the velocity of propagation of the wave, and f_0, g_0, h_0 quantities independent of x, y, z or t. Since

$$\frac{df}{dx} + \frac{dg}{dy} + \frac{dh}{dz} = 0,$$

we have $\qquad lf_0 + mg_0 + nh_0 = 0,$

and therefore $\qquad lf + mg + nh = 0.$

Thus the electric polarization is perpendicular to the direction of propagation of the wave.

By equation (2), Art. 266, we have

$$-\mu \frac{d\alpha}{dt} = \frac{dZ}{dy} - \frac{dY}{dz},$$

and
$$Z = \frac{4\pi}{K} h, \qquad Y = \frac{4\pi}{K} g.$$

Hence

$$\frac{d\alpha}{dt} = \frac{4\pi}{\mu K} \frac{2\pi}{\lambda} \{mh_0 - ng_0\} \sin \frac{2\pi}{\lambda} (lx + my + nz - Vt),$$

$$\alpha = \frac{4\pi}{\mu K V} (ng_0 - mh_0) \cos \frac{2\pi}{\lambda} (lx + my + nz - Vt);$$

or since
$$\mu K = \frac{1}{V^2},$$

$$\alpha = 4\pi V (ng - mh);$$

similarly
$$\beta = 4\pi V (lh - nf),$$

$$\gamma = 4\pi V (mf - lg).$$

Hence
$$l\alpha + m\beta + n\gamma = 0,$$

so that the magnetic force is at right angles to the direction of propagation of the wave, and since

$$f\alpha + g\beta + h\gamma = 0,$$

the magnetic force is perpendicular also to the electric polarization.

Since $\{\alpha^2 + \beta^2 + \gamma^2\}^{\frac{1}{2}} = 4\pi V \{f^2 + g^2 + h^2\}^{\frac{1}{2}},$

the resultant magnetic force is $4\pi V$ times the resultant electric polarization.

Hence in a plane electric wave, and therefore on Maxwell's Theory in a plane wave of light, there is in the front of the wave an electric polarization, and at right angles to this, and also in the wave front, there is a magnetic force bearing a constant ratio to the polarization. We shall see in Art. 270 that in a plane polarized light wave the electric polarization is at right angles to, and the magnetic force in, the plane of polarization.

In strong sunlight the maximum electric intensity is about 10 volts per centimetre, and the maximum magnetic force about one-fifth of the horizontal magnetic force due to the earth in England.

268. Propagation by the Motion of Faraday Tubes. The results obtained by the preceding analysis follow very simply from the view that the magnetic force

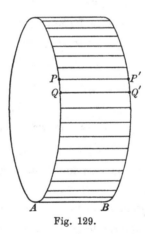

Fig. 129.

is due to the motion of the Faraday tubes. The electro-motive force round a circuit moving in a magnetic field is equal to the rate of diminution of the number of tubes

of magnetic induction passing through the circuit. Thus
let P, Q (Fig. 129) be two adjacent points on a circuit, P', Q'
the positions of these points after the lapse of a time δt.
Then the diminution in the time δt of the number of
tubes of magnetic induction passing through the circuit
of which PQ forms a part may, as in Art. 136, be shown
to be equal to the number of tubes which pass through
the sum of the areas $PP'Q'Q$. The number passing
through $PP'Q'Q$ is equal to

$$PQ \times PP' \times B \sin \phi \sin \theta,$$

where B is the magnetic induction, ϕ the angle it makes
with the plane $PP'Q'Q$, and θ the angle between PP' and
PQ. If V is the velocity with which the circuit is moving
$PP' = V\delta t$. Thus the rate of diminution in the number
of tubes passing through the circuit is

$$\Sigma PQ . VB \sin \phi \sin \theta.$$

Hence we may regard the electromotive force round
the circuit as equivalent to an electric intensity at each
point P of the circuit whose component along PQ is equal
to $VB \sin \phi \sin \theta$. As the component of this intensity
parallel to B and V vanishes, the resultant intensity is
at right angles to B and V and equal to

$$BV \sin \psi,$$

where ψ is the angle between B and V. In this case
the circuit was supposed to move, the tubes of induction
being at rest; we shall assume that the same expression
holds when the circuit is at rest and the tubes of magnetic induction move with the velocity V across an element
of the circuit at rest.

Let us now introduce the view that the magnetic force
is due to the motion of the Faraday tubes. Let OA (Fig.

130) represent the velocity of the Faraday tubes, OP the electric polarization, and OB the magnetic induction, which

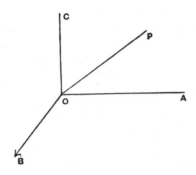

Fig. 130.

in a non-crystalline medium is parallel to the magnetic force and therefore (see page 483) at right angles to OP and OA. By what we have just proved the electric intensity is at right angles to OB and OA, and therefore along OC. Now in a non-crystalline medium the electric intensity is parallel to the electric polarization; hence OP and OC must coincide in direction; thus the Faraday tubes move at right angles to their length.

Again, if E is the electric intensity, by what we have just proved

$$E = BV \dots \dots \dots (1).$$

But if H is the magnetic force, μ the magnetic permeability,

$$B = \mu H,$$

and by Art. 265

$$H = 4\pi V P \dots \dots \dots (2),$$

where P is the electric polarization.

Hence by (1) and (2)

$$E = 4\pi\mu V^2 P.$$

If K is the specific inductive capacity of the dielectric

$$P = \frac{K}{4\pi} E;$$

hence we have $V^2 = 1/\mu K$. The tubes therefore move with the velocity $1/\sqrt{\mu K}$ at right angles to their length.

269. Evidence for Maxwell's Theory. We shall now consider the evidence furnished by experiment as to the truth of Maxwell's theory.

We have already seen that Maxwell's theory agrees with facts as far as the velocity of propagation through *air* is concerned. We now consider the case of other dielectrics.

The velocity of light through a non-magnetic dielectric whose specific inductive capacity is K is on Maxwell's theory equal to $1/\sqrt{K}$.

Hence

$$\frac{\text{velocity of light in this dielectric}}{\text{velocity of light in air}}$$

$$= \sqrt{\frac{\text{specific inductive capacity of air}}{\text{specific inductive capacity of dielectric}}}.$$

But by the theory of light this is also equal to

$$\frac{1}{n},$$

where n is the refractive index of the dielectric. Hence on Maxwell's theory

$n^2 =$ electrostatic measure of the specific inductive capacity.

In comparing the values of n^2 and K we have to remember that the electrical conditions under which these quantities are on Maxwell's theory equal to one another, are those which hold in a wave of light where the electric intensity is reversed millions of millions of times per second. We have at present no means of directly measuring K under these conditions.

To make a fair comparison between n^2 and K we ought to take the value of K determined for electrical oscillations of the same frequency as those of the vibrations of the light for which n is measured. As we cannot find K for vibrations as rapid as those of the visible rays, the other alternative is to use the value of n for waves of very great wave length; we shall call this value n_∞.

The process by which n_∞ is obtained is not however very satisfactory. Cauchy has given the formula

$$n = A + B\lambda^{-2} + C\lambda^{-4}$$

connecting n with the wave length λ, which holds accurately within the limits of the visible spectrum, unless the refracting substance is one which shows the phenomenon known as 'anomalous dispersion.' To find n_∞ we apply this empirical formula to determine the refractive index for waves millions of times the length of those used to determine the constants A, B, C which occur in the formula. For these reasons we should expect to find cases in which K is not equal to n_∞^2, but though these cases are numerous there are many others in which K is approximately equal to n_∞^2. A list of these is given in the following table:

Name of Substance	K	n_∞^2
Paraffin	2·29	2·022
Petroleum spirit	1·92	1·922
Petroleum oil	2·07	2·075
Ozokerite	2·13	2·086
Benzene	2·38	2·2614*
Carbon bisulphide	2·67	2·678*

....

As examples where the relation does not hold, we have

Glass (extra dense flint)	10·1	2·924*
Calcite (along axis)	7·5	2·197*
Quartz (along optic axis)	4·55	2·41*
Distilled water	76	1·779*

....

Sir James Dewar and Professor Fleming have shown that the abnormally high specific inductive capacities of liquids such as water, disappear at very low temperatures, the specific inductive capacities at such temperatures becoming comparable with the square of the refractive index.

Maxwell's Theory of Light has been developed to a considerable extent and the consequences are found to agree well with experiment. In fact the electromagnetic is the only theory of light yet advanced in which the difficulties of reconciling theory with experiment do not seem insuperable.

270. Hertz's Experiments. The experiments made by Hertz on the properties of electric waves, on their

---

* These are the values of n_0^2 where n_0 is the refractive index for sodium light.

reflection, refraction, and polarization, furnish perhaps the most striking evidence in support of Maxwell's theory, as it follows from these experiments that the properties of these electric waves are entirely analogous to those of light waves. We regret that we have only space for an exceedingly brief account of a few of Hertz's beautiful experiments; for a fuller description of these and other experiments on electric waves with their bearings on Maxwell's theory, we refer the reader to Hertz's own account in *Electrical Waves* and to *Recent Researches in Electricity and Magnetism* by J. J. Thomson.

We saw in Art. 245 that when a condenser is discharged by connecting its coatings by a conductor, electrical oscillations are produced, the period of which is approximately $2\pi\sqrt{LC}$ where C is the capacity of the condenser, and L the coefficient of self-induction of the circuit connecting its plates. This vibrating electrical system will, on Maxwell's theory, be the origin of electrical waves, which travel through the dielectric with the velocity V and whose wave length is $2\pi V\sqrt{LC}$. By using condensers of small capacity whose plates were connected by very short conductors Hertz was able to get electrical waves less than a metre long. This vibrating electrical system is called a vibrator.

Hertz used several forms of vibrators; the one used in the experiment we are about to describe consists of two equal brass cylinders placed so that their axes are coincident. The two cylinders are connected to the two terminals of an induction coil. When this is in action sparks pass between the cylinders. The cylinders correspond to the plates of the condenser, and the air between the cylinders (whose electric strength breaks down when the spark passes) to the conductor connecting the plates. The length of each of these cylinders is about 12 cm., and their diameters about 3 cm. ; their sparking ends are well polished.

To detect the presence of the electrical waves, Hertz used a very nearly closed metallic circuit, such as a piece of wire, bent into a circle, the ends of the wire being exceedingly close together. When the electric waves strike against this detector very minute sparks pass between the terminals; these sparks serve to detect the presence of the waves. Recently Sir Oliver Lodge has introduced a still more sensitive detector. It is founded on the fact discovered by Branly that the electrical resistance of a number of metal turnings, placed so as to be loosely in contact with each other, is greatly affected by the impact of electric waves, and that all that is necessary to detect these waves is to take a glass tube, fill it loosely with iron turnings, and place the tube in series with a battery and a galvanometer. When the waves fall on the tube its resistance, and therefore the deflection of the galvanometer, is altered.

The analogy between the electrical waves and light waves is very strikingly shown by Hertz's experiments with parabolic mirrors.

If the vibrator is placed in the focal line of a parabolic cylinder, and if the Faraday tubes emitted by it are parallel to this focal line; then if the laws of reflection of these electric waves are the same as for light waves, the waves emitted by the vibrator will, after reflection from the cylinder, emerge as a parallel beam, and will therefore not diminish in intensity as they recede from the mirror. When such a beam falls on another parabolic cylinder, the axis of whose cross section coincides with the axis of the beam, it will be brought to a focus on the focal line of the second mirror.

The parabolic mirrors used by Hertz were made of sheet zinc, and their focal length was about 12·5 cm. The vibrator was placed so that the axes of the cylinders coincided with the focal line of one of the mirrors. The detector, which was placed in the focal line of an equal parabolic mirror, consisted of two pieces of wire; each of these wires had a straight piece about 50 cm. long, and was then bent at right angles so as to pass through the back of the mirror, the length of the bent piece being about 15 cm. The ends of the two pieces coming through the mirror were bent so as to be exceedingly near to each other. The sparks passing between these ends were observed from behind the mirror. The mirrors are represented in Fig. 131.

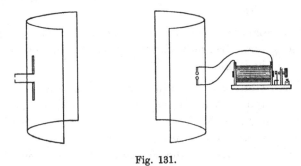

Fig. 131.

Reflection of Electric Waves.

To show the reflection of these waves the mirrors were placed side by side so that their openings looked in the same direction and their axes converged at a point distant about 3 metres from the mirrors. No sparks passed between the points of the detector when the vibrator was in action. If however a metal plate about 2 metres square was placed at the intersection of the axes of the mirrors, and at right angles to the line which bisects the angle between the axes, sparks appeared at the detector. These sparks however disappeared if the metal plate was turned through a small angle. This experiment shows that the electric waves are reflected and that, approximately at any rate, the angle of incidence is equal to the angle of reflection.

Refraction of Electric Waves.

To show the refraction of these waves Hertz used a large prism made of pitch. This was about 1·5 metres high, and it had a refracting angle of 30° and a slant side of 1·2 metres. When the electric waves from the mirror containing the vibrator passed through this prism, the sparks in the detector were not excited when the axes of the two mirrors were parallel, but sparks were produced when the axis of the mirror containing the detector made a suitable angle with that containing the vibrator. When the system was adjusted for minimum deviation, the sparks were most vigorous in the detector when the angle between the axes of the mirrors was equal to 22°. This would make the refractive index of pitch for these electrical waves equal to 1·69.

Electric Analogy to a plate of Tourmaline.

If a properly cut tourmaline plate is placed in the path of a plane polarized beam of light incident at right angles on the plate, the amount of light transmitted through the tourmaline plate depends upon its azimuth. For one particular azimuth all the light will be stopped, while for an azimuth at right angles to this the maximum amount of light will be transmitted.

If a screen be made by winding metal wire round a large rectangular framework so that the turns of the wire are parallel to one pair of sides of the frame, and if this screen be interposed between the mirrors when they are facing each other with their axes coincident, then it will stop the sparks in the detector when the turns of the wire are parallel to the focal lines of the mirrors, and thus to the Faraday tubes proceeding from the vibrator: the sparks will however recommence if the framework is turned through a right angle so that the wires are perpendicular to the focal lines of the mirror.

If this framework is substituted for the metal plate in the experiment on the reflection of waves, the sparks will appear in the detector when the wires are parallel to the focal lines of the cylinders and will disappear when they are at right angles to them. Thus this framework reflects but does not transmit Faraday tubes parallel to the wires, while it transmits but does not reflect Faraday tubes at right angles to them. It thus behaves towards the transmitted electrical waves as a plate of tourmaline does towards light waves. By using a framework wound with exceedingly fine wires placed very close together Du Bois and Rubens have recently succeeded in polarizing in this way radiant heat, whose wave length, though greater than that of the rays of the visible spectrum, is exceedingly small compared with that of electric waves.

Angle of Polarization.

When light polarized in a plane at right angles to the plane of incidence falls upon a plate of refracting substance, and the normal to the wave front makes with the normal to the refracting surface an angle $\tan^{-1}\mu$, where μ is the refractive index, all the light is refracted and none reflected. When light is polarized in the plane of incidence some of the light is always reflected.

Trouton has obtained a similar effect with electric waves. From a wall 3 feet thick reflection was obtained when the Faraday tubes proceeding from the vibrator were perpendicular to the plane of incidence, while there was no reflection when the vibrator was turned through a right angle so that the Faraday tubes were in the plane of incidence. This proves that on the electromagnetic theory of light we must suppose that the Faraday tubes are at right angles to the plane of polarization.

A very convenient arrangement for studying the properties of electric waves is described in a paper by Professor Bose in the *Philosophical Magazine* for January 1897.

CHAPTER XIV

THERMOELECTRIC CURRENTS

271. Seebeck discovered in 1821 that if in a closed circuit of two metals the two junctions of the metals are at different temperatures, an electric current will flow round the circuit. If, for example, the ends of an iron and of a copper wire are soldered together and one of the junctions is heated, a current of electricity will flow round the circuit; the direction of the current is such that the current flows from the copper to the iron across the hot junction, provided the mean temperature of the junctions is not greater than about 600° Centigrade.

The current flowing through the thermoelectric circuit represents a certain amount of energy, it heats the circuit and may be made to do mechanical work. The question at once arises, what is the source of this energy? A discovery made by Peltier in 1834 gives a clue to the answer to this question. Peltier found that when a current flows across the junction of two metals it gives rise to an absorption or liberation of heat. If it flows across the junction in one direction heat is absorbed, while if it flows in the opposite direction heat is liberated. If the current flows in the same direction as the current at the hot junction in a thermoelectric circuit of the two metals heat is absorbed; if it flows in the same direction as the current at the cold junction of the circuit heat is liberated.

Thus, for example, heat is absorbed when a current flows across an iron-copper junction from the copper to the iron.

The heat liberated or absorbed is proportional to the quantity of electricity which crosses the junction. The amount of heat liberated or absorbed when unit charge of electricity crosses the junction is called the Peltier Effect at the temperature of the junction.

Now suppose we place an iron-copper circuit with one junction in a hot chamber and the other junction in a cold chamber, a thermoelectric current will be produced flowing from the copper to the iron in the hot chamber, and from the iron to the copper in the cold chamber.

Now by Peltier's discovery this current will give rise to an absorption of heat in the hot chamber and a liberation of heat in the cold one. Heat will be thus taken from the hot chamber and given out in the cold. In this respect the thermoelectric couple behaves like an ordinary heat-engine.

272. The experiments made on thermoelectric currents are all consistent with the view that the energy of these currents is entirely derived from thermal energy, the current through the circuit causing the absorption of heat at places of high temperature and its liberation at places of lower temperature. We have no evidence that any energy is derived from any change in the molecular state

of the metals caused by the passage of the current or
from anything of the nature of chemical combination
going on at the junction of the two metals.

Many most important results have been arrived at
by treating the thermoelectric circuit as a perfectly reversible thermal engine, and applying to it the theorems
which are proved in the Theory of Thermodynamics to
apply to all such engines. The validity of this application
may be considered as established by the agreement between the facts and the result of this theory. There are
however thermal processes occurring in the thermoelectric
circuit which are not reversible, i.e. which are not reversed
when the direction of the current flowing through the
circuit is reversed. There is the conduction of heat along
the metals due to the difference of temperatures of the
junctions, and there is the heating effect of the current
flowing through the metal which, by Joule's law, is proportional to the square of the current and is not reversed
with the current. Inasmuch as the ordinary conduction
of heat is independent of the quantity of electricity passing
round the circuit, and the heat produced in accordance
with Joule's law is not directly proportional to this
quantity, it is probable that in estimating the connection
between the electromotive force of the circuit, which is
the work done when unit of electricity passes round the
circuit, and the thermal effects which occur in it, we
may leave out of account the conduction effect and the
Joule effect and treat the circuit as a reversible engine.
If this is the case, then, as Lord Kelvin has shown, the
Peltier effect cannot be the only reversible thermal effect
in the circuit. For let us assume for a moment that the
Peltier effect is the only reversible thermal effect in the

circuit. Let P_1 be the Peltier effect at the cold junction
whose absolute temperature is T_1, so that P_1 is the
mechanical equivalent of the heat liberated when unit of
electricity crosses the cold junction; let P_2 be the Peltier
effect at the hot junction whose absolute temperature is
T_2, so that P_2 is the mechanical equivalent of the heat
absorbed when unit of electricity crosses the hot junction.
Then since the circuit is a reversible heat-engine, we have
(see Maxwell's *Theory of Heat*)

$$\frac{P_1}{T_1} = \frac{P_2}{T_2}$$

$$= \frac{\text{work done when unit of electricity goes round the circuit}}{T_2 - T_1}.$$

But the work done when unit of electricity goes round
the circuit is equal to E, the electromotive force in the
circuit, and hence

$$E = (T_2 - T_1) \cdot \frac{P_1}{T_1}.$$

Thus on the supposition that the only reversible
thermal effects are the Peltier effects at the junctions,
the electromotive force round a circuit whose cold junction
is kept at a constant temperature should be proportional
to the difference between the temperatures of the hot
and cold junctions. Cumming, however, showed that
there were circuits where, when the temperature of the
hot junction is raised, the electromotive force diminishes
instead of increasing, until, when the hot junction is
hot enough, the electromotive force is reversed and the
current flows round the circuit in the reverse direction. This reasoning led Lord Kelvin to suspect that
besides the Peltier effects at the junction there were

reversible thermal effects produced when a current flows along an unequally heated conductor, and by a laborious series of experiments he succeeded in establishing the existence of these effects. He found that when a current of electricity flows along a copper wire whose temperature varies from point to point, heat is liberated at any point P when the current at P flows in the direction of the flow of heat at P, i.e. when the current is flowing from hot places to cold, while heat is absorbed at P when the current flows through it in the opposite direction. In iron, on the other hand, heat is absorbed at P when the current flows in the direction of the flow of heat at P, while heat is liberated when the current flows in the opposite direction. Thus when a current flows along an unequally heated copper wire it tends to diminish the differences of temperature, while when it flows along an iron wire it tends to increase those differences. This effect produced by a current flowing along an unequally heated conductor is called the Thomson effect.

Specific Heat of Electricity.

273. The laws of the Thomson effect can be conveniently expressed in terms of a quantity introduced by Lord Kelvin and called by him the 'specific heat of the electricity in the metal.' If σ is this 'specific heat of electricity,' A and B two points in a wire, the temperatures of A and B being respectively t_1 and t_2, and the difference between t_1 and t_2 being supposed small, then σ is defined by the relation,

$\sigma (t_1 - t_2) =$ heat developed in AB when unit of electricity

passes through AB from A to B.

The study of the thermoelectric properties of conductors is very much facilitated by the use of the thermoelectric diagrams introduced by Professor Tait. Before proceeding to describe them we shall enunciate two results of experiments made on thermoelectric circuits which are the foundation of the theory of these circuits.

The first of these is, that if E_1 is the electromotive force round a circuit when the temperature of the cold junction is t_0 and that of the hot junction t_1, E_2 the electromotive force round the same circuit when the temperature of the cold junction is t_1, and that of the hot junction t_2, then $E_1 + E_2$ will be the electromotive force round the circuit when the temperature of the cold junction is t_0, and that of the hot junction t_2. It follows from this result that E, the electromotive force round a circuit whose junctions are at the temperatures t_0 and t_1, is equal to

$$\int_{t_0}^{t_1} Q dt,$$

where $Q dt$ is the electromotive force round the circuit when the temperature of the cold junction is $t - \tfrac{1}{2}dt$, and the temperature of the hot junction is $t + \tfrac{1}{2}dt$. The quantity Q is called the thermoelectric power of the circuit at the temperature t.

The second result relates to the electromotive force round circuits made of different pairs of metals whose junctions are kept at assigned temperatures. It may be stated as follows: If E_{AC} is the electromotive force round a circuit formed of the metals A, C, E_{BC} that round a circuit formed of the metals B, C, then $E_{AC} - E_{BC}$ is the electromotive force acting round the circuit formed of the

metals A and B; all these circuits being supposed to work
between the same limits of temperature.

274. Thermoelectric Diagrams. The thermoelectric line for any metal (A) is a curve such that the
ordinate represents the thermoelectric power of a circuit
of that metal and some standard metal (usually lead) at a
temperature represented by the abscissa. The ordinate is
taken positive when for a small difference of temperature
the current flows from lead to the metal A across the
hot junction.

It follows from Art. 273, that if the curves α and β
represent the thermoelectric lines for two metals A and B,
then the thermoelectric power of a circuit made of the
metals A and B at an absolute temperature represented
by ON will be represented by RS, and the electromotive
force round a circuit formed of the two metals A and B

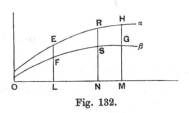

Fig. 132.

when the temperature of the cold junction is represented
by OL, that of the hot junction by OM, will be represented by the area $EFGH$.

Let us now consider a circuit of the two metals A and
B with the junctions at the absolute temperatures OL_1,
OL_2, Fig. 133, where OL_1 and OL_2 are nearly equal. Then

the electromotive force round the circuit (i.e. the work done when unit of electrical charge passes round the circuit) is represented by the area $EHGF$. Consider now the thermal effects in the circuit. We have Peltier effects

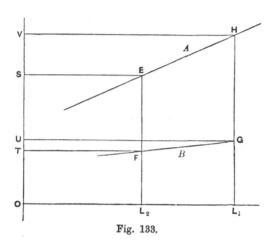

Fig. 133.

at the junctions; suppose that the mechanical equivalent of the heat absorbed at the hot junction when unit of electricity crosses from B to A it is represented by the area P_1, let the mechanical equivalent of the heat liberated at the cold junction be represented by the area P_2. There are also the Thomson effects in the unequally heated metals; suppose that the mechanical equivalent of the heat liberated when unit of electricity flows through the metal A from the hot to the cold junction is represented by the area K_1, and that the mechanical equivalent of the heat liberated when unit of electricity flows through B from the hot to the cold junction is represented by the area K_2. Then by the First Law of Thermodynamics, we have

$$\text{area } EFGH = P_1 - P_2 + K_2 - K_1 \ldots\ldots\ldots(1).$$

The Second Law of Thermodynamics may be expressed in the form that if H be the amount of heat absorbed in any reversible engine at the absolute temperature t, then

$$\Sigma \frac{H}{t} = 0.$$

In our circuit the two junctions are at nearly the same temperature, and we may suppose that the temperature at which the absorption of heat corresponding to the Thomson effect takes place is the mean of the temperatures of the junctions, i.e. $\frac{1}{2}(OL_1 + OL_2)$.

Hence by the Second Law of Thermodynamics, we have

$$0 = \frac{P_1}{OL_1} - \frac{P_2}{OL_2} + \frac{K_2 - K_1}{\frac{1}{2}(OL_1 + OL_2)} \ldots\ldots\ldots(2).$$

Hence from (1) and (2) we get

$$\text{area } EFGH = \frac{1}{2}\left\{\frac{P_1}{OL_1} + \frac{P_2}{OL_2}\right\}(OL_1 - OL_2),$$

or since OL_1 is very nearly equal to OL_2 and therefore P_1 is very nearly equal to P_2, this gives approximately

$$\text{area } EFGH = \frac{P_1}{OL_1}(OL_1 - OL_2).$$

But when OL_1 is very nearly equal to OL_2, the area

$$EFGH = GH(OL_1 - OL_2),$$

so that $P_1 = GH . OL_1,$

thus P_1 is represented by the area $GHVU$. Now P_1 is the Peltier effect at the temperature represented by OL_1, hence we see that at any temperature

Peltier effect = (thermoelectric power) (absolute

temperature),

or $$P = Qt,$$

where t is the absolute temperature.

By the definition of Art. 273 we see that if σ_1 is the specific heat of electricity for the metal A, σ_2 that for B, then

$$K_1 - K_2 = (\sigma_1 - \sigma_2) L_2 L_1.$$

But by (1)

$$\text{area } EFGH = P_1 - P_2 + K_2 - K_1,$$

and $$P_1 = \text{area } GHVU,$$

$$P_2 = \text{area } FEST.$$

Hence $$K_1 - K_2 = \text{area } SEHV - \text{area } TFGU$$

$$= (\tan \theta_1 - \tan \theta_2) OL_1 \times L_2 L_1,$$

where θ_1, θ_2 are the angles which the tangents at E and F to the thermoelectric lines for A and B make with the axis along which temperature is measured. Hence

$$\sigma_1 - \sigma_2 = (\tan \theta_1 - \tan \theta_2) OL_1 \ldots \ldots \ldots (3).$$

When the temperature interval $L_1 L_2$ is finite the areas $UGHV$ and $FEST$ will still represent the Peltier effects at the junctions, and the area $TFGU$ the heat absorbed when unit of electricity flows along the metal B from a place where the temperature is OL_2 to one where it is OL_1.

The preceding results are independent of any assumption as to the shape of the thermoelectric lines. The results of the experiments made by Professor Tait and others show, that over a considerable range of temperatures, these lines are straight for most metals and alloys, while Le Roux has shown that the 'specific heat of electricity' for lead is excessively small. Let us assume that it is zero and suppose that the diagram represents the thermoelectric lines of metals with respect to lead: then since these lines are straight, θ is constant for any metal and σ_2 vanishes when it refers to lead, the value of σ the 'specific heat of electricity' in the metal is by (3) given by the equation

$$\sigma = \tan \theta \cdot t,$$

where t denotes the absolute temperature.

The thermoelectric power Q of the metal with respect to lead at any temperature t is given by the equation

$$Q = \tan \theta \, (t - t_0),$$

where t_0 is the absolute temperature where the line of the metal cuts the lead-line; t_0 is defined as the neutral point of the metal and lead.

Let us consider two metals; let θ_1, θ_2 be the angles their lines make with the lead-line, and t_1 and t_2 their neutral temperatures, then Q_1 and Q_2 their thermoelectric powers with respect to lead are given by the equations

$$Q_1 = \tan \theta_1 \, (t - t_1),$$
$$Q_2 = \tan \theta_2 \, (t - t_2);$$

hence Q, the thermoelectric power of a circuit consisting of the two metals, is given by the equation

$$Q = (\tan \theta_1 - \tan \theta_2) \, (t - T_0),$$

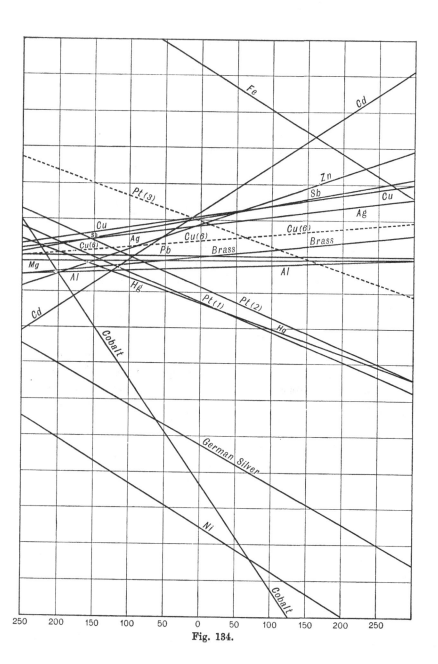

Fe

Cd

Zn

Sb

Cu

Pt (3)

Ag

Cu

Cu(6)

Sb

Ag

Cu(6)

Cu(6)

Brass

Pb

Brass

Cu(6)

Mg

Al

Al

Hg

Cd

Pt (1)

Pt (2)

Hg

Cobalt

German Silver

Ni

Cobalt

250   200   150   100   50   0   50   100   150   200   250

Fig. 134.

where T_0 is the neutral temperature for the two metals and is given by the equation

$$T_0 = \frac{t_1 \tan \theta_1 - t_2 \tan \theta_2}{\tan \theta_1 - \tan \theta_2}.$$

The electromotive force round a circuit formed of these metals, the temperatures of the hot and cold junctions being T_1, T_2, respectively, is equal to

$$\int_{T_2}^{T_1} Q dt = (\tan \theta_1 - \tan \theta_2)(T_1 - T_2)(\tfrac{1}{2}(T_1 + T_2) - T_0).$$

This vanishes when the mean of the temperatures of the junctions is equal to the neutral temperature. If the temperature of one junction is kept constant the electromotive force has a maximum or minimum value when the other junction is at the neutral temperature.

In Fig. 134 the thermoelectric lines for a number of metals are given. The figure is taken from a paper by Noll, *Wiedemann's Annalen*, vol. 53, p. 874. The abscissæ represent temperatures, each division being 50° C., the ordinates represent the E.M.F. for a temperature difference of 1° C., each division representing 2·5 microvolts. To find the E.M.F. round a circuit whose junctions are at t_1 and t_2 degrees we multiply the ordinate for $\tfrac{1}{2}(t_1 + t_2)$ degrees by $(t_2 - t_1)$.

CHAPTER XV

THE PROPERTIES OF MOVING ELECTRIC CHARGES

275. As the properties of moving electric charges are of great importance in the explanation of many physical phenomena, we shall consider briefly some of the simpler properties of a moving charge and other closely allied questions.

Magnetic Force due to a Moving Charged Sphere.

The first problem we shall discuss is that of a uniformly charged sphere moving with uniform velocity along a straight line. Let e be the charge on the sphere, a its radius, and v its velocity; let us suppose that it is moving along the axis of z, then when things have settled down into a steady state the sphere will carry its Faraday tubes along with it. If we neglect the forces due to electromagnetic induction, the Faraday tubes will be uniformly distributed round the sphere and the number passing normally through unit area at a point P will be $e/4\pi OP^2$, O being the centre of the charged sphere. These tubes are radial and are moving with a velocity v parallel to the axis of z, hence the component of the velocity at right angles to their direction is $v \sin \theta$, where θ is the angle OP

makes with the axis of z; by Art. 265 these moving tubes will produce a magnetic force at P equal to

$$4\pi\,(e/4\pi\,.\,OP^2)\,v\sin\theta = ev\sin\theta/OP^2.$$

The direction of this force is at right angles to the tubes, i.e. at right angles to OP; at right angles also to their direction of motion, i.e. at right angles to the axis of z; thus the lines of magnetic force will be circles whose planes are at right angles to the axis of z and whose centres lie along this axis. Thus we see that the magnetic field outside the charged sphere is the same as that given by Ampère's rule for an element of current ids, parallel to the axis of z, placed at the centre of the sphere, provided $ev = ids$.

276. As the sphere moves, the magnetic force at P changes, so that in addition to the electrostatic forces there will be forces due to electromagnetic induction, these will be proportional to the intensity of the magnetic induction multiplied by the velocity of the lines of magnetic induction, i.e. the force due to electromagnetic induction at a point P will be proportional to $\mu\,(ev\sin\theta/OP^2)\times v$, where μ is the magnetic permeability of the medium; while the electrostatic force will be $e/K\,.\,OP^2$, where K is the specific inductive capacity of the medium. The ratio of the force due to electromagnetic induction to the electrostatic force is $\mu K v^2\sin\theta$ or $\sin\theta v^2/V^2$, where V is the velocity of light through the medium surrounding the sphere; hence in neglecting the electromagnetic induction we are neglecting quantities of the order v^2/V^2. The direction of the force due to electromagnetic induction at P is along NP, if PN is the normal drawn from P to the axis of z; this force tends to make the Faraday tubes congregate in the plane through the centre of the sphere at right angles to its direction of motion; when the sphere is moving with the velocity of light it can be shown that all the Faraday tubes are driven into this plane.

Increase of Mass due to the Charge on the Sphere.

277.   Returning to the case when the sphere is moving so slowly that we may neglect v^2/V^2; we see that since H, the magnetic force at P, is $ev \sin\theta/OP^2$, and at P there is kinetic energy equal to $\mu H^2/8\pi$ per unit volume (see Art. 163), the kinetic energy per unit volume at P is

$$\mu e^2 v^2 \sin^2\theta/8\pi . OP^4.$$

Integrating this for the volume outside the sphere, we find that the kinetic energy outside the sphere is $\dfrac{\mu e^2 v^2}{3a}$, where a is the radius of the sphere.   Thus if m be the mass of the uncharged sphere the kinetic energy when it has a charge e is equal to

$$\frac{1}{2}\left(m + \frac{2}{3}\frac{\mu e^2}{a}\right)v^2.$$

Thus the effect of the charge is to increase the mass of the sphere by $2\mu e^2/3a$.   It is instructive to compare this case with another, in which there is a similar increase in the effective mass of a body; the case we refer to is that of a body moving through a liquid.   Thus when a sphere moves through a liquid it behaves as if its mass were $m + \frac{1}{2}m'$, where m is the mass of the sphere, and m' the mass of liquid displaced by it.   Again when a cylinder moves at right angles to its axis through a liquid its apparent mass is $m + m'$, where m' is the mass of the liquid displaced by the cylinder.   In the case of an elongated body like a cylinder, the increase in mass is much greater when it moves sideways than when it moves point foremost, indeed in the case of an infinite cylinder the increase in the latter case vanishes in comparison with that in the former; the increase in mass being $m' \sin^2 \theta$, where θ is the angle the direction of motion of the cylinder makes with its axis. In the case of bodies moving through liquids the increase in mass is due to the motion of the body setting in motion the liquid around it, the site of the increased mass is not the body itself but the space around it where the liquid is moving. In the electrical problem we may regard the increased mass as due to the Faraday tubes setting in motion the ether as they move through it. From the expression for the energy per unit volume we see that the increase in mass is the same as if a mass $4\pi\mu N^2$ were bound by the tubes, and had a velocity given to it equal to the velocity of the tubes at right angles to themselves, the motion of the tubes along their length not setting this mass in motion. Thus on this view the increased mass due to the charge is the mass of ether set in motion by the tubes. If we regard atoms as made up of exceedingly small particles charged with negative electricity, embedded in a much larger sphere of positive electricity, the positive charge on this sphere being equal to the sum of the negative charges embedded in it, it is possible to regard all mass as electrical in its origin, and as arising from the ether set in motion by the Faraday tubes connecting the electrical charges of which the atoms are supposed to be made up. For a development of this view the reader is referred to the author's *Conduction of Electricity through Gases and Electricity and Matter*.

Momentum in the Electric Field.

278. The view indicated above, that the Faraday tubes set the ether moving at right angles to the direction of these tubes, suggests that at each point in the field there is momentum whose direction is at right angles to the tubes, and by symmetry in the plane through the tube and the line along which the centre of the charged sphere moves. As the mass of the ether moved per unit volume at P is $4\pi\mu N^2$ where N is the density of the Faraday tubes at P, the momentum per unit volume would, on this view, be $4\pi\mu N^2 v \sin\theta$. This is equal to BN where B is the magnetic induction and N the density of the Faraday tubes at P, the direction of the momentum being at right angles to B and N. We shall now prove that this expression for the momentum is general and is not limited to the case when the field is produced by a moving charged sphere.

279. Since the magnetic force due to moving Faraday tubes is (Art. 265) equal to 4π times the density of the tubes multiplied by the components of the velocity of the tubes at right angles to their direction, and is at right angles both to the direction of the tubes and to their velocity; we see if α, β, γ are the components of the magnetic force parallel to axes of x, y, z at a place where the densities of the Faraday tubes parallel to x, y, z are f, g, h, and where u, v, w are the components of the velocity of the tubes, α, β, γ are given by the equations

$$\alpha = 4\pi\,(hv - gw), \quad \beta = 4\pi\,(fw - hu), \quad \gamma = 4\pi\,(gu - fv).$$

If all the tubes are not moving with the same velocity we shall have

$$\alpha = 4\pi \ (h_1v_1 - g_1w_1 + h_2v_2 - g_2w_2 + h_3v_3 - g_3w_3 + \dots)$$

with similar expressions for β, γ. Here u_1, v_1, w_1 are the components of the velocity of the tubes f_1, g_1, h_1; u_2, v_2, w_2 those of the tubes f_2, g_2, h_2 and so on.

Now T the kinetic energy per unit volume at P is equal to

$$\frac{\mu}{8\pi}(\alpha^2 + \beta^2 + \gamma^2) = \frac{\mu}{8\pi} \times 16\pi^2 . (\{\Sigma\ (hv - gw)\}^2$$
$$+ \{\Sigma\ (fw - hu)\}^2 + \{\Sigma\ (gu - fv)\}^2)$$
$$= 2\pi\mu . \{(\Sigma\ (hv - gw))^2 + (\Sigma\ (fw - hu))^2 + (\Sigma\ (gu - fv))^2\} ;$$

the momentum per unit volume parallel to x due to the tubes f_1, g_1, h_1 is equal to $\dfrac{dT}{du_1}$, i.e. to

$$- 4\pi\mu\ \{h_1\Sigma\ (fw - hu) - g_1\Sigma\ (gu - fv)\}$$
$$= \mu\ (g_1\gamma - h_1\beta).$$

Similarly that due to the tubes f_2, g_2, h_2 is equal to

$$\mu\ (g_2\gamma - h_2\beta),$$

and so on, thus P the total momentum parallel to x per unit volume is given by the equation

$$P = \mu\ (\gamma\Sigma g - \beta\Sigma h)$$
$$= \mu\ (\gamma g - \beta h),$$

where f, g, h are the densities parallel to x, y, z of the whole assemblage of Faraday tubes. Similarly Q, R, the components of the momentum parallel to y and z, are given respectively by the equations

$$Q = \mu\ (\alpha h - \gamma f),$$
$$R = \mu\ (\beta f - \alpha g).$$

Thus we see that the vector P, Q, R is perpendicular to the vectors α, β, γ, f, g, h, and its magnitude is $BN \sin \theta$ where B is the magnetic induction at the point, N the density of the Faraday tubes and θ the angle between B and N; hence we see that each portion of the field possesses an amount of momentum equal to the vector product of the magnetic induction and the dielectric polarization.

280. Before considering the consequences of this result, it will be of interest to consider the connection between the momentum and the stresses which we have supposed to exist in the field. We have seen (Arts. 45, 46) that the electric and magnetic forces in the field could be explained by the existence of the following stresses:

$$\alpha \begin{cases} (1) \text{ a tension } \dfrac{KR^2}{8\pi} \text{ along the lines of electric force;} \\ (2) \text{ a pressure } \dfrac{KR^2}{8\pi} \text{ at right angles to these lines;} \end{cases}$$

here K is the specific inductive capacity, and R the electric force;

$$\beta \begin{cases} (1) \text{ a tension } \dfrac{\mu H^2}{8\pi} \text{ along the lines of magnetic force;} \\ (2) \text{ a pressure } \dfrac{\mu H^2}{8\pi} \text{ at right angles to these lines;} \end{cases}$$

here μ is the magnetic permeability of the medium and H the magnetic force.

Let us consider the effect of these tensions on an element of volume bounded by plane faces perpendicular to the axes of x, y, z. The stresses α are equivalent to a hydrostatic pressure $KR^2/8\pi$ and a tension $KR^2/4\pi$ along the lines of force. The effect of the hydrostatic pressure on the element of volume is equivalent to forces

$$-\frac{d}{dx}\left(\frac{KR^2}{8\pi}\right)\Delta x\,\Delta y\,\Delta z, \quad -\frac{d}{dy}\left(\frac{KR^2}{8\pi}\right)\Delta x\,\Delta y\,\Delta z,$$

$$-\frac{d}{dz}\left(\frac{KR^2}{8\pi}\right)\Delta x\,\Delta y\,\Delta z,$$

parallel to the axes of x, y, z respectively, Δx, Δy, Δz being the sides of the element of volume.

Let us now consider the tension $KR^2/4\pi$. We know that a stress N in a direction whose direction cosines are l, m, n is equivalent to the following stresses:

$$\begin{cases} Nl^2 \text{ acting on the face } \Delta y\,\Delta z \text{ parallel to } x, \\ Nlm \quad\quad,, \quad\quad ,, \quad\quad ,, \quad\quad ,, \quad y, \\ Nln \quad\quad,, \quad\quad ,, \quad\quad ,, \quad\quad ,, \quad z, \end{cases}$$

$$\begin{cases} Nlm \quad\quad,, \quad\quad ,, \quad \Delta x\,\Delta z \quad ,, \quad x, \\ Nm^2 \quad\quad,, \quad\quad ,, \quad\quad ,, \quad\quad ,, \quad y, \\ Nmn \quad\quad,, \quad\quad ,, \quad\quad ,, \quad\quad ,, \quad z, \end{cases}$$

$$\begin{cases} Nln \quad\quad,, \quad\quad ,, \quad \Delta x\,\Delta y \quad ,, \quad x, \\ Nmn \quad\quad,, \quad\quad ,, \quad\quad ,, \quad\quad ,, \quad y, \\ Nn^2 \quad\quad,, \quad\quad ,, \quad\quad ,, \quad\quad ,, \quad z. \end{cases}$$

Thus the effect of these stresses on the element of volume is equivalent to a force parallel to x equal to

$$\left\{\frac{d}{dx}\left(Nl^2\right)+\frac{d}{dy}\left(Nlm\right)+\frac{d}{dz}\left(Nln\right)\right\}\Delta x\,\Delta y\,\Delta z;$$

the forces parallel to y and z are given by symmetrical expressions.

In our case the tension is along the lines of force, hence $l=\dfrac{X}{R}$, $m=\dfrac{Y}{R}$, $n=\dfrac{Z}{R}$, where X, Y, Z are the components of the electric force, hence substituting these values for l, m, n and putting $N = \dfrac{KR^2}{4\pi}$, we see that the tension produces a force parallel to x equal to

$$\left(\frac{d}{dx}\frac{KX^2}{4\pi} + \frac{d}{dy}\frac{KXY}{4\pi} + \frac{d}{dz}\frac{KXZ}{4\pi}\right)\Delta x\,\Delta y\,\Delta z.$$

The force parallel to x due to the hydrostatic pressure and this tension is equal to

$$\left(-\frac{d}{dx}\frac{K(X^2+Y^2+Z^2)}{8\pi} + \frac{d}{dx}\frac{KX^2}{4\pi}\right.$$
$$\left. + \frac{d}{dy}\frac{KXY}{4\pi} + \frac{d}{dz}\frac{KXZ}{4\pi}\right)\Delta x\,\Delta y\,\Delta z;$$

when the medium is uniform, this may be written

$$\frac{K}{4\pi}\left\{Y\left(\frac{dX}{dy}-\frac{dY}{dx}\right) - Z\left(\frac{dZ}{dx}-\frac{dX}{dz}\right)\right.$$
$$\left. + X\left(\frac{dX}{dx}+\frac{dY}{dy}+\frac{dZ}{dz}\right)\right\}\Delta x\,\Delta y\,\Delta z.$$

Now $KX,\ KY,\ KZ = 4\pi f,\ 4\pi g,\ 4\pi h,$

and by equation (4) Art. 234,

$$\frac{dX}{dy}-\frac{dY}{dx} = \frac{dc}{dt},\quad \frac{dZ}{dx}-\frac{dX}{dz} = \frac{db}{dt},\quad \frac{dY}{dz}-\frac{dZ}{dy} = \frac{da}{dt},$$

while $K\left(\dfrac{dX}{dx}+\dfrac{dY}{dy}+\dfrac{dZ}{dz}\right) = 4\pi\rho;$

thus the force parallel to x due to the electric stresses may be written

$$\left(g\frac{dc}{dt} - h\frac{db}{dt} + X\rho\right)\Delta x\,\Delta y\,\Delta z.$$

In the same way the magnetic stresses may be shown to give a force parallel to x equal to

$$\frac{\mu}{4\pi}\left\{\beta\left(\frac{d\alpha}{dy}-\frac{d\beta}{dx}\right)-\gamma\left(\frac{d\gamma}{dx}-\frac{d\alpha}{dz}\right)\right.$$
$$\left.+\alpha\left(\frac{d\alpha}{dx}+\frac{d\beta}{dy}+\frac{d\gamma}{dz}\right)\right\}\Delta x\,\Delta y\,\Delta z\,;$$

since by Art. 234

$$\frac{d\gamma}{dy}-\frac{d\beta}{dz}=4\pi\frac{df}{dt}\,,\quad\frac{d\alpha}{dz}-\frac{d\gamma}{dx}=4\pi\frac{dg}{dt}\,,\quad\frac{d\beta}{dx}-\frac{d\alpha}{dy}=4\pi\frac{dh}{dt}\,,$$

and
$$\mu\left(\frac{d\alpha}{dx}+\frac{d\beta}{dy}+\frac{d\gamma}{dz}\right)=4\pi\sigma,$$

where σ is the density of the magnetism, the magnetic stresses give rise to a force parallel to x equal to

$$\left(c\frac{dg}{dt}-b\frac{dh}{dt}+\alpha\sigma\right)\Delta x\,\Delta y\,\Delta z\,;$$

hence the system of electric and magnetic stresses together gives rise to a force parallel to x equal to

$$\left(\frac{d}{dt}\left(cg-bh\right)+X\rho+\alpha\sigma\right)\Delta x\,\Delta y\,\Delta z.$$

The terms $X\rho$ and $\alpha\sigma$ represent the forces acting on the charged bodies and the magnets in the element of volume, and are equal to the rate of increase of momentum parallel to x of these bodies, the remaining term

$$\frac{d}{dt}\left(cg-bh\right)\Delta x\,\Delta y\,\Delta z$$

equals the rate of increase of the x momentum in the ether in the element of volume. This agrees with our previous investigation; for we have seen (p. 524) that the momentum parallel to x per unit volume is equal to $gc-hb$.

281. A system of charged bodies, magnets, circuits carrying electric currents &c. *and the ether* forms a self-contained system subject to the laws of dynamics; in such a system, since action and reaction are equal and opposite, the whole momentum of the system must be constant in magnitude and direction, if any one part of the system gains momentum some other part or parts must lose an equal amount. If we take the incomplete system got by leaving out the ether, this is not true. Thus take the case of a charged body struck by an electric wave, the electric force in the wave acts on the body and imparts momentum to it, no other material body loses momentum, so that if we leave out of account the ether we have something in contradiction to the third law of motion. If we take into account the momentum in the ether there is no such contradiction, as the momentum in the electric waves after passing the charged body is diminished as much as the momentum of that body is increased.

282. Another interesting example of the transference of momentum from the ether to ordinary matter is afforded by the pressure exerted by electric waves, including light waves, when they fall on a slab of a substance by which they are absorbed. Take the case when the waves are advancing normally to the slab. In each unit of volume of the waves there is a momentum equal to the product of the magnetic induction B and the dielectric polarization N; B and N are at right angles to each other, and are both in the wave front; the momentum which is at right angles to both B and N is therefore in the direction of propagation of the wave. In the wave $B = 4\pi\mu NV$, so that $BN = \dfrac{1}{4\pi} \dfrac{B^2}{\mu V}$, V being the velocity of light; B is a periodic function, and may be represented by an expression of the form $B_0 \cos{(pt - nx)}$, x being the direction of propagation of the wave; the mean value of B^2 is therefore $\frac{1}{2}B_0^2$. Thus the average value of the momentum per unit volume of the wave is $\dfrac{1}{8\pi} \dfrac{B_0^2}{\mu V}$, the amount of momentum that crosses unit area of the face of the absorbing substance per unit time is therefore $\dfrac{1}{8\pi\mu} \dfrac{B_0^2}{V} \times V$, or $\dfrac{1}{8\pi\mu} B_0^2$. As the wave is supposed to be absorbed by the slab no momentum leaves the slab through the ether, so that in each unit of time $\dfrac{B_0^2}{8\pi\mu}$ units of momentum are communicated to the slab for each unit area of its face exposed to the light: the effect on the slab is the same therefore as if the face were acted upon by a pressure $B_0^2/8\pi\mu$. It should be noticed that μ is the magnetic permeability of the dielectric through which the waves are advancing, and not of the absorbing medium.

If the slab instead of absorbing the light were to reflect it, then if the reflection were perfect each unit area of the face would in unit time be receiving $B_0^2/8\pi\mu$ units of momentum in one direction, and giving out an equal amount of momentum in the opposite direction; the effect then on the reflecting surface would be as if a pressure $2 \times B_0^2/8\pi\mu$ or $B_0^2/4\pi\mu$ were to act on the surface. This pressure of radiation as it is called was predicted on other grounds by Maxwell; it has recently been detected and measured by Lebedew and by Nichols and Hull by some very beautiful experiments.

283. If the incidence is oblique and not direct, then if the reflection is not perfect there will be a tangential force as well as a normal pressure acting on the surface. For suppose i is the angle of incidence, B_0 the maximum magnetic induction in the incident light, B_0' that in the reflected light, then across each unit of wave front in the incident light $B_0^2/8\pi\mu$ units of momentum in the direction of the incident light pass per unit time, therefore each unit of surface receives per unit time $\cos i B_0^2/8\pi\mu$ units of momentum in the direction of the incident light, or $\cos i \sin i B_0^2/8\pi\mu$ units of momentum parallel to the reflecting surface. In consequence of reflection

$$\cos i \sin i B_0'^2/8\pi\mu$$

units of momentum in this direction leave unit area of the surface in unit time, thus in unit time

$$\cos i \sin i (B_0^2 - B_0'^2)/8\pi\mu$$

units of momentum parallel to the surface are communicated to the reflecting slab per unit time, so that the slab will be acted on by a tangential force of this amount. Professor Poynting has recently succeeded in detecting this tangential force.

Since the direction of the stream of momentum is changed when light is refracted, there will be forces acting on a refracting surface, also when in consequence of varying refractivity the path of a ray of light is not straight the refracting medium will be acted upon by forces at right angles to the paths of the ray; the determination of these forces, which can easily be accomplished by the principle of the Conservation of Momentum, we shall leave as an exercise for the student.

532 PROPERTIES OF MOVING ELECTRIC CHARGES [CH. XV

284. We shall now proceed to illustrate the distribution of momentum in some simple cases.

Case of a Single Magnetic Pole and an Electrified Point.

Let A be the magnetic pole, B the charged point, m the strength of the pole, e the charge on the point, then at a point P the magnetic induction is m/AP^2 and is directed along AP, the dielectric polarization is $e/4\pi BP^2$ and is along BP, hence the momentum at P is

$$\frac{me \sin APB}{4\pi . AP^2 . BP^2}$$

and its direction is the line through P at right angles to the plane APB. The lines of momentum are therefore circles with their centres along AB and their planes at right angles to it, the resultant momentum in any direction evidently vanishes. There will however be a finite *moment* of momentum about AB: this we can easily show by integration to be equal to *em*. Thus in this case the distribution of momentum is equivalent to a moment of momentum *em* about AB. The distribution of momentum is similar in some respects to that in a top spinning about AB as axis. Since the moment of momentum of the ether does not depend upon the distance between A and B it will not change either in magnitude or direction when A or B moves in the direction of the line joining them. If however the motion of A or B is not along this line, the direction of the line AB and therefore the direction of the axis of the moment of momentum of the ether, changes. But the moment of momentum of the system consisting of the ether, the charge point, and the pole must remain constant; hence when the momentum in the ether changes, the momentum of the system consisting of the pole and the charge must change so as to compensate for the change in the momentum of the ether. Thus suppose the charged point moves from B to B' in the time δt, then in that time the moment of momentum

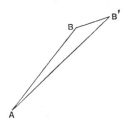

Fig. 135.

in the ether changes from em along AB to em along AB'; this change in the moment of momentum of the ether is equivalent to a moment of momentum whose magnitude is $em\delta\theta$, where $\delta\theta = \angle BAB'$, and whose axis is at right angles to AB in the plane BAB'. The change in the moment of momentum of the pole and point must be equal and opposite to this. Since the resultant momentum of the ether vanishes in any direction, the change in the momentum of the pole must be equal and opposite to the change in momentum of the point, and these two changes must have a moment of momentum equal to $em\delta\theta$: we see that this will be the case if δI the change in momentum of the point is at right angles to the plane BAB' and equal to $\dfrac{em\delta\theta}{AB}$, while the change in momentum in the pole is equal and opposite to this. This change in momentum $\dfrac{em\delta\theta}{AB}$ occurring in the time δt may be regarded as produced by a force F acting on the point at right angles to the plane BAB' and given by the equation

$$F = \frac{em}{AB} \cdot \frac{\delta\theta}{\delta t}.$$

Now

$$\delta\theta = \frac{BB' \sin ABB'}{AB},$$

or if v be the velocity of the point,

$$\delta\theta = \frac{v\delta t \sin ABB'}{AB},$$

or

$$\frac{\delta\theta}{\delta t} = \frac{v \sin ABB'}{AB};$$

thus

$$F = \frac{emv}{AB^2} \sin ABB'$$

$$= evH \sin \phi,$$

where H is the magnetic force at the point and ϕ the angle between H and the direction in which the point is moving; from this we see that a moving charged point in a magnetic field is acted on by a force at right angles to the velocity of the point, at right angles also to the magnetic force at the point, and equal to the product of the charge, the magnetic force and the velocity of the point at right angles to the magnetic force. Thus we see that we can deduce the expression for the force acting on a charged point moving across the lines of magnetic force directly from the principle of the Conservation of Momentum. We should have got an exactly similar expression if we had supposed the charge at rest and the pole in motion; in this case we must take v to be the velocity of the pole and ϕ the angle between v and AB.

285. From the expression given on page 524 for the momentum in the field we can prove that the momentum in the ether due to a charged point at P and the magnetic force produced by a current flowing round a small closed circuit, is equivalent to a momentum passing through P whose components F, G, H parallel to the axes of x, y, z respectively are given by the equations

$$F = \mu i \alpha \left(m \frac{d}{dz} \frac{1}{r} - n \frac{d}{dy} \frac{1}{r} \right),$$

$$G = \mu i \alpha \left(n \frac{d}{dx} \frac{1}{r} - l \frac{d}{dz} \frac{1}{r} \right),$$

$$H = \mu i \alpha \left(l \frac{d}{dy} \frac{1}{r} - m \frac{d}{dx} \frac{1}{r} \right),$$

where i is the current flowing round the circuit, α the area of the circuit and l, m, n the direction cosines of the normal to its plane, x, y, z are the coordinates of P and r the distance of P from the centre of the circuit, the charge at P is supposed to be the unit charge. We see that

$$\frac{dF}{dy} - \frac{dG}{dx} = \mu i \alpha \left\{ m \frac{d^2}{dydz} \frac{1}{r} - n \left(\frac{d^2}{dy^2} \frac{1}{r} + \frac{d^2}{dx^2} \frac{1}{r} \right) + l \frac{d^2}{dxdz} \frac{1}{r} \right\},$$

or since
$$\frac{d^2}{dx^2} \frac{1}{r} + \frac{d^2}{dy^2} \frac{1}{r} + \frac{d^2}{dz^2} \frac{1}{r} = 0,$$

$$\frac{dF}{dy} - \frac{dG}{dx} = \mu i \alpha \left(l \frac{d^2}{dxdz} \frac{1}{r} + m \frac{d^2}{dydz} \frac{1}{r} + n \frac{d^2}{dz^2} \frac{1}{r} \right)$$

$$= \mu i \alpha \frac{d}{dz} \left(l \frac{d}{dx} \frac{1}{r} + m \frac{d}{dy} \frac{1}{r} + n \frac{d}{dz} \frac{1}{r} \right)$$

$$= c,$$

c being the z component of the magnetic induction at P

due to the small circuit. We have similarly if a and b are the x and y components respectively of this induction

$$\frac{dH}{dx} - \frac{dF}{dz} = b,$$

$$\frac{dG}{dz} - \frac{dH}{dy} = a.$$

The usual expression for the electromotive force due to induction follows at once from the principle of the Conservation of Momentum. For the momentum in the ether is equivalent to a momentum through P whose components are F, G, H. Suppose that in consequence of the motion of the circuit or the alteration of the current through it, F, G, H become $F + \delta F$, $G + \delta G$, $H + \delta H$, then the momentum in the ether still passes through P but has now components $F + \delta F$, $G + \delta G$, $H + \delta H$ instead of F, G, H; but the momentum of the whole system, point circuit and ether must remain constant; thus to counter-balance the changes in momentum δF, δG, δH at P due to the ether, we must have changes in momentum of the unit charge at P equal to $-\delta F$, $-\delta G$, $-\delta H$. Suppose that the time taken by the changes δF, δG, δH is δt, then in the time δt the x momentum of the unit charge at P must change by $-\delta F$, i.e. the unit charge must be acted on by force $-\dfrac{dF}{dt}$. Thus there is at P an electric force whose component parallel to x is $-\dfrac{dF}{dt}$, similarly the components parallel to y and z are $-\dfrac{dG}{dt}$, $-\dfrac{dH}{dt}$. The electric force whose components we have just found is the force due to electromagnetic induction, and its magnitude is that given by Faraday's law. To prove this we notice that the line integral of the electric force round a fixed circuit of which ds is an element is equal to

$$-\int\left(\frac{dF}{dt}\frac{dx}{ds}+\frac{dG}{dt}\frac{dy}{ds}+\frac{dH}{dt}\frac{dz}{ds}\right)ds$$

$$=-\frac{d}{dt}\int\left(F\frac{dx}{ds}+G\frac{dy}{ds}+H\frac{dz}{ds}\right)ds$$

$$=-\frac{d}{dt}\int\left\{l\left(\frac{dG}{dz}-\frac{dH}{dy}\right)+m\left(\frac{dH}{dx}-\frac{dF}{dz}\right)+n\left(\frac{dF}{dy}-\frac{dG}{dx}\right)\right\}dS$$

by Stokes' theorem; here l, m, n are the direction cosines of the normal to a surface filling up the closed curve, dS is an element of this surface. Substituting the values already given for $\frac{dG}{dz}-\frac{dH}{dy}$, &c. the preceding expression becomes

$$-\frac{d}{dt}\int(la+mb+nc)\,dS;$$

the integral in this expression is the number of lines of magnetic induction passing through the closed circuit, hence we see that the line integral of the electric force due to induction round a closed circuit equals the rate of diminution in the number of lines of magnetic induction passing through the circuit; this however is exactly Faraday's law of induction (see Art. 229).

286. When a charged particle is moving so rapidly that v^2/V^2 cannot be neglected, the distribution of the Faraday tubes round the particle is no longer uniform and the expression $2\mu e^2 v/a$ given in Art. 277 for the momentum of the charged sphere has to be modified. For an investigation of this case we refer the reader to

Recent Researches in Electricity and Magnetism, where, page 21, it is shown that when the velocity of the charged sphere is w, R the momentum parallel to z is in the general case given by the equation

$$R = \frac{1}{2}\frac{\mu e^2}{a}\frac{V^2}{w(V^2 - w^2)^{\frac{1}{2}}}\left\{ \vartheta\left(1 - \frac{1}{4}\frac{V^2}{w^2}\right) \right.$$
$$\left. + \frac{1}{2}\sin 2\vartheta\left(1 + \frac{1}{4}\frac{V^2}{w^2}\cos 2\vartheta\right)\right\},$$

where $$\sin\vartheta = \frac{w}{V}.$$

From this value of R we see that when w approaches V, the value of R/w, the apparent mass, increases rapidly with w; thus if an appreciable amount of the mass of a body is due to electric charge, the mass of the body will increase with the velocity, it is only however when the velocity of the body approaches that of light that this increase becomes appreciable, in the limiting case where the velocity is that of light the apparent mass would be infinite. The influence of velocity on the apparent mass of particles travelling with great velocities has been detected by Kaufmann by some very interesting experiments, a short account of which will be found in the author's *Conduction of Electricity through Gases*, page 533. Kaufmann found that a particle moving with a velocity about five per cent. less than the velocity of light, had a mass about three times that with small velocities.

The increase in the mass of a slowly moving charged sphere is $2\mu e^2/3a$, i.e. 4 (potential energy of the sphere)$/3V^2$, thus if this mass were to move with the velocity of light its kinetic energy would be two-thirds of the electrical potential energy. The same proportion between the in-

crease in the mass due to electrification and the electrical
potential energy can be shown to hold for any system of
electrified bodies as well as for the simple case of the
charged sphere.

**287. Effects due to changes in the velocity of
the moving charged body.** We shall take first the
case of a charged sphere moving so slowly that the lines of
force are symmetrically distributed around it, and consider

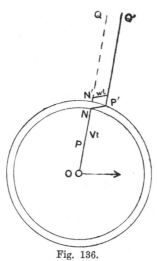

Fig. 136.

what will happen when the sphere is suddenly stopped.
The Faraday tubes associated with the sphere have inertia
and are in a state of tension, thus any disturbance communicated to one end of a tube will travel along the tube
with a finite and constant velocity—the velocity of light.
Let us suppose that the stoppage of the particle takes a
finite small time τ. We can find the configuration of the
tubes, after a time t has elapsed since the sphere began

to be stopped, in the following way. Describe with the centre of the charged sphere as centre two spheres, one having the radius Vt, the other the radius $V(t-\tau)$. Then since no disturbance can have reached the portions of the Faraday tubes situated outside the surface of the outer sphere these tubes will be in the positions they would have occupied if the sphere had not been stopped, while since the disturbance has passed over the tubes within the inner sphere, these tubes will be in their final position. Thus consider a tube which when the particle was stopped was along the line OPQ, O being the centre of the charged sphere, this will be the final position of the tube; hence at the time t the portion of this tube inside the inner sphere will be in the position OP, the portion $P'Q'$ outside the outer sphere will be in the position it would have occupied if the sphere had not been stopped, i.e. if O' is the position to which O would have come if the sphere had not been stopped, $P'Q'$ will be a straight line passing through O'. Thus to preserve its continuity the tube must bend round in the shell between the surfaces of the two spheres, and take the position $OPP'Q'$. Thus the tube which before the sphere was stopped was radial, has now, in the shell, a tangential component, and this implies a tangential electric force; this tangential force is, as the following calculation shows, much greater than the radial force at P before the sphere was brought to rest.

Let us suppose that δ, the thickness of the shell, is so small that the portion of the Faraday tube inside it may be regarded as straight, then, if T is the tangential force inside the pulse, R the radial force, we have

$$\frac{T}{R} = \frac{P'N'}{PN'} = \frac{OO' \sin \theta}{\delta} = \frac{wt \sin \theta}{\delta} \quad \dots\dots\dots(1),$$

where w is the velocity with which the sphere was moving before it was stopped, and θ the angle OP makes with the direction of motion of the sphere; t is the time since the sphere was stopped. Since $OP = Vt$ and $R = e/K.OP^2$, K being the specific inductive capacity of the medium, we have, writing r for OP,

$$T = \frac{ew \sin \theta}{KV.r\delta}.$$

Thus the tangential force varies inversely as the distance and not as the square of the distance.

The tangential Faraday tubes move radially outwards with the velocity V, they will therefore produce a magnetic force at right angles to the plane of the pulse and in the opposite direction to the magnetic force at P before the sphere was stopped; this force is equal to

$$V \times 4\pi.\frac{KT}{4\pi} = \frac{ew \sin \theta}{r\delta};$$

the magnetic force before the sphere was stopped was $ew \sin \theta/r^2$, thus the magnetic force in the pulse, which however only lasts for a very short time, exceeds that in the steady field in the proportion of r to δ.

Thus the pulse produced by the stoppage of the sphere is the seat of very intense electric and magnetic forces; the pulses formed by the stoppage of the regularly electrified particles of the cathode rays form, in my opinion, the well-known Röntgen rays.

288. Energy in the Pulse. The energy due to the magnetic force in the field is per unit volume

$$\frac{\mu}{8\pi} \frac{e^2w^2 \sin^2 \theta}{\delta^2 r^2};$$

integrating this through the pulse we find that the energy
due to the magnetic force in the pulse is

$$\frac{\mu e^2 w^2}{3\delta^2}.$$

The energy due to the tangential electric force in the
pulse is per unit volume

$$\frac{KT^2}{8\pi} = \frac{e^2 w^2 \sin^2\theta}{8\pi . K V^2 \delta^2 r^2};$$

integrating this through the pulse we find that this energy
is equal to $\frac{\mu e^2 w^2}{3\delta}$, since $\mu K = \frac{1}{V^2}$.

Thus the total energy in the pulse is $\frac{2}{3}\frac{\mu e^2 w^2}{\delta}$; and this
energy radiates away into space. The energy in the field
before the sphere was stopped was $\frac{1}{3}\mu e^2 w^2/a$, where a is
the radius of the sphere (see Art. 277). Thus if δ is not
much greater than the diameter of the sphere a very considerable fraction of the kinetic energy is radiated away
when the particle is stopped.

289. Distribution of Momentum in the Field.
There is no momentum inside the surface of the sphere
whose radius is $b(t - T)$, there is a certain amount of
momentum in the pulse, and momentum in the opposite
direction in the region outside the pulse; we shall leave
it as an exercise for the student to show that the momentum in the pulse is equal and opposite to that outside
it, so that as soon as the sphere is reduced to rest the
whole momentum in the field is zero.

290. Case of an Accelerated Charged Body.
The preceding method can be applied to the case when
the charged body has its velocity altered in any way, not

necessarily reduced to zero. Thus if the velocity instead
of being reduced to zero is diminished by δw, we can show
in just the same way as before that the magnetic force H
in the pulse is given by the equation

$$H = \frac{e\Delta w \cdot \sin \theta}{r\delta},$$

and the tangential electric force T by

$$T = \frac{e\Delta w \sin \theta}{KVr\delta}.$$

Now $\delta = V\delta t$ if δt is the time required to change the
velocity by Δw, hence we have

$$H = \frac{e}{V}\frac{\Delta w}{\delta t}\frac{\sin \theta}{r}, \qquad T = \frac{e}{KV^2}\frac{\Delta w}{\delta t}\frac{\sin \theta}{r};$$

but $\Delta w/\delta t = -f$, where f is the acceleration of the particle,
hence

$$H = -\frac{e}{V}\frac{f\sin \theta}{r}, \qquad T = -\frac{e}{KV^2}\frac{f\sin \theta}{r}.$$

It must be remembered that f is not the acceleration
of the sphere at the time when H and T are estimated
but at the time r/V before this. We see that when the
velocity of the sphere is not uniform, part of the magnetic
and electric force will vary inversely as the distance from
the centre of the sphere, while the other part will vary
inversely as the square of this distance; at great distances
from the sphere the former part will be the most important.

The energy in the pulse emitted whilst the velocity is
changing is equal to

$$\frac{2}{3}\frac{e^2f^2}{V^2}d,$$

where d is the thickness of the pulse; since $d = V\delta t$, where

δt is the time the acceleration lasts, the energy emitted in the time δt is

$$\frac{2}{3}\frac{e^2 f^2}{V}\,\delta t,$$

thus the rate of emission of energy is $2e^2 f^2/3V$.

291. Magnetic and Electric Forces due to a charged particle vibrating harmonically through a small distance. The magnetic force proportional to the acceleration which we have just investigated arises from the motion of the tangential part of the Faraday tubes—the portion $P'N'$ of Fig. 134; the radial tubes are however also in motion, their velocity at right angles to their length being $w \sin \theta$, where w is the velocity of the particle when its acceleration is f, i.e. at a time r/V before the force is estimated. This motion of the radial tubes produces a magnetic force $ew \sin \theta/r^2$ in the same direction as that due to the acceleration. Thus H the magnetic force at P is equal to

$$\frac{ew \sin \theta}{r^2} + \frac{ef \sin \theta}{Vr},$$

and is at right angles to OP and to the axis of z along which the particle is supposed to be moving. Let the velocity of the particle along this line be $\omega \sin pt$ and its acceleration therefore $\omega p \cos pt$. The magnetic force at P at the time t will depend upon the velocity and acceleration of the particle at the time $t - \dfrac{r}{V}$, these are respectively

$\omega \sin p \left(t - \dfrac{r}{V} \right)$ and $\omega p \cos p \left(t - \dfrac{r}{V} \right)$, thus H the magnetic force at P is given by the equation

$$H = \frac{ew \sin \theta \sin p \left(t - \dfrac{r}{V} \right)}{r^2} + \frac{e\omega \sin \theta p \cos p \left(t - \dfrac{r}{V} \right)}{Vr}.$$

If a, β, γ are the components of this force parallel to the axes of x, y, z, then

$$\alpha = -\frac{y}{r\sin\theta}H, \qquad \beta = \frac{x}{r\sin\theta}H, \qquad \gamma = 0.$$

Hence

$$\alpha = \frac{d}{dy}\frac{e\omega\sin p\left(t-\frac{r}{V}\right)}{r}, \qquad \beta = -\frac{d}{dx}\frac{e\omega\sin p\left(t-\frac{r}{V}\right)}{r}.$$

If X, Y, Z are the components of the electric force, we have by equation (1), page 489,

$$K\frac{dX}{dt} = \frac{d\gamma}{dy} - \frac{d\beta}{dz} = \frac{d^2}{dxdz}\frac{e\omega\sin p\left(t-\frac{r}{V}\right)}{r},$$

$$K\frac{dY}{dt} = \frac{d\alpha}{dz} - \frac{d\gamma}{dx} = \frac{d^2}{dydz}\frac{e\omega\sin p\left(t-\frac{r}{V}\right)}{r},$$

$$K\frac{dZ}{dt} = \frac{d\beta}{dx} - \frac{d\alpha}{dy} = -\left(\frac{d^2}{dx^2} + \frac{d^2}{dy^2}\right)\frac{e\omega\sin p\left(t-\frac{r}{V}\right)}{r}.$$

Hence the periodic parts of X, Y, Z are given by the equations

$$KX = -\frac{1}{p}\frac{d^2}{dxdz}\frac{e\omega\cos p\left(t-\frac{r}{V}\right)}{r},$$

$$KY = -\frac{1}{p}\frac{d^2}{dydz}\frac{e\omega\cos p\left(t-\frac{r}{V}\right)}{r},$$

$$KZ = \frac{1}{p}\left(\frac{d^2}{dx^2} + \frac{d^2}{dy^2}\right)\frac{e\omega\cos p\left(t-\frac{r}{V}\right)}{r}.$$

In addition to these there are the components

$$-\frac{e}{K}\frac{d}{dx}\frac{1}{r}, \quad -\frac{e}{K}\frac{d}{dy}\frac{1}{r}, \quad -\frac{e}{K}\frac{d}{dz}\frac{1}{r},$$

of the electrostatic force due to the charge at O. In this investigation ω is supposed to be so small compared with V that ω^2/V^2 may be neglected.

INDEX

(The numbers refer to the pages)

Absolute measurement
    of a resistance 467, 470, 473
    of a current 475
Alternating currents, distribution
    of 424, 428
Ampère's law 330, 355
Angle, Solid 218
Anode 285
Axis of a magnet 197

Ballistic galvanometer 382
Boundary conditions
    for two dielectrics 128
    for two magnetizable sub-
      stances 262
    for two conductors carrying
      currents 326
Bunsen's Cell 303

Cadmium Cell 304
Capacity
    of a condenser 84
    of a sphere 84
    of two concentric spheres 85,
      141
    of two parallel plates 90, 135
    of two coaxial cylinders 93,
      141
    specific inductive 120
Capacities, comparison of two 108
    determination of, in electro-
      magnetic measure 476
Cathode 285
Cavendish experiment 31
Charge of electricity 8
Charge, unit 12

Circuit, magnetic 352
Circular currents
    magnetic force due to 356
    force between two 358
Clark's Cell 303, 475
Coefficients
    of capacity 43
    of induction 43
    of potential 42
    of self-induction 365, 448
    of mutual induction 365, 450
Condensers 84
    comparison of two 109
    in parallel 116
    in cascade 117
    parallel plate 90
Condenser in an alternating current
    circuit 443
Conductors 9
Conjugate conductors 314
Coulomb's Law 36, 127
Couples
    between two magnets 205
    on a current in a magnetic
      field 360
Currents
    electric 283
    strength of 285
    magnetic force due to 329, 356
    distribution of steady 310, 320
    distribution of alternating 424,
     428
    dielectric 480
Cylinder
    electric intensity due to 21
    capacity of 93, 141

Daniell's Cell 301
Declination, magnetic 233
Dielectric
    currents 480
    plane and an electrified point
      169
    sphere in an electric field 172
Dimensions of electrical quantities
    461
Dip 235
Discharge of Leyden jar 436
Dissipation function 317
Distribution of
    steady currents 310, 320
    alternating currents 424, 429
    currents due to an impulse 404
Diurnal variation 241
Doublet, electric field due to 158
Duperrey's lines 238
Dynamical system illustrating in-
    duction 397

Electric
    intensity 13
    potential 25
    screens 51
    images 145
    currents 283
Electrification
    by friction 1
    positive and negative 2
    by induction 4
Electrolysis 285, 287
Electrolyte, E.M.F. required to
    liberate ions of 305
Electromagnetic
    induction 387
      Faraday's law of 393
      Neumann's law of 393
    screening 434
    wave, plane 492
Electrometers 97
    quadrant 98
Electromotive force of a cell 296
Electroscope 5
Element, Rational current 372
Ellipsoids in magnetic field 274
Energy
    in the electric field 37, 70, 127
    of a shell in a magnetic field
      222

Energy
    in the magnetic field 270
    due to a system of currents 369
    in a pulse 541
    due to a moving charged sphere
      521
Equipotential surface 29

Faraday's
    laws of Electrolysis 287
    laws of electromagnetic induc-
      tion 393
    tubes 67, 480, 494
    tubes, tension in 73
    tubes, pressure perpendicular
      to 74
Force
    lines of 60
    tubes of 65
    on an uncharged conductor 82
    on an electrified system 53
    between electrified bodies 12
    on charged conductor 56
    between bodies in a dielectric
      134
    on a dielectric 142
    between magnets 192
    due to a magnet 202
    between two small magnets 208
    on a shell in a magnetic field
      222
    on a current in a magnetic
      field 359

Galvanometer
    tangent 375
    sine 379
    ballistic 382
    Desprez-D'Arsonval 380
    resistance of 386
Gauss's proof of law of force be-
    tween poles 211
Gauss's theorem 14
Grove's Cell 303

Heat produced by a current 293,
    316
Hertz's experiments 499
Hysteresis 257

Impedance 413

Impulse, distribution of currents induced by 404
Induction
  magnetic 247
  electromagnetic 387
  total normal electric 13
Insulators 9
Intensity
  electric 13
  of magnetization 198
Inversion 175
Ions 286
Isoclinic lines 236
Isogonic lines 236

Jar, Leyden 114
  discharge of, 436
Joule's Law 293

Kirchhoff's Laws 310

Law of force between electrified bodies 12, 31
Lenz's Law 445
Leyden jar 114
  in parallel 116
  in cascade 117
  discharge of 436
Light, Maxwell's Theory of 480 *et seq.*
Lines of force 25, 60
  refraction of 131
Lorenz's Method 470

Magnet
  pole of 196
  axis of 197
  moment of 197, 215
  potential due to 198
  resolution of 199
  force due to 200, 202
  couple on 204
Magnetic
  force 195
  disturbances 243
  potential 195
  shell 216
  shell, force due to 226
  shell, force acting on 222
  shielding 266
  induction 246

Magnetic
  induction, tubes of 249
  permeability 252
  retentiveness 257
  susceptibility 251
  declination 233
  dip 235
Magnetic field
  energy in 270
  due to current 329
  due to two straight currents 342
  due to circular current 356
Magnetization, intensity of 198
Magnetized sphere, field due to 227
Magnets 191
  action between two small 205
Mass due to electric charge 516
Maxwell's Theory 480 *et seq.*
Model illustrating magnetic induction 397
Moment of magnet, determination of 215
Momentum in electric field 523
Moving electric charges 519 *et seq.*
Mutual induction, coefficient of 365
  determination of 450
  comparison of 452

Neumann's Law of electromagnetic induction 393
Neutral temperature 518

Ohm, determination of 470, 474
Ohm's Law 289
Oscillating electric charge 543

Parallel plate condenser 90
Peltier Effect 507
Periodic electromotive force 411
Permeability, magnetic 252
  affected by temperature 256
Plane uniformly electrified 22
Plane and electrified point 145
Planes
  parallel, separated by dielectric 135
  two parallel and electrified point 181

Planes
  magnetic force due to currents
    in parallel 346
Polarization
  in a dielectric 125
  of a battery 304
Pole, unit 194
Poles of a magnet 196
Potential
  electric 25
  of charged sphere 27
  of a magnet 198
Propagation of electromagnetic disturbance 488 *et seq.*
Pulse due to stopping or starting charge 539

Ratio of units 475
Refraction of lines of force 131
Resistance
  electric 289
  of conductors in series 290
  of conductors in parallel 291
  specific 293
  measurement of 374
  absolute 467, 473
Resolution of a magnet 199
Retentiveness, magnetic 257
Rotating circuit 414

Saturation, magnetic 255
Screening
  electric 51
  electromagnetic 434
  magnetic 266
Secondary circuit, effect of on apparent self-induction and resistance 419
Self-induction
  coefficient of 365
  coefficient of, of a solenoid 367
  coefficient of, of two parallel circuits 368
  determination of 448
  comparison of 454
Shell, magnetic 216
Sine galvanometer 379
Solenoid 349
Solid angle 218

Specific inductive capacity 120
  determination of 143
Specific resistance 293
Specific Heat of Electricity 510
Sphere
  electric intensity due to 20
  potential due to 27
  capacity of 84, 85
  and an electrified point 149
  in a uniform electric field 157
  inversion of 176
  magnetic field due to 227
  in a uniform magnetic field 265
Spheres
  intersecting at right angles 162
  in contact 183
Surface density 36
Susceptibility 252

Tangent galvanometer 375
Temperature, effect of, on magnetic permeability 256
Terrestrial magnetism 232
Thermoelectric
  currents 506
  diagrams 512
Thomson Effect 510
Transformers 417
Tubes
  of electric force 65
  Faraday 67, 480, 494
  Faraday, tension in 73
  Faraday, pressure perpendicular to 74
  of magnetic induction 249

Units
  electrostatic system 465
  electromagnetic system 465

Variation
  in magnetic elements 240
  diurnal 241
Voltaic cell 295

Wave, Electromagnetic 492 *et seq.*
Wheatstone's Bridge 310, 384
Work done when unit pole is taken round a circuit 332

CAMBRIDGE: PRINTED BY JOHN CLAY, M.A. AT THE UNIVERSITY PRESS.

Printed in the United States
By Bookmasters